D0936819

STATISTICAL ECOLOGY   VOLUME 3

MANY SPECIES POPULATIONS,

ECOSYSTEMS, AND SYSTEMS ANALYSIS

STATISTICAL ECOLOGY

Based on the Proceedings of the International Symposium on Statisti-
cal Ecology, New Haven, Connecticut, August 1969, sponsored by the
Ford Foundation, U.S. Forest Service, The Pennsylvania State
University, and Yale University under the auspices of the Interna-
tional Statistical Institute and the International Association of
Ecology.

ORGANIZING COMMITTEE

Director: G. P. Patil; Co-Chairmen: G. P. Patil (U.S.A.); E. C.
Pielou (Canada); W. E. Waters (U.S.A.). Members: E. Batschelet
(U.S.A.); D. R. Cox (England); J. Gani (England); D. W. Goodall
(U.S.A.); J. Gulland (Italy); C. S. Holling (Canada); H. Klomp
(Netherlands); V. Labeyrie (France); B. Matérn (Sweden); C. R. Rao
(India); V. Schultz (U.S.A.); J. G. Skellam (England); L. R. Taylor
(England); E. J. Williams (Australia).

LOCAL ARRANGEMENTS COMMITTEE

Chairman: Carroll B. Williams; Members: Z. White; K. D'Antonio;
M. Feldman.

THE PENN STATE STATISTICS SERIES

An International Series in Statistics and Applications
General Editor: G. P. Patil, Professor of Statistics
                  The Pennsylvania State University
Published by The Pennsylvania State University Press, University
Park, Pennsylvania, 16802, U.S.A.

RANDOM COUNTS IN SCIENTIFIC WORK   (Edited by G. P. Patil)
Volume 1:   Random Counts in Models and Structures
Volume 2:   Random Counts in Biomedical and Social Sciences
Volume 3:   Random Counts in Physical Science, Geo Science, and
            Business

STATISTICAL ECOLOGY   (Edited by G. P. Patil, E. C. Pielou and
                       W. E. Waters)

Volume 1:   Spatial Patterns and Statistical Distributions
Volume 2:   Sampling and Modeling Biological Populations and
            Population Dynamics

STATISTICAL ECOLOGY VOLUME 3

MANY SPECIES POPULATIONS, ECOSYSTEMS, AND SYSTEMS ANALYSIS

Edited by

G. P. PATIL
Professor of Mathematical Statistics
The Pennsylvania State University, University Park, Pa.

E. C. PIELOU
Professor of Mathematical Biology
Queen's University, Kingston, Ontario

W. E. WATERS
Chief of Forest Insect Research
Forest Service, USDA, Washington, D.C.

The Pennsylvania State University Press
University Park and London

Library of Congress Catalogue Card Number 77-114352

Copyright © 1971 by The Pennsylvania State University

International Standard Book Number (ISBN) 0-271-00113-5

Printed in the United States of America

The Pennsylvania State University Press
University Park, Pennsylvania 16802

The Pennsylvania State University Press, Ltd.
London W. 1

CONTENTS

PREFACE

It is becoming increasingly evident that the interrelated problems
of human population growth, environmental contamination, and
depletion of vital natural resources are both serious and complex.
The dilemmas involved pose a challenge to the ecologist, as scientist and practitioner, that cannot be met by the traditional,
subjective approaches that largely characterize ecology today.
New, more sophisticated concepts and methodology are required to
probe into the form and substance of man-environment relationships
at all levels of organization. Since the variables of concern are
quantifiable, generally, and the relationships dynamic with respect
to time and place, the concepts and techniques of mathematical and
statistical analysis must be drawn upon for insight and prediction.
The primary objective of the present Symposium was to provide
opportunity for contact and communication between quantitatively-
oriented ecologists and mathematicians-statisticians-systems
analysts familiar with biological systems. The emphasis was on
analytical approaches and techniques applicable to the solution
of man-environment problems, but the exchange of ideas and information in all aspects of quantitative ecology was encouraged.

Mission-oriented research projects, undertaken to solve particular
problems, must draw upon an underlying body of knowledge and theoretical concepts that only basic research can provide. Pure science
and applied science depend on each other, as advances in the
physical sciences have shown; they are like the two legs of a biped.
We therefore planned the Symposium in the belief that it would also
further the consolidation of "pure" ecology and thus be beneficial
to applied ecology as well.

The Symposium brought together a large group of research workers:
ecologists who use mathematical and statistical methods, and statisticians whose work has ecological applications. We hope that the
ecologists will have learned of new statistical methods adapted to
their requirements; and that the statisticians will have learned of
new ecological work in which their expertise is needed. We also
hope that readers of these Proceedings will benefit similarly.

Because of their length, the Proceedings have been published in
three volumes. The three volumes bring together the papers relating
to the three areas of statistical ecology designated by the respective titles: Volume 1, Spatial Patterns and Statistical
Distributions; Volume 2, Sampling and Modeling Biological

<u>Populations and Population Dynamics</u>; Volume 3, <u>Many-Species</u>
<u>Populations, Ecosystems, and Systems Analysis</u>. The individual
papers represent but a small scattering of knowledge and experience
in the respective subject matter areas. However, taking them all
together, they cover a very wide range of points of view and
methodology. Many of the papers were followed by discussions
prepared in advance by participants who had seen pre-prints; all
were open to spontaneous discussion from the floor and the speakers
immediately wrote down their comments and questions for inclusion
in the Proceedings. The reader can distinguish preplanned discussions from spontaneous ones by the underlines they carry with
the names of the speakers. If only the private discussions could
have been magically transcribed and reproduced here too!

The first session of the Symposium opened with a welcoming
address by Dr. F. Mergen, Dean of the Yale School of Forestry, and
a keynote address by Dr. F. J. Anscombe, Chairman of the Statistics
Department at Yale. Near the close of the Symposium, a banquet was
held at the New Haven Lawn Club at which Professor G. E. Hutchinson
of the Biology Department at Yale gave an address.

Acknowledgments always constitute a risk--of omission or
commission--but we would like to record our special thanks to the
sponsoring organizations and supporting institutions and to the
members of the organizing committee and the local arrangements
committee. We are very grateful to Dr. C. I. Noll, Dean of the
College of Science of The Pennsylvania State University for his
support of the Symposium and the publication of the Proceedings in
the Penn State Statistics Series, an International Series in Statistics and Applications, published by The Pennsylvania State
University Press. We also wish to record our appreciation to the
staff of the University Press for their interest and attention.

Finally, we wish to express the hope that this Symposium will
stand as a prototype for further gatherings of like personalities
with a common concern for the ecological dilemma of man's survival
in a deteriorating environment and a scientific dedication to
finding solutions.

New Haven   G. P. Patil
                                                      E. C. Pielou
                                                      W. E. Waters

SIMULATION STUDIES AND ECOLOGY:
A SIMPLE DEFINED SYSTEM MODEL

K. A. KERSHAW
G. P. HARRIS*
Imperial College
London S. W. 7

SUMMARY

The major environmental variables and their interactions with
the physiology of the lichen Parmelia caperata, are discussed
in relation to a simulation of the growth of the lichen on oak
trees in S. W. England. The results from the initial model show
water availability and its interactions with the physiology of
the lichen, largely controls the vertical zonation of this
species.

---

* Now at McMaster University, Hamilton, Ontario.

1.   INTRODUCTION

Over the last few years a number of simulation studies of biological systems have been made (Holling [11], Paulik & Greenough
[15], Waggoner & Reifsnyder [21], Duncan, Loomis, Williams, &
Hanau [8], etc.) In common, all have used digital simulation
rather an analog approach. Furthermore, each model also illustrates three other features common to, and fundamental to
simulation studies: Each model is a simplification of the real
system it is simulating - all irrelevant detail is omitted.
Secondly each model potentially represents a 'building block'
which eventually can be fitted by minor adjustments, into a
more complete model of a complex system. Thirdly the models
attempt to answer a number of specific questions to which an
experimental answer is either difficult or impossible - there
is little merit in a model which is constructed purely for its
own sake.

    At the present level of our knowledge it is essential to take
a defined part of a complete ecosystem, which is then simplified
down to its essential components. To attempt to model all the
interactions of a complex system as a primary goal is completely
out of the question. It is desirable to construct a number of
'unit models' which are eventually fitted together. In a recent
investigation into the lichen zonation on oak trees in S.W.
England it became obvious that a simulation study might materi-
ally advance our knowledge of the causal mechanisms involved.
A number of ecological factors had been investigated which
offered a reasonable explanation of the observed distribution
of Parmelia caperata and the hypothesis accordingly required
an experimental approach to test whether or not the growth was
controlled by different light and water regimes. Unfortunately
lichens are very difficult to grow under laboratory conditions
and their very slow growth rates does not make them the best of
experimental material. Transplant methods [4] were also rejected due to the time factor and a simulation study seemed to
offer the best line of approach to the problem.

    The system can be simplified in a number of ways, both by
intent, and intrinsically by the structure of a lichen thallus:
The complete absence of roots and the corticolous habitat eliminates one of the most complex and intractable parts of an
ecosystem model--the soil. Water uptake is extremely rapid

and takes place over the entire area of the thallus. Mineral
nutrition can be ignored (at least in the initial models) since
we are considering the zonation from the top of a single (average)
tree to the bottom. Mineral nutrition is probably of importance
when the epiphytic lichen flora of different tree species is to
be considered ([3], [9], [12]) but is of doubtful significance
in the vertical zonation of any one tree species. Finally and
somewhat surprisingly, lichens turned out to be ideal experimental material for experimentation to determine in detail those
aspects of lichen metabolism of ecological significance. What

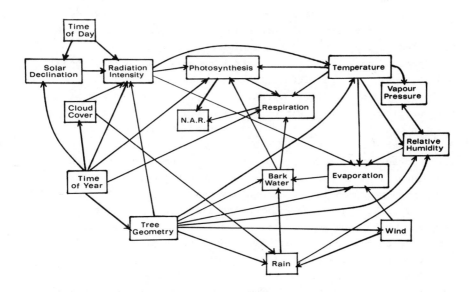

Figure 1.   A diagram of the basic interactions between parts of
            the lichen/microenvironment system

little is known about the ecology and physiology of corticolous lichens suggests that light and water are the two factors of extreme importance, and immediately a simple hypothesis can be set up for testing. This hypothesis comprises radiation intensity, water availability, the physiology of the lichen, and their interactions at different heights in an 'average' oak tree. Admittedly this simple system when broken down into detail required observational data on a large number of variables and their interactions (Fig. 1), but it should be emphasized the inclusion of both nutrition and the soil complex in such a model would present some unsolvable problems.

2.   THE MODEL

2.1   Water Availability When Raining

The amount of water arriving at any one height in a tree is a function of the proportion of the rainfall arriving directly, plus drips (if any) arriving from higher levels in the tree. This wetting-up cycle (see Fig. 1) thus reduces to the geometry of the tree, i.e. the relative proportion of branches, twigs, and leaves (if present) at each height zone, cutting off direct rain, and also storing a fixed minimum quantity of water per unit area of bark. A simple approach to this problem has been made by measuring the diffuse light intensity at the sample zones on each of the 25 trees sampled [2], as a percentage of full daylight, for both winter and summer. The light gradient is then equated directly to the proportion of rain arriving directly at each height, and equally, the proportion of the total 'drips' that actually fall on each height from above when the higher zones are saturated. The water storage of the bark was simply measured in the laboratory by using constant diameter discs of bark, again sampled from the six height of those trees sampled, which were soaked and weighed. Thus over simulated time, the bottom zone of bark receives a fixed proportion of the total rainfall during a time interval, the proportion being directly related to the percentage of full daylight at the bottom zone. As the top zone becomes saturated, 'drip' starts and the same fixed proportion of the total drips is added into the bark storage of the lowest zone. Equally, proportions of

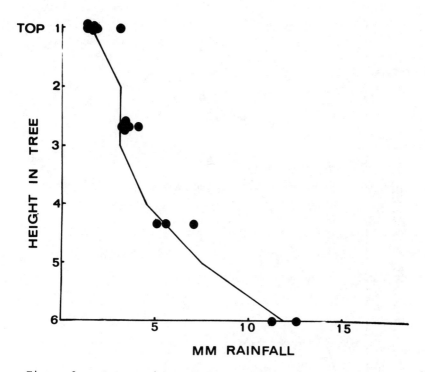

Figure 2a. A comparison of the observed and simulated rainfall amounts required to produce .02 mls cm^2 on the bark surface at each height in the tree

the drip from the top zone are added into the unit area at the other sample heights of the model tree. The final stage is when all the upper zones are dripping on to the lowest zone which is thus receiving a proportion of direct rain, plus the same proportion of the total amounts dripping from the other zones. For summer conditions the canopy is treated as fully expanded on the 1st May up to the end of October and a second light zonation used to compute the wetting-up cycle for the

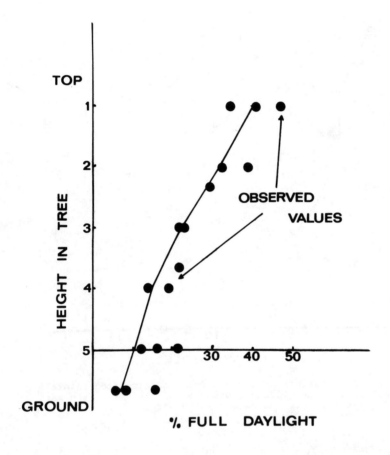

Figure 2b. A comparison of the observed summer light profiles
            for three oak trees with the interference profile
            from the simulation model of Figure 2a

summer months. This simple model of wetting-up has been tested
against observed data, monitored from a number of storms by
electrical conductivity probes mounted in sample trees at four
different heights. The results (Fig. 2a,2b) show very good agreement between the simulated and observed wetting rates for summer
conditions. There is little difference between the summer and
winter rates, due simply to the constant total capacity of the
canopy, which when saturated, and with rain still falling,
merely releases a quantity of 'drips' equal to the incoming
rain. Thus although the proportion of direct rain arriving at
any one point is much less in the summer, this is fully compen-
sated for by an increased quantity of drips arriving from the
canopy after it has reached saturation, and the overall 'wetting
rate' remains similar to the winter rate.

2.2   Radiation Levels at the Different Sample Heights and Evaporation

The relative light intensity required for the calculation of
photosynthesis levels (see above) as a first approximation is
taken directly from the light profile used to simulate the
wetting rates in the model tree. The percentage drop in diffuse
radiation inside the canopy is used to apportion the total incoming radiation [2]. However, the rate of evaporation has
also to be calculated at the different sample heights, from
parameters that are normally recorded at meteorological stations throughout the British Isles.

It is this aspect of the model that has in fact been the most
difficult to achieve and is also the part most open to criticism.
Very little information is available on the rates of evaporation
inside a forest canopy although the evaporation from open water
surfaces and grass cover is well understood ([14], [16], etc.).
The turbulence effects inside a canopy are completely unknown.
Denmead [6] has suggested that the evaporation rates at different heights in a canopy are proportional to the mass of
foliage at those heights. The wetting-up model (see above) has
demonstrated that the light intensity gradient gives a good
measure of foliage distribution with height, and accordingly
this has been used to scale evaporation rates in the way Denmead
has suggested. Input radiation and back radiation are calculated from time of year, time of day, cloud cover, temperature
and latitude of the site, and the resultant net radiation above

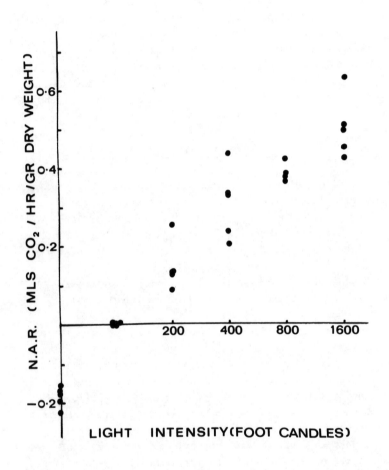

Figure 3.   The observed net assimilation rate of saturated
Parmelia caperata thalli at six light intensities
and 20°C, from five replicate experiments

the canopy is used to calculate total evaporation. The total
evaporation is then partitioned to give an evaporation gradient
in the canopy.

It is obvious that a steep evaporation gradient exists in
most tree canopies, the crown of a tree being completely dry
long before the lower levels. The slope of this evaporation
gradient we have taken as a straight function of foliage mass.
It may be necessary to adjust the present model in this respect
when a clearer understanding of canopy evaporation is gained.

2.3 The Physiology of Parmelia caperata

Stalfelt [19] gives Q_{10} for photosynthesis ranging from 2.5 at
$0^{\circ}C$ up to 0.5 at $25^{\circ}C$. These have been used in the model. The
effect of light intensity on photosynthesis rate has been determined experimentally for saturated Parmelia caperata, under
constant temperature conditions and at different light levels
(Fig. 3), using an Infra Red Gas Analyser to monitor the CO_2
level in the experimental 'leaf chamber'. Smith [18] suggests
that the respiration rate of Peltigera in March, April, May,
October, and November is considerably greater than at other
times of year, and similarly Harris [10] has found increased
rates in Parmelia for the same periods. This seasonal effect
is included in the model.

The most difficult experimental problem is the well-known
relationship between metabolic activity and water content of
the thallus. This has been clearly demonstrated by Ried [17],
Stalfelt [20], Butin [5], and others, for a variety of species,
but there is considerable discrepancy between the optimum levels
given by different workers for the same and different species.
Equally some of the experimental techniques used are open to
criticism. On the basis that the optimum level of photosynthesis might be closely controlled by percentage saturation of
the thallus for any one species, this was experimentally determined with cosiderable accuracy. Harris [10] has developed a
technique for simultaneously measuring both the CO_2 changes in
dry air circulated over a wet illuminated lichen thallus, and
also the weight of the lichen. This utilizes the hygrometer
principle; the experimental material is mounted on a glass float
which can freely rise into the experimental air chamber as water
is lost by evaporation. The CO_2 concentration in the chamber
is monitored by an Infra Red Gas Analyser, the bottom of the

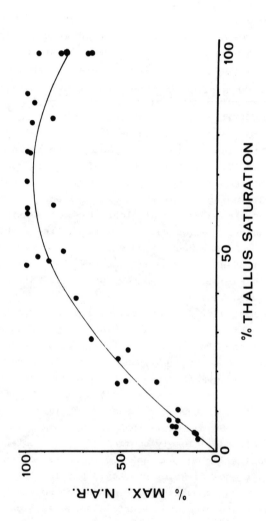

Figure 4.   The interaction between thallus water content and net assimilation rate, from five replicate experiments

experimental chamber being below water level and the surface of
the water sealed with liquid paraffin to prevent any CO_2 trans-
ference from the water. The movement of the float is measured
by a travelling microscope thus allowing the weight at any point
to be known accurately and expressed as a percentage saturation
by comparison with the final dry weight of the thallus. The
results for replicate experiments are given in Figure 4.
together with a fitted curve (fitted by the standard I.B.M.
POLYFT routine). This general equation of net assimilation
against water content is subsequently used for generating the
absolute level of photosynthesis in the model by correcting the
net assimilation curve for the dark respiration rate, at a
particular thallus water content. The interaction between dark
respiration and percentage saturation of the thallus (Fig. 5) was
investigated in a similar manner by simply removing the light
source and covering the experimental chamber with a black cloth.

Throughout these experiments light and temperature were held
constant and although the effect of temperature change on net
assimilation was predictable with the known Q_{10} of 2.0 for
respiration, there seemed little likelihood of a third order
interaction of net assimilation rate/water content/light intensity, other than the normal relationship existing between light
intensity and rate of photosynthesis. However it became evident
that in fact the respiration rate in the light, assumed to be
simply equal to the dark rate, was in fact directly proportional
to light intensity. This phenomenon will be dealt with in detail in the second paper of this series but briefly, the normal
respiration rate of a lichen in the dark is severely substrate
limited.

A lichen consists of two components, an alga and a fungus.
The fungal partner obtains its carbohydrate supplies directly
from its algal partner and thus in a high light intensity
(2000 foot candles) its respiration rate is optimal. As the
light intensity decreases the photosynthesis rate of the algal
partner declines and with it the supply of carbohydrates to the
fungal partner. (This result has been confirmed experimentally
and will be reviewed in the second paper of this series.) Thus
a further correction to the absolute photosynthesis rate is
necessary, to correct for this increased respiration rate in
the light.

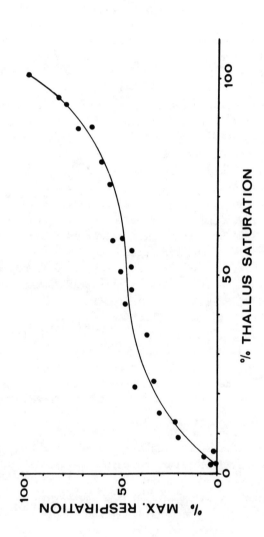

Figure 5.   The interaction between thallus water content and respiration rate
in the dark, from five replicate experiments

3. THE COMPUTER PROGRAMMING

A large number of computer languages are available at present
to handle a variety of simulation problems. Several simulation
languages are specifically designed to handle events, facilities,
and queuing problems but were eliminated by their lack of function routines. DYNAMO has served as a standard simulation
language in the past [7] but has the major disadvantage that
it is not compiled by all computer installations. FORTRAN IV
although lacking timing routines has a very extensive range of
function routines and is available as a language on most larger
machines. Accordingly all the programming utilized this language, the computing being carried out on an IBM 7094. The
meteorological data was obtained as standard hourly records on
punched cards from the Meteorological Office at Bracknell and
a tape prepared for two years data. Each data record included
the year, month, day, hour, cloud cover, relative humidity,
air temperature, wind speed, and rain fall. Each hourly record
was broken down into four fifteen-minute periods, and water
availability, net radiation, and net assimilation rate calculated
for six heights in the tree. Time increments were achieved by
simple nested loop counts with standard IF and GO TO statements.
The output is in the form of weekly net assimilation totals at
six heights expresses as mgm's carbon fixed per gm dry weight
of thallus, plus grand totals at six heights of the two years
simulated time.

4. RESULTS AND DISCUSSION

The results of simulated net assimilation rate expressed as
mg's of carbon fixed, per gm dry weight of thallus are given
in Figure 6A for a simulated time period of two years.
This result shows an optimum at height 3-4 which when directly
compared with the observed distribution of Parmelia caperata
(Fig. 7), with a mean maximum percentage cover in zone 4-5,
suggests that the model in its present state is fairly satisfactory. However a direct comparison of the observed and
simulated data is not strictly valid since the lichen thallus
itself will interact with the system. Thus as increments of
carbon are converted to mm's of thallus growth, the amount of

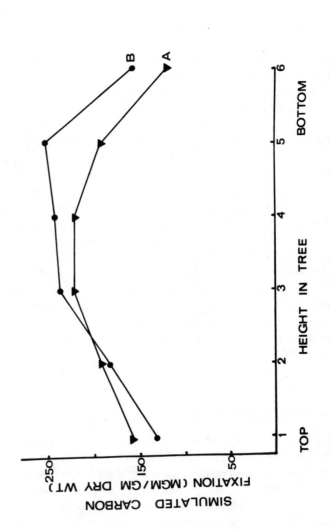

Figure 6   A.   Simulated mgrs. of Carbon fixed per gram dry weight of thallus at six
                  heights in the model tree, using meteorological data for Plymouth
            B.   Simulated carbon fixation under an increased evaporation gradient

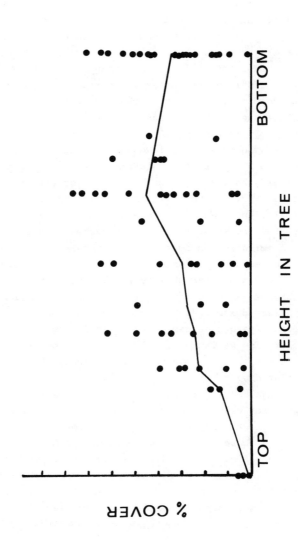

Figure 7. The percentage cover of Parmelia caperata on the 25 sampled oak trees, at different heights

water held at each height zone in the tree is altered slightly.
This in turn alters the wetting-up rates of the model tree,
and of course the relative increments of carbon per unit of
thallus. Although the simulated pattern of net assimilation
strongly suggests the present observed cover distribution of
Parmelia caperata, the system is a dynamic one which will follow
a compound interest law rather than a straight arithmetic sequence, and it now requires an upgrading of the existing model
to a dynamic form. This at the moment is being attempted.

The evaporation gradient used in the model was arrived at
simply by partitioning the total evaporation in the same proportions as the mass of foliage at each height as Denmead [6] has
suggested. The observed distribution of Parmelia caperata in
the field shows a considerable degree of variation from tree
to tree (Fig. 7) and it is interesting to compare the simulated
assimilation rates with those obtained using a steeper evaporation gradient (Fig. 6B). This produces a shift of the optimum
net assimilation to zone 5, and strongly emphasizes the importance of accurate information on this aspect of the model, and
points to water availability as the dominant factor in the
ecology of Parmelia caperata. A close examination of the individual trees sampled reveals a considerable variation in
'exposure' and tree geometry and it is suggested that the
similar wide variation of lichen cover from tree to tree, reflects the range of evaporation gradients which is closely
correlated with exposure.

We have been aware that the approach to the problem of
evaporation used in the current model only represents one of
the approximations available at the moment, and clearly it is
necessary to re-examine this part of the model in greater detail
before any final conclusions can be drawn. The results from
this initial model are sufficiently encouraging however, to
necessitate an upgrading to a 'dynamic' version, and at the
same time some sophistication of environmental measurements
and reprogramming of the relevant subroutines can be conveniently done.

In addition to the approximation used in the simulation of
evaporation in the model, the problem of direct sunlight and
sun-flecks has been initially ignored, and light values have
been calculated at each height from the total incoming diffuse

light only. Again upgrading of the existing program is possible,
but since individual tree geometry controls the pattern of
direct sunlight as well as evaporation gradient, a general model
is essential in the first instance. It is possible that the
wide variation of percentage cover of Parmelia caperata on the
sample trees is in part, also a product of sun-fleck distribution as well as variation of evaporation gradient. Clearly an
explanation of the variation of lichen cover from tree to tree
will involve careful analysis of direct sunlight throughout the
year as well as evaporation rates. The former could be achieved
by use of hemispherical photography at different heights in the
tree at different times of the year (see [1]).

The wetting-up cycle and the thallus physiology have turned
out to be more straightforward than was originally thought.
The effect of the level of saturation of the thallus, on net
assimilation, initially suggested the dominant role of water
availability in the ecology of the species and it is significant
that the effect of altering the evaporation gradient in the
model (cf Fig. 6A & B) has confirmed this.

The primary model has served two purposes. It has continually directed research into specific areas and at the same
time thrown up new concepts fundamental to the understanding of
both the ecology and physiology of lichens. This 'feedback'
from the model will be discussed in more detail in the second
paper of this series, but as the simulation study developed
environmental variables, previously overlooked, were examined
in detail as relevant parts of the system. The interaction in
a physical sense, of the lichen thallus with the remainder of
the system was not previously appreciated. It is now of obvious
importance and probably plays a dominant role in any concept of
plant succession in a corticolous habitat, gradually leading to
more mesic conditions and the successful establishment of other
migrules. This in turn leads to the possibility of competition
between lichen species; the evidence at the moment suggests
that the early stages of succession leading to the appearance
of Parmelia caperata represents an open community structure
with little competition. Equally the interaction of the lichen
thallus with the physical system and the necessity of a dynamic
approach, raises the importance of the early colonization of
Parmelia caperata. The physiology of the asexual propagules
of Parmelia caperata may be the key to the whole zonation.

Both competition, and colonization of migrules necessitate an experimental approach in the first instance, to provide the numerical data on which to base a model. Kershaw and Millbank [13] have demonstrated that lichens can be successfully grown under simulated natural conditions in the laboratory, and provisional growth experiments have shown sufficient promise to enable colonization studies to be made at least.

The second function of the model has been to confirm the dominant role of water availability in any understanding of the ecology of Parmelia caperata. Previously light has been suggested as the major environmental variable and those species at the base of the trunk thought to be adapted to low light intensities. This is not the case.

This primary model, with its simplifications has served as an informative and powerful research method.

ACKNOWLEDGMENTS

It gives us great pleasure to thank Professor A. J. Rutter and Dr. A. Goldsworthy of this department respectively, for their helpful discussions on the evaporation and net assimilation sections of the model. This paper covers some of the material included in a Ph.D. thesis and one of us (G. P. Harris) wishes to acknowledge the financial support received from N.E.R.C.

REFERENCES

[1]   Anderson, M. 1964. Studies of the woodland light climate.
      I. The photographic computation of light conditions.
      J. Ecol. 52:27-41.

[2]   ————. 1964. Studies of the woodland light climate.
      II. Seasonal variation in the light climate. J. Ecol.
      52:643-663.

[3]   Barkmann, J. J. 1958. Phytosociology and ecology of crytogamic epiphytes. Assen. Netherlands.

[4]   Brodo, I. M. 1961. Transplant experiments with corticolous lichens using a new technique. Ecology. 42:838-841.

[5]   Butin, H. 1954. Physiologisch-okologische. Untersuchungen uber den Wasserhaushalt und die Photosynthese bei Flechten. Biol. Zbl. 73:459-502.

[6]   Denmead, O. T. 1964.   Evaporation sources and apparent
      diffusivities in a forest canopy.   J. Appl. Meteor.
      3:383-389.

[7]   De Witt, C. T., and Brower, R. 1968.   Uber ein Dynamisches
      Modall des Vegativen Wachstums von Pflanzenbestanden.
      Zeitsch. Ang. Bot. 42:1-12.

[8]   Duncan, W. G., Loomis, R. S., Williams, W. A., and Hanau,
      R. 1967.   A model for simulating photosynthesis in plant
      communities.   Hilgardia. 38:181-205.

[9]   Hale, M. E. 1952.   Vertical distribution of cryptogams in
      a virgin forest of Wisconsin.   Ecology. 33:398.

[10] Harris, G. P. 1969.   A study of the ecology of corticolous
      lichens.   Ph.D. Thesis.   London.

[11] Holling, C. S. 1966.   The Strategy of Building Models of
      complex Ecological Systems.   Systems Analysis in Ecology.
      Acad. Press.   195-214.

[12] Kershaw, K. A. 1964.   Preliminary observations on the dis-
      tribution and ecology of epiphytic lichens in Wales.
      Lichenologist. 2:263-276.

[13] ───────, and Millbank, J. W. 1969.   A controlled
      environment lichen growth chamber.   Lichenologist.
      4:83-87.

[14] Monteith, J. L. 1965.   Evaporation and environment.
      Symp. Soc. Exp. Biol. 19:205-234.

[15] Paulik, G. J. and Greenough, J. W. 1966.   Management
      analysis for a Salmon Resource System.   Systems Analysis
      in Ecology.   Acad. Press.   215-252.

[16] Penman, H. L. and Schofield, R. K. 1951.   Some physical
      aspects of assimilation and transpiration.   Symp. Soc.
      Exp. Biol. 5:115-129.

[17] Ried, A. 1960.   Thallusbau und Assimilationshaushalt von
      Lamb-und Krustenflechten.   Biol. Zbl. 79:129-151.

[18] Smith, D. C. 1961.   The physiology of Pettigera polydactyla
      (Neck.) Hoffm.   Lichenologist 1:209-226.

[19] Stalfelt, M. G. 1939.   Der Gasaustausch der Flechten.
      Planta. 29:11-31.

[20] ───────. 1939.   Vom System der Wasserversorgurnfabhangiger
      Stoffwechselcharaktere.   Bot. Notiser. 176-192.

[21] Waggoner, P. E. and Reifsnyder, W. E. 1967.   Simulation of
      the temperature, humidity and evaporation profiles in a
      leaf canopy.   Journ. Appl. Met. 7:400-409.

RECORD OF PREPLANNED AND SPONTANEOUS DISCUSSIONS

R.A. PARKER (Washington State University)

This paper, the first of two to be presented by these authors,
reflects many of the current problems associated with ecosystem
simulation. Although a relatively simple system has been selected,
it is obvious that the complexities involved in the interaction
between the physical and biological components are immense. Attention is forceably directed to the need for adequate physiological
and field observations. Furthermore, it is particularly important
that observations on functional characteristics in the laboratory
be made under conditions as close as possible to the ones prevail-
ing in the field. In fact, where experimental techniques allow,
better values may be obtained in the field. The use of this
simulation study to other workers in the field would be enhanced
if the exact equations utilized were included.

The authors note that they have used FORTRAN IV to program their
system rather than using one of the many existing simulation languages, primarily because of its cosmopolitan nature and function
subroutine capability. Although most of the simulation languages
are easy to use there is often inadequate flexibility available to
handle specifics that arise in framing a particular ecological
system. Too often the user of a simulation language is required
to make further assumptions beyond those necessary for system formulation in order that his work be compatible with the computational
techniques available. Merits of any particular language certainly
vary with the problem attacked so that one must keep an open mind
to insure proper selection.

I feel that the most significant conclusion in this paper is
recognition of the need for a dynamic model. Clearly, the simplest
approach is to assume that the biological component in question is
influenced by certain physical features of the system but that it
in turn does not affect those features. This, of course, is often
not the case and points to the need for examining the simultaneous
interaction that occurs among many components. The role of model
variation (i.e., modification) and its potential effect on simulation results has been emphasized. On the other hand, it should be
noted that one can often obtain results of apparently different
functional form (as well as magnitude) by altering some of the
system parameters (e.g., rate constants). Unfortunately it may be
difficult to predict the effects in advance, and a systematic

examination of model behavior over a suitable range of values of system parameters is required.

SIMULATION STUDIES AND ECOLOGY: USE OF THE MODEL

K. A. KERSHAW
G. P. HARRIS*
Imperial College
London S. W. 7

SUMMARY

The use of simulation as an ecological research tool is discussed in relation to two fundamental aspects of the physiology and ecology of <u>Parmelia</u> <u>caperata</u>. The existence of the light dependence of the respiration of the lichen, and the wide range of physiological rariability were both infered originally from simulation studies and subsequently experimentally confirmed.

*Now at McMaster University, Hamilton, Ontario

1. INTRODUCTION

In the first paper of this series it has been shown that the vertical zonation of _Parmelia caperata_ can be partly explained in terms of the light and water availability at different heights on the trunk, and the innate physiology of lichen thallus. However in addition to the final simulation result, the construction of the model has forcibly drawn attention to a number of factors and interactions which would have certainly been otherwise completely overlooked. There has been a constant feedback from the incomplete model which has led certainly to the realization of several fundamental aspects of lichen physiology and ecology.

2. LIGHT RESPIRATION RATE IN PARMELIA CAPERATA

Considerable effort has been devoted to measuring respiration in higher plants in the light and apart from the occurrence of photorespiration in some species [4], it is often necessary to assume that the light respiration is active at more or less the same rate in the dark. The same assumption was made in the development of the net assimilation subroutine used in the model described previously. Equally, having demonstrated the independent effects of light on photosynthetic rates, and percentage saturation of the thallus on net assimilation, the possibility of an interaction between light and water on net assimilation was not seriously considered. The net assimilation subroutine was written, utilizing the observed dark respiration rate against water content of the thallus (Fig. 1a) which was added to the observed net assimilation/water content results (Fig. 1b) to give the 'observed' photosynthesis/water content relationship. Similarly the net assimilation/light observed result was corrected for dark respiration rate (Fig. 2). The simulated net assimilation for any temperature, light, and thallus water content was then simply calculated from the series of fitted curves.

The resultant subroutine was tested against known conditions with a complete lack of agreement between the observed and predicted net assimilation rates (Fig. 3).

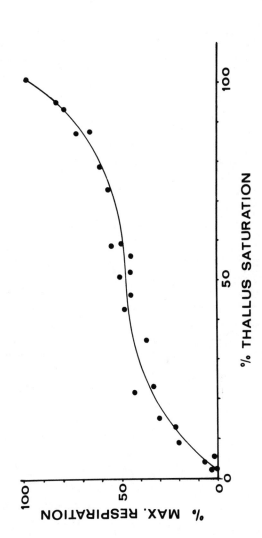

Figure 1a.   The observed interaction between thallus water content and respiration rate

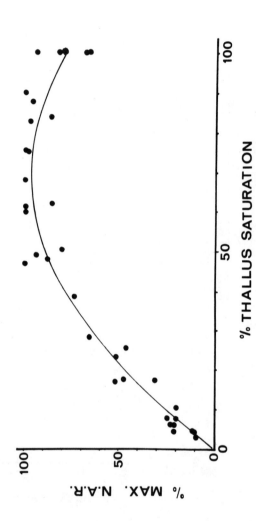

Figure 1b.   The observed interaction between net assimilation rate and water
content of the thallus

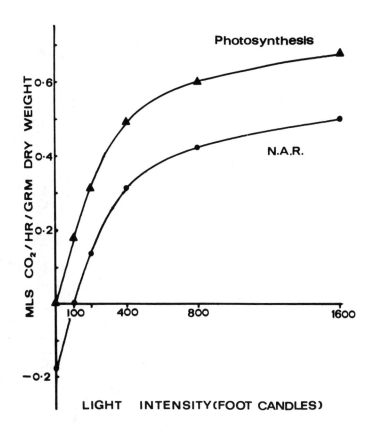

Figure 2.   The net assimilation rate at six light intensities and
corrected 'absolute' photosynthesis rate

Thus at 800 feet candles and $20^{\circ}C$ the optimum net assimilation rate is observed to be at about 70% saturation of the
thallus.   The predicted result shows a steady increase up to
100% thallus saturation.   Outside these limits the predicted
results for low light intensity and high temperature shows an
optimum at 40% thallus saturation, falling to a negative value
at 100% saturation.   Conversely at high light and low temperature there is a very steep rise in net assimilation to 100%
thallus saturation.   Clearly this merely reflects the expected
balance of CO_2 fixation in the light, with temperature controlled
respiration.   However, the observed net assimilation at 1600

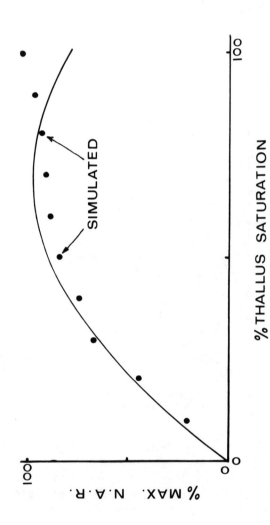

Figure 3.   The simulated and observed interaction between net assimilation rate
and thallus water content, at 800 f.c., and 20°C

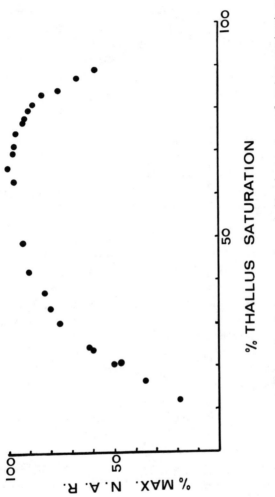

Figure 4.   The interaction between net assimilation rate and water content of
the thallus at 1600 f.c., and 10°C

foot candles and $10^{\circ}C$ (Fig. 4) gives a very similar result to that obtained previously and suggests that the shape of the net assimilation curve is more or less constant over a wide range of temperature and light conditions. In other words the respiration is linked directly to the rate of carbon fixation.

Smith [3] has demonstrated the rapid movement of carbohydrates fixed by the algal component in Peltigera polydactyla, to the fungal component, where it eventually appears as the sugar alcohol mannitol. On the basis of this evidence it appears then that the discrepancy between the observed and predicted net assimilation results could be a function of the amount of substrate made available by the algal component, controlling the respiration rate of the fungal symbiont. This would lead to a permanent substrate limitation in the dark respiration rates. The other possible explanation would be a high level of photorespiration in the algal component. This latter possibility was tested experimentally and gave a negative result.

The effect of substrate availability on net assimilation rate was tested by exposing saturated lichen thallus to four levels of illumination (plus a dark treatment): The CO_2 concentration in the experimental loop was monitored with an Infra Red Gas Analyzer, and when the net assimilation rate had reached a steady value, 5 ml of 1% (w/v) mannitol was injected on to the lichen material. The results are shown diagrammatically in Figure 5. The addition of mannitol in the dark boosts the respiration rate to a considerable extent, this effect being steadily reduced in the light to an almost imperceptible amount at 1600 foot candles, when presumably the abundant supply of carbohydrates from the algal component at this light intensity virtually removes the substrate limitation on the fungal respiration rate. (Fig. 6)

The resultant modification of the net assimilation subroutine by introducing a substrate limiting equation gives an extremely good fit of the observed with the predicted results (Fig. 7). In retrospect it is by no means surprising that the metabolism of each component in the lichen thallus is very closely linked to the other. It is equally true however that this marked interaction between the respiration and photosynthetic rates in Parmelia caperata would have escaped attention in the current

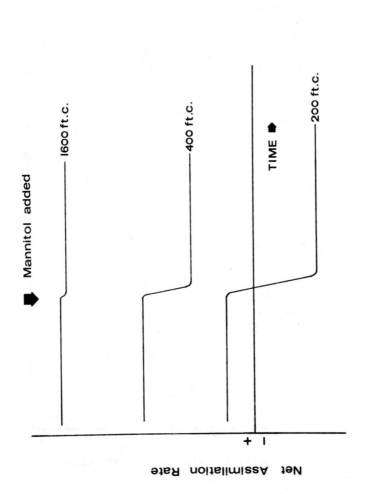

Figure 5.   The effect of mannitol on the net assimilation rate at 20°C, as measured by an Infra Red Gas Analyzer.

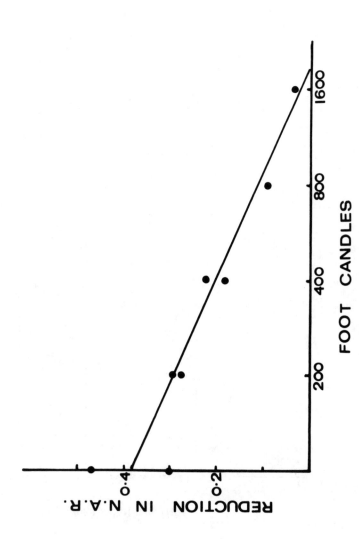

Figure 6.   The reduction in net assimilation rate of Parmelia caperata produced
by the addition of mannitol at five light intensities

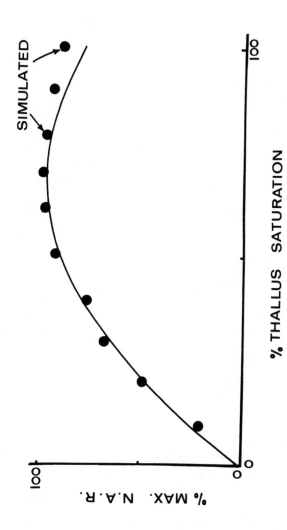

Figure 7. The simulated and observed interaction between net assimilation rate and thallus water content at 800 f.c., and 20°C

ecological investigation, without the simulation study defining the marked discrepancy between the predicted and observed net assimilation.

3. THE DISTRIBUTION OF PARMELIA CAPERATA IN S. ENGLAND

The present model has been used to test the hypothesis that light and water control the vertical distribution of Parmelia caperata on oak in South West England. The results have been encouraging, and it is interesting to speculate on the validity of a simulation using meteorological data from the S. E. of England, to predict the potential growth of the lichen in the southeast. Parmelia caperata is completely absent from areas close to urbanization or industrial development reflecting the susceptibility of most lichens to air pollution. However it occurs frequently in rural areas of Kent and Surrey, but it is only markedly frequent on the bases of old trees in open hedgerows or very open woodland. This contrasts with the much wider vertical distribution in the South West and its abundance in closed canopy woodland.

The predicted zonation from the model using meteorological data from the Boscombe Down meteorological station which is fairly typical of much of the London area, shows an extremely reduced carbon fixstion compared with the results using Plymouth meteorological data (Fig. 8). The vertical distribution is also somewhat different and very much at variance with the observed occurence of Parmelia caperata would be of extremely low percentage cover. One is forced immediately to the conclusion that the model is incomplete and cannot be extrapolated outside the original defined system.

However the model basically consists of the inter-relations between net radiation, and water availability at different heights, with the physiology of the lichen. Extrapolating from the Plymouth model would certainly not alter the physical laws governing net radiation, and water availability but the physiology of the lichen could possibly be adapted to the less mesic conditions in the southeast. This latter possibility has been investigated by examining the optimal net assimilation percentage thallus saturation relationship in a series of replicates collected from different parts of Southern England. The results for Wiltshire Surrey and Norfolk compared with those for Devon,

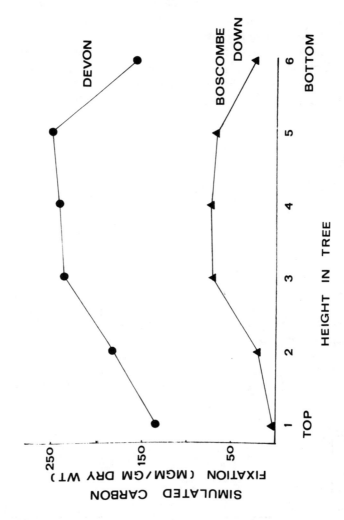

Figure 8.   Simulated carbon fixation using meteorological data for Plymouth compared with Boscombe Down

are given below (Fig. 9a, b, c & d) and show striking differences
in the position of their net assimilation optima. A comparison
of the rainfall distribution in Southern England (Fig. 10) shows
an obvious correlation of rainfall with the level of thallus
saturation at which the net assimilation rate reaches an optimum. Clearly Parmelia caperata as a species either consists
of a number of physiological strains each of which fills a
suitable ecological niche, or becomes fully adapted physiologi-
cally to less mesic habitats than its norm.

It should be emphasized that this physiological variation in

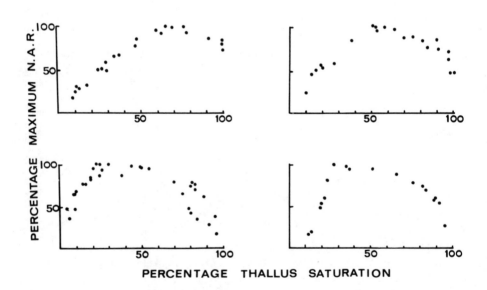

Figure 9.   The interaction between net assimilation rate and
             thallus water content for Parmelia caperata collected
             from a) Devon, b) Wiltshire, c) Surrey, d) Norfolk

OVER 40 INS. RAIN

—— **30 INS. ISOHYET**

Figure 10.   Rainfall distribution in southern England and optimal
water contents for maximum net assimilation for
Parmelia caperata

Parmelia caperata related only to net assimilation rate as
monitored under constant light and temperature conditions.
Since this work was completed we have found a similar variation
in the response of respiration rate to water content of the
thallus, in different replicate gatherings of material. Before
the model can be used to simulate growth under different levels
of rainfall in different parts of the country, more information
is required on possible physiological variations of the complete
metabolism of the lichen. This should obviously include responses to both temperature and substrate limitation in the

dark and at low light intensities, as well as the Q_{10} of photosynthesis.

4. DISCUSSION

Both the metabolic variation present in Parmelia caperata and the relationship between substrate availability and respiration rate were completely unexpected complications to what was initially envisaged as a simple model.

The importance of fully testing the model at all stages of its development cannot be emphasized enough. Accurate observational data is obviously required to test the model, and any departure of the model from the observed data should be vigorously examined as a potential incomplete section of the total model. An understanding of the light dependence of the respiration rate was only gained in this way, coupled with a refusal to insert a 'cooks factor' to make the simulated results fit the observed.

As has been discussed in the previous paper of this series the present model represents the first step in what will eventually be a much more complex and complete simulation of the system. Through the model and its development a much fuller understanding of the ecology of lichens has already been gained. The physical act of converting observational data into a series of FORTRAN IV statements forces a different mental approach to the original sampling methods and more especially gives a completely different evaluation to environmental variables. This in turn has resulted in a feedback of information from a model which is still in its developmental stage.

The analogy of a series of building blocks being fitted together to give an increasingly more complete picture is valid. Implicit in this is the concept of simplicity leading to final complexity. This is comparable with the approach of Greig-Smith ([1],[2] etc.) to pattern analysis as an ecological tool. In visually homogeneous vegetation any detected pattern would be controlled by the interactions of a minimum number of environmental factors, which might possibly be soluble. Pattern at the community level is clearly controlled by a large number of variables amongst which at the present level of our ecological knowledge, are a number of unknowns. A similar approach

is inherent in simulation studies. We have chosen a simple
plant group and although growing in a specialized habitat,
represent early colonizers in a situation where competition is
not of importance. The system is also 'simple' in that the host
of environmental variables always present in the soil and inter-
acting with plant roots are not applicable. Nutrition can be
conveniently ignored at least initially, and from the remaining
variables a simple model can be built with some hope of a
positive outcome.

Already some further blocks are partially complete and will
be fitted into the model, thus including in the system the
other two most abundant lichen species characteristic of oak
in S. W. England.

REFERENCES

[1] Greig-Smith, P. 1952. The use of random and contiguous
     quadrats in the study of the structure of plant com-
     munities. Ann. Bot. Lond. N.S. 16:293-316.

[2] _____. 1957. Quantitative Plant Ecology. Butter-
     worths, London.

[3] Smith, D. C. 1963. Studies in the physiology of lichens.
     4. Carbohydrates in Peltigera polydactyla and the
     utilization of absorbed glucose. New. Phytol. 62:
     205-216.

[4] Tregunna, E. B., Krotkov, G., and Nelson, C. D. 1964.
     Further evidence on the effects of light on respiration
     during photosynthesis. Canad. J. Bot. 42:989-997.

RECORD OF PREPLANNED AND SPONTANEOUS DISCUSSIONS

R. A. PARKER (Washington State University)

As I mentioned after the first paper in this series, I believe that
the importance of good estimates of system parameters applicable to
the particular natural ecological system under consideration cannot
be overemphasized. For example, are the Q_{10} values obtained under
constant temperature and light conditions really close enough to
those exhibited by the same organism in nature? Further insight
into questions of this kind could be gained in the laboratory if
one also experimented under simulated diurnal cycles of temperature
and light. I certainly do not mean to underestimate the necessity

of looking at physiological parameters under fixed conditions, but
I do think that one might be able to explain some additional discrepancies between laboratory and field observations if a dual
approach were adopted.

In this paper the authors have modified their original model by
incorporating an effect of substrate availability on net assimilation rate. Although they claim "complete lack of agreement" in
Figure 3 without this addition and "an extremely good fit" in Figure 7 after the addition, it appears to me that there may be
something more fundamental at stake. Needless to say the actual
deviations observed in Figure 7 are less than those observed in
Figure 3. Equally apparent, however, is that the simulated form
of the response is the same, and that this form is not reflected
in the real data. One might ask whether the difference is due to
an inadequacy of the observational methods or in the model. It
might be that a more refined observational technique would, in fact,
confirm the form of the model results. On the other hand, if this
is not the case, the authors would be forced to look more carefully
at the model as it now exists. Significant, I believe, is their
assessment of the observed data in other parts of England relative
to simulated results since it has pointed toward modifications required for more general applicability. Additional laboratory and
field investigations also have been suggested to provide a better
understanding of the physiological parameters involved.

As a whole, I believe that this series of papers has brought
out very well the tremendous difficulties encountered in simulating
any ecological system, even a "simple" one. The quality of simulated results can only be measured by comparison with actual
observations on all important variables preconditioned by proper
formulation of the model (i.e., based on known biological and
physiochemical responses). I do not believe that success can ever
really be claimed until short-term regulation of a natural system
is possible by altering system components in a manner pretested by
a study of simulated results.

P. GREIG-SMITH (University College of North Wales, Bangor, U.K.)

These comments bear on the papers in this session generally, rather
than specifically on the last one.

The analyses discussed all represent ecologically relatively
simple situations. With animals, at least at the higher trophic
levels, the nature of interactions is fairly clear e.g., one

predator attacks a limited number of known prey species. Even
herbivores, at least the large ones, eat certain plant species and
not others. Plant examples of such analyses generally concern
ecosystems in which one species is predominant. Moreover, the
species are generally ones for which obvious descriptive parameters
of individuals are available e.g., forest trees. There also seems
to be an assumption in analyses that the ecosystem discussed can
be regarded in practice as environmentally uniform. This may perhaps apply to animals, with their power of movement, but it does
not apply generally to plants. In spite of the ecologically simple
nature of the systems discussed, any workable models of them become,
as has been demonstrated, very complex. Consider the plant portion
of a more typical, complex ecosystem. The competition for re-
sources--water, light, nutrients--is very much more complex; we
scarcely begin to know, for example, anything about the mechanism
by which one individual takes up more of a nutrient than another
individual growing alongside it. Moreover, many plant species
present no satisfactory descriptive parameters. The question I
would ask of systems ecologists is "How practical is the systems
approach in this situation?"

I think I know the answer that will be made. "It is not practical in terms of individual species, but that does not matter. We
can predict in terms of energy flow, of nutrient cycles, etc."
This may serve in terms of applied work of resource management,
but it will not help very much towards an understanding of the
mechanisms of ecosystems, one of the principal objectives of ecology. It may not even serve the first purpose. At such a level
of generalization many different combinations of species may give
the same result in terms of energy flow or nutrient cycles, but
the actual specific composition may be important in relation to
amenity value or to utilization.

D. W. GOODALL (Utah State University)

Greig-Smith's comments that models of ecosystems are directed only
to separating trophic levels are unjustified. Ecosystem models are
under development in which each important species is treated separately, and I agree that the principle of niche separation makes it
unlikely that a satisfactory model will result if a measure of
specific distinction is not included. One-species models such as
that described in this paper will provide the necessary building
blocks for more comprehensive ecosystem models.

Admittedly models for entire ecosystems will necessarily be of a complexity which is daunting at the present stage. But techniques and equipment for simulation work are still developing very rapidly, and there seems no reason to doubt that they will keep up with the development of understanding and the accumulation of the data they will require. Model building is a slow process, requiring constant interaction with the real world, and this puts a premium on an early start.

SYSTEMS OF EQUATIONS FOR PREDICTING FOREST GROWTH AND YIELD

G. M. FURNIVAL
R. W. WILSON, JR.
School of Forestry
Yale University
New Haven, Connecticut

U. S. D. A., Forest Service
Northeastern Forest Experiment Station

SUMMARY

Mensurational studies of growth and yield often employ a system
or a set of equations to describe stand development. In the
past, the individual equations in the system have been fitted
one at a time by the method of least squares, but this method
of estimation is not entirely satisfactory. It is often necessary to treat the same variable as dependent in one equation
and independent in another. In addition, the model may specify
that certain coefficients in one equation are functionally
related to coefficients appearing in another equation, yet
it may be impossible to utilize this information when the equations are fitted separately. Furthermore, no cognizance is
taken of the fact that the residuals from the several equations
are almost certain to be correlated. Consequently, estimates
of the variances of predicted growth and yield are suspect and,
in fact, are generally not computed.

The purpose of this paper is to outline a method of treating
a growth and yield model as a system of equations with simultaneous estimation of all the coefficients in the system. The
techniques to be described are well-known and widely used in
Econometrics, but have never, so far as we know, been applied
to a mensurational problem.

1.   ESTIMATION IN SIMULTANEOUS SYSTEMS

The theory of simultaneous systems of linear euqations is
thoroughly covered in the econometric literature ([1], [2]).
We will attempt to develop some of the more important concepts
in an intuitive fashion by means of some simple examples. Our
treatment will be completely lacking in rigor and formality.

As our first example, let us consider the problems involved
in estimating the coefficients of the two equation system

$$E(Y_1 \mid X_1, X_2) = \beta_{11} X_1 + \beta_{12} X_2 \tag{1}$$

$$E(Y_2 \mid X_1, X_2) = \beta_{21} E(Y_1 \mid X_1, X_2) + \beta_{22} X_2 \tag{2}$$

where the Y's are "endogenous" variables determined within the
system and the X's are "exogenous" variables determined outside
the system. There appears to be no barrier to estimating the
coefficients of (1) by least squares. The difficulty arises in
(2) where the expected value of Y_1 appears as an "independent"
variable. A straightforward approach would be to replace the
expected values of Y_2 with observed values and apply ordinary
least squares. However, there is at least an intuitive objection to this procedure. The observed values will in general
differ from the expected values; hence, an independent variable
subject to error will be brought into the system.

Equation (1) suggests a way out of our difficulty. Given
least squares estimates of β_{11} and β_{12}, we might estimate the
expected values of Y_1 from the corresponding values of the X's
with the hope that the calculated values would be close approximations to the expected values. We would then substitute
calculated values for the expected values of Y_2 and apply a
second stage of least squares to estimate the coefficients of (2).

Alternatively and equivalently, we might substitute the right
hand side of (1) for $E(Y_1 \mid X_1, X_2)$ in (2) and rewrite our system
as

$$E(Y_1 \mid X_1, X_2) = \alpha_{11} X_1 + \alpha_{12} X_2 \tag{3}$$

$$E(Y_2 \mid X_1, X_2) = \alpha_{21} X_1 + \alpha_{22} X_2 \tag{4}$$

where

$$\alpha_{11} = \beta_{11} \tag{5}$$

$$\alpha_{12} = \beta_{12} \tag{6}$$

$$\alpha_{21} = \beta_{21} \, \alpha_{11} \tag{7}$$

$$\alpha_{22} = \beta_{21} \, \alpha_{12} + \beta_{22} \tag{8}$$

Equations (3) and (4) collectively are called the "reduced form" of our system and the α_{ij} may be estimated by least squares. The coefficients of (1) and (2) which make up the "structural form" in which we are interested can then be estimated from (5) through (8).

The two methods of estimation which we have just described are called, respectively, two-stage least squares and indirect least squares. For our particular structural form, the two methods give identical results and are as efficient as any known alternative. The estimates of the coefficients are biased but consistent.

Unfortunately, for other structural forms, estimation may be more difficult and in some cases impossible. If we replace (2) by

$$E(Y_2 \mid X_1, X_2) = \beta_{21} \, E(Y_1 \mid X_1, X_2) + \beta_{22} X_1 + \beta_{23} X_2 \tag{9}$$

two-stage least squares is no longer possible. The calculated values of Y_1 are, of course, a linear function of X_1 and X_2 and therefore the matrix of sums of squares and products generated in the second stage will be singular. Similarly, we reach an impasse if we attempt to estimate the structural coefficients of (9) from the coefficients of the equations in the reduced form. Equations (7) and (8) are replaced by

$$\alpha_{21} = \beta_{21} \, \alpha_{11} + \beta_{22} \tag{10}$$

$$\alpha_{22} = \beta_{21} \, \alpha_{12} + \beta_{23} \tag{11}$$

and we have two equations in three unknowns.

On the other hand, if (2) is replaced by

$$E(Y_2 \mid X_1, X_2) = \beta_{21} \, E(Y_1 \mid X_1, X_2) \tag{12}$$

still another kind of problem arises when we attempt to estimate from the reduced form. Equations (7) and (8) now become

$$\alpha_{21} = \beta_{21} \, \alpha_{11} \tag{13}$$

$$\alpha_{22} = \beta_{21} \, \alpha_{12} \tag{14}$$

and clearly two estimates of β_{21} are possible--one from (13) and another from (14). The root of the difficulty here is that acceptance of (12) as a valid structural equation imposes a constraint

$$\alpha_{21} \, / \, \alpha_{22} = \alpha_{11} \, / \, \alpha_{12} \tag{15}$$

on the coefficients of the system and this constraint is ignored in estimating by indirect least squares.

There is no obvious barrier to the application of two-stage least squares to (1) and (12); a unique estimate of β_{21} can clearly be obtained. However, the constraint given by (15) is applied only in the second stage of the estimation procedure. Hence, we might suspect--and it turns out to be true--that there is some loss of efficiency.

We have in (1), (9) and (12) examples of just-identified, under-identified and over-identified structural equations. When a system contains only just-identified equations, two-stage and indirect least squares give identical results and either may be employed. If a system contains an under-identified equation, consistent estimates of the coefficients of that equation are not obtainable. When an over-identified equation occurs in the system, two-stage least squares gives consistent estimates but more efficient methods are available and will be discussed later.

We clearly need some criteria for determining if an equation is under, over, or just-identified. A simple rule which is not infallible can be stated as follows:

(i) If the number of unknown coefficients in an equation is greater than the number of exogenous variables in the system, the equation is under-identified.

(ii) If the converse is true, the equation is over-identified.

(iii) If the number of unknown coefficients is equal to the number of exogenous variables, the equation is just-identified. There are two exogenous variables in our system. Thus, equation (2) with two coefficients to be estimated is just-identified. Equation (9) with three coefficients is under-identified and equation (12) with one coefficient is over-identified.

The preferred method of estimation for over-identified systems is the least generalized residual variance (LGRV). We rewrite (1) and (12) in terms of the observed values of the Y's

$$Y_1 - \beta_{11} X_1 - \beta_{12} X_2 = e_1 \qquad (16)$$

$$Y_2 - \beta_{21} Y_1 = e_2 \qquad (17)$$

where the e_i are random disturbances. We then attempt to
minimize the determinant of the variance-covariance matrix of
the disturbances. Unfortunately, the computations are quite
complex, involving the solution of a set of non-linear equations. Therefore, an alternative method of estimation, called
three-stage least squares, is often employed. This method of
estimation takes advantage of the fact that the non-linear
terms in the LGRV equations are the sample variances and covariances of the disturbances. We simply substitute the two-
stage least squares estimates of these variances and covariances
in the LGRV equations. The result is a linear system that can
be easily solved.

2. DATA

The data used were obtained from measurements made on white
pine stands in a growth study conducted cooperatively by the
U. S. Forest Service and the agricultural experiment stations
of Maine, New Hampshire and Massachusetts.

One of the hypotheses to be tested in the study concerned
the relative influence of stand age and basal area on growth.
For this reason, a serious effort was made to obtain a balanced
distribution of plots with respect to stand age and basal area
per acre (Fig. 1). In addition, an attempt was made to obtain
a uniform distribution over classes of soil, parent material
and soil drainage.

Nominal plot size was 1/10-acre but some plots 1/5- and
1/20-acre in size were also used. The data consist of observations from 175 plots at the beginning and the end of a three
year period on the following set of variables:

(i) Main stand age (at breast height).

(ii) Average total height (H) of dominants and codominant
trees in the main stand.

(iii) Total number of trees (N) per acre in the main stand.

(iv) Average diameter breast high (D) corresponding to the
average basal area per tree in the main stand.

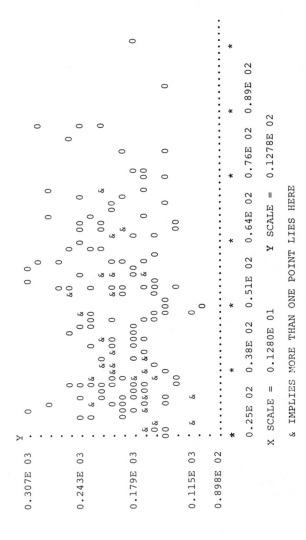

Figure 1. Distribution of field plots by basal area per acre (Y) and stand age at breast height (X).

(v)   Basal area (B) per acre in the main stand.

(vi)   Total cubic foot volume (v) in stems of the main stand.

3. THE YIELD MODEL

The system of equations employed for our yield model is

$$\log H = \beta_{11} + \beta_{12} \log A + e_1 \qquad (18)$$

$$\log N = \beta_{21} + \beta_{22} \log D + e_2 \qquad (19)$$

$$\log B = \beta_{31} + \beta_{32} \log H + e_3 \qquad (20)$$

$$\log F = \beta_{41} + \beta_{42} \log D + e_4 \qquad (21)$$

$$\log B = \log K + \log N + 2 \log D \qquad (22)$$

$$\log V = \log F + \log H + \log B \qquad (23)$$

where

V = volume in cubic feet

H = average stand height in feet

A = average stand age at breast height

B = basal area in square feet per acre

D = the diameter of the tree of average basal area

N = number of trees per acre

F = cylindrical form factor

K = $\pi/576$

The form of equation (18), our height or site index equation, was chosen by trial and error after a number of the more commonly employed S-shaped curves failed to fit our data. These failures were probably due to the fact that our age was measured from breast height rather than from seed.

All of the remaining equations, with the exception of (21), come directly from the literature of forest mensuration. Equation (19) is the Reineke stand density equation [3] and (20) is a stocking equation employed by Coile and Schumacher [4]. Equation (22) is an identity relating basal area to number of trees and their average diameter, and equation (23) is another identity which defines F as the cylindrical form factor for the stand.

Equation (21) allows the form factor to be a function of average stand diameter and, when employed in conjunction with equation (23), gives an expression for volume in terms of number of trees, average diameter and stand height. We would have

preferred to use a single logarithmic stand volume equation
(Spurr [5]) of the form

$$\log V = \beta_{41} + \beta_{42} \log N + \beta_{43} \log B + \beta_{44} \log H + e_4$$

but identification becomes a problem. There are four unknown
coefficients and only two exogenous variables, age and the
"dummy" or constant, in the system.

The density and stocking equations form a truly simultaneous
subset within our yield model. This fact becomes apparent if
we substitute for log B from (22) and rewrite (19) and (20) as

$$\log N - \beta_{22} \log D = \beta_{21} + e_2 \qquad (24)$$

$$\log N + 2 \log D = \beta_{31} + \beta_{32} \log H - \log K + e_3 \qquad (25)$$

It seems biologically reasonable that average diameter and
number of trees should be jointly and simultaneously determined.
However, so far as we know this concept has not been employed
in other yield models.

The coefficients of the model were estimated by direct least
squares and by two-stage least squares from the data taken at
the second measurement of the plots. The results are given in
Table 1.

4. THE GROWTH MODEL

Our growth model was derived from the yield model by differencing
equations (18) through (20). We assume that the disturbance in
the height equation, (18), is largely a function of site and
hence should remain essentially constant over time. Therefore,
we write

$$\log H_1 - \log H_0 = \beta_{12} (\log A_1 - \log A_0) + r_1$$

However, the disturbances in equations (19) and (20) are,
respectively, the logarithms of density and stocking when density is defined as the ratio of actual to expected number of
trees and stocking is defined as the ratio of actual to expected
basal area.

Schumacher [4] has proposed that the relationship of future
stocking, S_1, to past stocking, S_0, is given by

$$\log S_1 = G \log S_0 + r$$

Table 1. Estimates of the coefficients and their standard
errors obtained from the yield and the growth models
by the several methods of estimation.

ESTIMATE	INTERCEPT			SLOPE		
	Coef.	Std.Err.	100CV	Coef.	Std.Err.	100CV
	b_{11}			b_{12}		
** Yield Model						
DLS	0.5590	0.05499	9.84	0.7266	0.03227	4.44
2SLS	0.5590	0.05467	9.78	0.7266	0.03208	4.42
** Growth Model						
DLS				1.015	0.03137	3.09
2SLS				1.015	0.03128	3.08
3SLS				1.009	0.03114	3.09
	b_{21}			b_{22}		
** Yield Model						
DLS	4.529	0.05912	1.31	-1.985	0.05726	2.88
2SLS	4.349	0.08208	1.89	-1.809	0.07979	4.41
** Growth Model						
DLS	4.128	0.2061	4.99	-1.463	0.1472	10.06
2SLS	4.271	0.3905	9.14	-1.566	0.2803	17.90
3SLS	4.297	0.1699	3.95	-1.585	0.1208	7.63
	b_{31}			b_{32}		
** Yield Model						
DLS	1.821	0.1183	6.50	0.2564	0.06586	25.68
2SLS	1.932	0.1366	7.07	0.1945	0.07603	39.09
** Growth Model						
DLS	2.086	0.2465	11.82	0.2300	0.1074	46.69
2SLS	1.429	0.7432	51.99	0.5174	0.3252	62.85
3SLS	1.506	0.3176	21.10	0.4849	0.1386	28.59
	b_{41}			b_{42}		
** Yield Model						
DLS	-0.1194	0.01562	13.07	-0.2484	0.01513	6.09
2SLS	-0.1958	0.02255	11.52	-0.1738	0.02192	12.61
** Growth Model						
DLS	-0.1194	0.01562	13.07	-0.2484	0.01513	6.09
2SLS	-0.1194	0.01555	13.03	-0.2484	0.01507	6.06
3SLS	-0.1192	0.01482	12.43	-0.2486	0.01430	5.75

where G is $A_0 A_1^{-1}$. We assume that essentially the same relationship holds for density and write

$$\log N_1 - \beta_{21} - \beta_{22} \log D_1 = G(\log N_0 - \beta_{21} - \beta_{22} \log D_0) + r_2$$

$$\log B_1 - \beta_{31} - \beta_{32} \log H_1 = G(\log B_0 - \beta_{31} - \beta_{32} \log H_0) + r_3$$

or

$$(\log N_1 - G \log N_0) = \beta_{21} (1-G) + \beta_{22} (\log D_1 - G \log D_0) + r_2$$

$$(\log B_1 - G \log B_0) = \beta_{31} (1-G) + \beta_{32} (\log H_1 - G \log H_0) + r_3$$

In many yield studies volume is measured on only one occasion.
We have therefore retained in the growth model the form factor
equation

$$\log F = \beta_{41} + \beta_{42} \log D_1 + r_4$$

used in the yield model. The identities given by (22) and (23)
are, of course, also utilized.

Since lagged values of endogenous variables can be treated
as exogenous, the growth model is obviously over-identified.
Therefore, the coefficients were estimated by three-stage least
squares as well as by direct and two-stage least squares. The
results are included in Table 1.

5. RESULTS

Table 1 shows that the coefficients estimated from the yield
model are quite different from those obtained for the growth
model. We are almost certain that these discrepancies are due
to biases in the estimates of the coefficients of the yield
model caused by correlation between the residuals and the exogenous variable, stand age.

The residuals from the first three equations in the yield
model are, as we have stated before, closely associated with
site, density and stocking. It is quite likely that site is
negatively correlated with age. Such a correlation could be
brought on by the reversion of abandoned farmlands to forest
since the poorer lands are probably the first to be dropped
from cultivation. Another factor in areas with an active timber market is the tendency to cut stands shortly after they
reach merchantable size. Stands on the better sites will

reach a suitable size and be harvested at an early age; hence older stands will be associated with poorer sites.

Correlations between age, on the one hand, and stocking and density, on the other, are also likely in data gathered for yield-table construction. The tendency is to omit low-density, under-stocked stands in the relatively abundant younger age classes but to take almost anything in the older age classes which are rare and hard to find. In our data, in particular, we feel very sure that such a correlation exists. The plots were selected with the intention of obtaining a balanced distribution of basal area classes within age classes; therefore, basal area and stocking are probably too high for the younger ages and too low for the older stands.

We believe that the correlations just discussed are greatly reduced and are much less of a problem in the growth model. The residuals in this model involve changes in site, stocking and density over time plus, of course, other disturbances such as measurement errors. There is no obvious reason why measurement errors should be correlated with the exogenous variables and changes in site with time should be non-existent or at least very small. Greater changes in density and stocking are to be expected for young stands as compared to old, but most of this effect should be eliminated by the use of the Schumacher formula for change in stocking.

The estimates produced by the several methods of estimation are quite different for many of the coefficients. Of particular interest is the difference between the direct and two-stage least squares estimates of β_{22} for the yield model. The direct least squares estimate of -1.985 makes the simultaneous subset formed by equations (24) and (25) essentially colinear; hence, a solution for average diameter and number of trees is effectively impossible. The two-stage least squares estimate, on the other hand, is -1.809 and a solution can be obtained.

The standard errors for the three-stage least squares estimates of the coefficients in the growth model are surprisingly small. In most cases, the errors are not appreciably larger than those for the same coefficient in the yield model. We had anticipated that the short interval between remeasurements would make the com-parison much less favorable.

The estimated variances and covariances of the three-stage

Table 2.   Estimated variance-covariance matrix for the coefficients in the growth model obtained by the three-stage least squares method of estimation.

	b_{12}	b_{21}	b_{22}	b_{31}	b_{32}	b_{41}	b_{42}
b_{12}	0.000970	-0.000118	0.000264	-0.000010	-0.000133	0.000030	-0.000048
b_{21}	-0.000118	0.014605	-0.020313	0.015056	-0.034364	0.000003	-0.000002
b_{22}	0.000264	-0.020313	0.028883	-0.020929	0.048381	-0.000004	0.000002
b_{31}	-0.000010	0.015056	-0.020929	0.019213	-0.043876	0.000054	-0.000054
b_{32}	-0.000133	-0.034364	0.048381	-0.043876	0.100880	-0.000131	0.000138
b_{41}	0.000030	0.000003	-0.000004	0.000054	-0.000131	0.000205	-0.000210
b_{42}	-0.000048	-0.000002	0.000002	-0.000054	0.000138	-0.000210	0.000220

least squares coefficients for the growth model are given in Table 2. These statistics are important by-products of a systems method of estimation since they can be used to estimate the standard errors of predicted growth and yield. Table 3 gives some examples of calculated prediction errors for a stand with a site-index of 65. The estimates of basal area are least precise; the error here reaches 20 percent. However, the changes in height, number of trees, average diameter and volume are almost always predicted with an error of less than ten percent.

Table 3. Estimated standard errors of predicted change for stands with a site index of 65 and an age of 90 years at the end of the prediction period.

From Age	Yield Parameter				
	H	N	D	B	V
10	7.65	0.17	6.14	8.83	11.29
20	5.95	0.59	5.54	10.24	9.53
30	5.03	1.32	5.92	11.94	8.90
40	4.41	2.37	6.53	13.74	8.68
50	3.94	3.76	7.16	15.57	8.63
60	3.54	5.46	7.77	17.40	8.61
70	3.15	7.45	8.34	19.21	8.49
80	2.62	9.73	8.86	21.00	7.84

REFERENCES

[1] Goldberger, Arthur S. 1964. Econometric Theory. John Wiley and Sons, Inc., New York.

[2] Klein, L. R. 1962. An Introduction to Econometrics. Prentice-Hall, Englewood Cliffs, N. J.

[3] Reineke, L. H. 1933. Perfecting a stand density index for ever-aged stands. Journal of Agricultural Research, 46:627-638.

[4] Schumacher, F. X. and Coile, T. S. 1960. Growth and Yields of Natural Stands of the Southern Pines. T. S. Coile, Inc., Durham, N. C.

[5] Spurr, Steven H. 1952. Forest Inventory. Ronald Press, New York.

RECORD OF PREPLANNED AND SPONTANEOUS DISCUSSIONS

W. G. WARREN   (Forest Products Laboratory)

I have always deplored the piecemeal use of regression as commonly practiced by forest mensurationists and have spoken against it whenever the opportunity afforded. Therefore, I must thank the authors for their further drawing attention to this question. However, I am not convinced that the methods they have put forward in this paper really overcome the problem.

I must confess to a suspicion, bred largely of ignorance, of the techniques used by econometricians, so that although there may be formal justification for these arguments they are somewhat outside the realm of my experience.

It seems to me to be more natural to divide the variables into two sets--the dependent and the independent, the predicated and the predictor, the endogenous and the exogenous, or whatever one wants to call them--and to structure these in the conventional manner of a multivariate linear model. In this form one should be able to see clearly what is estimable and to construct the estimates and their precisions in a straightforward manner. If the purpose is the estimation of some of the various measures of growth or yield, for specified forest units or periods, I do not see that anything is gained by the "confounding" of predicated and predictor variables. In fact, this would seem to create the potential for misinterpretation or misevaluation.

In other words, the real solution to the problem is not to create it.

On the other hand, I can see a role for this sort of thing in the development of, say, simulation models for forest management decisions, but, in these circumstances, accurate values of the parameters and their standard errors are, to some extent, irrelevant. Essentially all that is needed are first-order approximations to the parameters plus a study of the robustness of the decision against changes in the values of the parameters or changes in the model.

Since I received the paper only yesterday, I have not had time to go through the examples in detail, but I did experience some difficulty in relating the examples to the theory. I feel that it is not presumptious of me to suggest that if I had difficulty so also will many forest mensurationists. Thus I would like to see the link spelled out more explicitly.

One final point, I do not think that a linear model is properly specified unless the variance-covariance matrix is also given. In the paper we are forced to assume that this is standard, but, in view of the various manipulations of the variables, it would be reassuring to have this laid out specifically.

To sum up, I must thank the authors for their attempt to draw attention to a situation about which I feel there has been a lot of naive thinking. If my criticism seems adverse it is not so much that I disagree with what they have done but that I have a strong personal preference for a different approach.

E. J. WILLIAMS (University of Melbourne, Australia)

If the object of the study is to predict height and other characteristics from measurements of properties of the site, the appropriate technique is the method of least squares, leading to estimated regression relations. Two-stage and three-stage least squares analyses, as used in econometrics, are appropriate for determining underlying functional relations. Such relations are not necessarily suitable for prediction.

The biologist is therefore fortunate in that simple least squares techniques are usually adequate for his purposes.

B. B. STOUT (Department of Hort.-Forestry, Rutgers University)

This paper contains an exciting idea. In a forest we have plants and animals which influence and are influenced by others. In addition, the exogenous edaphic and climatic variables are operating. Can we by examining the coefficients of both the endogenous and exogenous variables gain some idea of which variables might be amendable to manipulation for the benefit of man?

COMPARATIVE ANALYTICAL STUDIES
OF SITE FACTOR EQUATIONS

R. G. WRIGHT
G. M. VAN DYNE
College of Forestry and Natural Resources
Colorado State University
Fort Collins

SUMMARY

This paper examines the models used for predicting forest productivity as measured by site index or tree height. These models normally take the form of multiple regression equations that relate various attributes of the site to direct measurements of productivity. A portion of the existing published equations are examined and conclusions are drawn as to the accuracy and relevance of these equations. New equations are developed and tested on original data to provide better estimates of productivity. The use of linear programming in determining the optimum site is explored and an application of this method is described. The problem of accurately recording original data is examined and the use of forms of available data to provide assessments of site productivity is covered with a case example.

1.   INTRODUCTION

The problem of accurately estimating forest productivity as a
function of a complex of environmental conditions, commonly called
site factors, is an important aspect of forest management. Productivity estimates originate from a need to have some quantitative
index of site quality on lands where timber is absent or where
conversion from one tree species to another would be economically
profitable. Forest growth is the reflection of the response to a
combination of factors. Those site variables which can be defined
quantitatively and which operate independently from the existence
of trees appear to provide the best means of predicting the productivity of a site.

The most common method of evaluating the quality of a forest
site is determining the height growth of trees. The height of
free-growing trees of a given age is defined as site index, where
age usually is designated as 25, 50, or 100 years. In the eastern
and southern United States, the base age is usually 50 years and
the site index is the expected height growth at this age. For many
western tree species, 100 years is the index age. Site index is
determined by averaging the total height and age of the dominant
and codominant trees in a stand and using these measurements as
coordinates for determining site index from specially prepared sets
of curves. Such curves, based on height-against-age regressions
of the form:

$$Ht = a + b \frac{1}{A}$$

are available for most American species [2]. To be a valid
estimate of productivity, site index requires that the stands
sampled be even-aged and homogeneous in historical and genetic
characteristics. The extent to which these criteria are met
will greatly affect the accuracy with which site productivity
can be determined.

Site factor equations commonly use site index or height, or
some other measure of tree growth, as the dependent variable.
Independent variables include many edaphic, climatic, and
physiographic variables. Height, when used as a dependent
variable, is commonly transformed to the logarithm, and the

reciprocal of age is then one of the independent variables.
The use of this latter set is necessary when there are few
stands growing near the index age. However, the inclusion
of an age term is not desirable for it absorbs a large amount
of the variation about the dependent variable inflating the
accuracy of the equation. As noted by Hodgkins [22], this
often leads to an incorrect assessment of the role played by
the independent site variables in the equation which, compared
with the age term, may be of little value. It is, therefore,
preferable to use site index as the dependent variable whenever
possible.

Site factor equations will often include independent variables which are dependent on the existence of trees on the site,
e.g. basal area. While this and other similar terms provide
some measure of stand density, like age, they have similar
drawbacks in that they are often highly correlated with the
dependent variable. The inclusion of such terms also negates
the use of the site equation for predicting the productivity
of barren land. In the equations developed here, such terms
were omitted. A more complete discussion of the applicability
of site index as a measure of productivity is given by Spurr
[34].

1.1 Background

The data analyzed in this study were acquired directly or in-
directly from the investigators conducting the original study,
and most have been used in published works. We acknowledge the
many researchers, agencies, and institutions who provided
original plot data utilized in these studies. These data sets
provide a valuable resource for evaluation of the analytical
methodology utilized in an important phase of natural resource
management. Our criticisms of published results and methods
are offered in a constructive sense in the desire to promote
better analyses in resource management.

A complete summarization of these data is given by Van Dyne,
Wright, and Dollar [38]. Most of these data have been collected
for the purpose of quantifying the site factor relationship in
the past years, but many data have not been subjected to intensive analysis. It has only been comparatively recently that
adequate computation facilities have become widely available.

Even when such facilities have been available, the biologist has often lacked an adequate understanding of the computational processes and results. Thus, the interchange of ideas between the biometrician and the biologist was reduced. Too often the validity of many equations derived in this manner has been left unquestioned.

1.2 Objectives

The objectives of this paper are: (i) to evaluate the accuracy and techniques used in the derivation of the published equations based on examination of the original data; (ii) to develop more accurate models defining the site factor-productivity relationship; (iii) to evaluate the importance of different site factors over different geographical provinces; (iv) to combine the techniques of linear programming with multiple regression procedures to examine multiple species relations in evaluating site quality; and (v) to show how other sources of available data, such as soil surveys, can be used to provide indirect quantitative assessments of site productivity.

2.   METHODS

2.1 Regression studies

The model most frequently used to define the relationship of site factors to site productivity is the standard multiple regression equation:

$$Y_i = \sum_j b_j X_{ij} + e_i$$

where Y_i is commonly the mean value of tree height or of site index on plot i, b_j is the partial regression coefficient for the jth site variable, X_{ij} is the value of the jth site variable on the ith plot, e_i is a random error, and i=1,2,...,n plots and j=1,2,...,m site factors. This is the model used in this paper, and all of the analyses were performed on a general multiple regression and correlation program [36] using a CDC 6400 computer.

The examination of the plots of the data plays an important role in any regression analysis. However, the information content of a plot will vary with the plotting procedure used.

Historically, the most common method has been to plot the residuals of the regression equation against each independent variable contained in the equation. This is a valid method but has a tendency to mask any interactions that might exist between independent variables. We sought to overcome this deficiency by plotting each independent variable adjusted for the variation from their means of all of the independent variables except the one being plotted by the following formula:

$$Y_{ij}^{*} = Y_{ij} - \sum_{\substack{j=i \\ j \neq k}} (X_{ij} - \bar{X}_{j})\, b_{j}$$

where k is the independent variable being plotted.

Plot examination, when used in connection with the respective partial correlation coefficients, gives some hint of the action of the independent variables, the nature of the correlation with the dependent variable, and an idea of whether some transformation based on the shape of the scatter and prior knowledge of the variable will be applicable. It is through an iterative procedure, using plots and regression statistics at each step, that one hopes to build a good model while at the same time containing a selective number of easy-to-measure terms.

The criteria used for testing the accuracy of fit of an equation were the coefficient of determination values expressed as a percentage, $100R^{2}$, and the standard error of estimate (SEE). The equations were expressed in standard partial regression coefficients, which operate as if the data were normalized. These coefficients give a further measure of the importance of the respective independent variables based on the magnitude of the coefficients alone; and, due to their standardized variance, they provide an absolute unit for comparing the relative importance of a variable throughout a range of studies. All of the equations in this paper appear in this form. In equations where the dependent variable has been transformed logarithmically, the SEE of the equation appears in the logarithmic form. In least squares analysis, a straight anti-logarithmic transformation of this parameter is not a technically valid conversion (Furnival [16]). Therefore, the conversion method of Furnival, given below, was used in this paper:

$$\text{SEE} = (\text{antilog } \bar{y})\ (\text{SEE})\ (1/\log_{10} e).$$

2.2 Linear programming studies

Linear programming was used herein for the maximization of a linear objective function of m variables subject to linear constraints. The objective function is:

$$Q_{max} = \sum_i^m c_i X_i$$

A series of constraints, linear in the X_i's are imposed on the objective function:

$$\sum_{j=1}^m a_{ij} X_j = d_i \text{ for } i = 1,2,\ldots,n$$

$$\underset{\sim}{A} = \left\{ a_{ij} \right\} = \text{a matrix of constraint coefficients}$$

$$\underset{\sim}{X} = \text{a vector of } X_i\text{'s}$$

$$\underset{\sim}{d} = \text{a vector of constraint levels}$$

Additional constraints are that the $X_j \geq 0$ for $j=1,2,\ldots,m$. The problem was solved using an iterative procedure until a feasible and realistic solution was obtained, giving the X vector which will maximize the objective function.

The solution of the linear programming model to maximize the growth of a given species on a particular site requires means of determining what parameters best control the growth on that site. For this purpose, the multiple regression equations developed from Carmean's [6] data were used.

White oak, on the well-drained sites, was considered to have the most potential in terms of market value in this analysis. The equation that best predicted the growth of this species on these sites was used as the objective function. The constant term was dropped from the equation. Site index was the parameter being maximized. Multiple regression equations, i.e., site factor equations, for chestnut, black, and scarlet oak were utilized in the constraint inequalities.

More details of methods and results from these analytical studies are given by Wright [40].

3. COMPARATIVE ANALYSES OF SITE FACTOR EQUATIONS

This section is divided into three parts based on the form of
the analysis conducted in the original investigation: 3.1
standard multiple regression, 3.2 step-wise regression, and
3.3 all possible regressions. In each subsection, the meaningful aspects of each study are discussed and the comparison
between our analysis and that conducted in the literature is
presented in tabular form. A complete set of the equations
can be found in Wright [40].

3.1 Multiple regression

The studies using standard multiple regression techniques are
the most common form found in the literature and will be discussed first. The study conducted by Carmean [6] on four
species of upland oaks in Southeastern Ohio best illustrates
some components of a well-planned site factor study. The
plots for each species were stratified according to two drainage
patterns--reducing much of the between-plot variation. The
value of this can be seen in the lists of important variables
for the different sites (Table 1). While the mean site index
was naturally higher on the better, well-drained sites, the
R^2 values were better for the species growing on the imperfectly
drained sites. The latter effect is because all factors limiting growth were related to poor drainage. It is easier to
relate site index to factors that directly measure one limiting
factor than to a complex of factors which may or may not be
limiting for various periods.

Each of the eight data sets contained the same 26 variables,
greatly facilitating comparative analyses. Carmean only presented equations for black oak, both expressed in terms of
height. We developed eight equations, two for each species on
both sites (Table 1). Carmean used height as the dependent
variable and we used site index. Age as an independent variable
often accounted for 50% of the variation of height, thus our
prediction by site factors alone was lower.

Trimble and Weitzman [35], studying five species of upland
oaks in West Virginia, also stratified plots to reduce unwanted
variation. Their published equation was derived from data based
on 57 plots of a homogeneous soil type. We used the data from
all 70 plots seeking a more generalized expression. It took

Table 1. Comparative results of multiple regression analyses. The ordered three most important variables and the R^2 of each respective equation is given where available by investigator and species.

Investigator	Species	From literature Important variables	R^2	Our analysis Imp. variables	R^2	SEE
Carmean ([4], [6])	black oak (poorly drained)	age A depth % silt	74	A depth % gravel % silt	54	5.8
Carmean	black oak (well-drained)	age aspect	74	A depth ridge distance	39	8.3
Carmean	chestnut oak (poorly drained)	-	-	length slope % gravel % Clay C	48	6.4
Carmean	chestnut oak (well-drained)	-	-	A depth % slope aspect	39	8.4
Carmean	scarlet oak (poorly drained)	-	-	% gravel A % gravel B moisture equivalent	63	4.6
Carmean	scarlet oak (well-drained)	-	-	A depth moisture equivalent ridge distance	36	6.4

Table 1 (continued).

Author	Species					
Carmean	white oak (poorly drained)	—	—	% clay B % clay A length slope	51	6.1
Carmean	white oak (well-drained)	—	—	Depth A ridge distance % silt B	35	6.8
Doolittle [12]	scarlet oak	A depth slope position % sand A	86	A depth slope position	88	4.5
Trimble & Weitzman [35]	chestnut oak	aspect total soil depth ridge distance	67	total soil depth aspect ridge distance	68	7.1
Allen [1]	loblolly pine	% clay A A depth	11	% slope % clay B A depth	21	8.1
Allen	shortleaf pine	% clay B % sand A	04	% slope % clay B A depth	11	6.5

an extra term to account for the variation induced by the addition of 13 plots to give a similar result (Table 1). Agreement with the published equation was exact.

Doolittle [12] also studies oaks, but in North Carolina. He found no significant difference between the site indices of scarlet and black oak growing on the same plots and the data from both were combined. Our model of this data produced a good R^2 of 88 (Table 1). Duplication of the published equation gave an R^2 of 86. However, he reported a value of 96 for this equation. Summary statistics, i.e., means were not given in the literature.

Allen's [1] and Goggans and Schultz's [18] studies on the Coastal Plains of Alabama and Mississippi provide good examples of the influence of prior land use practices on prediction equations based on such areas. The effects of fire and logging are difficult to measure, but they have an important bearing on the present status of a stand. Both areas sampled had a past history of such insults. Allen neglected to consider this as evidenced by the low R^2 values of the equations derived from his data (Table 1). These equations also show a wide discrepancy between the published and duplicated results we obtained. Again no summary statistics were available.

Goggans and Schultz [18] examined the growth of trees on areas that were farmed or in pasture at one time. The old fields on which the plantations studied were established had received a wide range of fertilizer treatments and they hoped to correlate nutrient concentration with tree height. All species were analyzed separately with height the dependent variable. Since the loblolly and slash pine equations were identical and the height of long-leaf pine was estimated from age alone, we developed one model with the combined data (Table 2). This study was unusual for it considered trees from 5 to 16 years of age and, because of this, the age term over-shadowed all others. This effect and the fact that nutrient competition may not be noticed in young stands destroyed their nutrient-height correlation. Working from the more general base of the pooled data, we found some indication of a nutrient-height relationship.

The three studies conducted by Linnartz ([26], [27]) and the two by Zahner [41] in Louisiana on pines all proved difficult

Table 2. Comparative results of multiple regression analyses. The ordered three most important variables and the R2 of each respective equation is given where available by investigator and species.

Investigator	Species	From literature Important variables	R^2	Our analysis Imp. variables	R^2	SEE
Goggans & Schultz [18]	mixed pines	age A depth	78	age phosphorus potassium	82	6.0
Linnartz ([26], [27])	loblolly pine	% sand B % sand A depth least permeable layer	23	internal drainage pH depth least permeable layer	23	7.2
Linnartz	longleaf pine	surface drainage % sand % silt	—	same equation	27	5.8
Linnartz	slash pine	least permeable layer internal drainage % slope	—	same equation	42	4.8
Zahner [41]	loblolly pine	A depth slope position % silt	60*	surface drainage % sand % silt	20	6.6
Zahner	shortleaf pine	A depth % clay % silt	75*	slope position % silt surface drainage	26	5.6

Table 2 (continued).

Dickenson [11]	slash pine	age fine texture depth mottling	84	age % sand % clay	82	6.0
Carmean [3]	Douglas-fir basalts	—	—	elevation total soil depth % gravel	76	13.6
Carmean	Douglas-fir gravel	—	—	% gravel B elevation soil texture	75	17.1
Carmean	Douglas-fir sand	—	—	elevation moisture equivalent % gravel C_1	40	20.2
Carmean	Douglas-fir shales	—	—	total soil depth moisture equivalent % gravel C_2	58	8.0
Carmean	Douglas-fir terraces	—	—	% gravel B % gravel A_2 moisture equivalent	52	21.0
Myers & Van Duesen [32]	Ponderosa pine	total soil depth % to ridge % slope	65 & 26	consistency B aspect water retention	52	7.2

* From literature; could not be verified.

to work with due, apparently, to a lack of important variables.
We were able to accurately duplicate Linnartz's three equations
and these were equal to the best models developed here (Table 2).
Zahner reported favorable R^2 values for his two equations, but
these equations could not be verified due to inadequate data
received. Due to these inadequacies, our models could not contain certain important variables and were correspondingly lower
in predictability (Table 2).

One further study done on Southeastern pines was that conducted by Dickenson [11] in Florida. We verified the accuracy
of his equation. However, because this equation contained two
variables with only three non-zero data points for each, we
feel these were of little importance. In our model these variables were not included and a slightly lower R^2 resulted (Table 2).

Carmean's [3] work in Washington is the only study presented
in this paper for Douglas fir. Plots were divided among the
five major soil types of the area. No equations were derived
in the literature and the results in Table 2 are from our equations for each of the soil types. Contrary to the findings of
Carmean's study on upland oaks above, no relationship was
found between the mean site index and R^2 values for the respective soil series.

Myers and Van duesen [32] divided their Ponderosa pine data
among two physiographic types in their study conducted in
Wyoming. In duplicating the two equations they derived from
their data, we found very large discrepancies. On about half
of the plots, we found that the measurement of total soil depth
was lacking. When these data points were omitted, the R^2 increased from 13 to 64 and from 5 to 26 as compared to respective
reported values of 82 and 67. Our final model (Table 2) was
derived by combining the data from both areas.

3.2 Step-wise regression

Two studies covered in this paper used step-wise regression as
an exclusive procedure in their analysis. Both studies were
analyzed on an IBM 1620 computer at the University of Georgia.
The very extensive study by Ike and Huppuch [23] concerned six
hardwood and four pine species in Northeastern Georgia. The
published presentation of the quantitative results of this
study is excellent and we recommend it as a format for future
studies.

The published equations were those that gave the best prediction with a small number of easy-to-use terms as independent variables. More precise equations, including more terms, are contained in an unpublished report by Ike. We had access to both sets of equations, but as our analysis revealed many similarities between the two, only the published set will be included here. All equations in this report were expressed as the logarithm of height. The results of our analysis are presented in Table 3.

This table shows relatively high R^2 values for all equations indicating a strict adherence to accepted sampling methods. In all but one case, we were able to develop more precise models than those given in the literature. The table shows the close agreement with some of the results in the literature and the large divergencies with others. This last feature was an unexplanable anomaly for the peripheral statistics published, such as means of the variables all check with our findings. The complete summary of this study can be found in Table 4.

The three equations derived from the study by Kormanik [24] conducted in the Piedmont area of Virginia all proved to be very accurate and give further illustration of the problem encountered when age is used as an independent variable. In all three equations, the age variable was at least three times as important as any other variable in the equation. Our equations utilized site index and the predictability was lower than when height was the dependent variable (Table 4).

3.3 All possible regressions

Considerable use has been made of Grosenbaugh's [20] and Furnival's [17] computer programs of all possible regressions in soil analyses. Both Hebb [21] and McClurkin [29] used this approach singularly and Della-Bianca and Olsen [10] used it as one step in an extensive analysis. Many discrepancies were found in the results of these studies. This, we feel, is due in part to the difficulty in properly interpreting the output of such programs, especially Grosenbaugh's which is quite complex. Screening programs as these can be misleading and preferably should be an initial, and not an all-inclusive, step in developing site factor equations.

Hebb's [21] study was done on lowland hardwoods in Tennessee. Only R^2 values were given; no equations were presented in his

Table 3. Summary of comparative analysis of Ike and Huppuch's [23] study.

Species	Dependent variable used here	R^2 reported in literature	R^2 found here for same equation	R^2 of best equation developed here
pitch pine	site index	79	79	84
shortleaf pine	log height	65	63	71
eastern white pine	log height	68	66	76
Virginia pine	log height	65	36	56
black oak	site index	86	73	66
chestnut oak	site index	85	85	84
northern red oak	log height	90	84	90
scarlet oak	site index	78	50	82
white oak	log height	74	69	73
yellow-poplar	site index	71	55	64

Table 4. Comparative results of step-wise regression analyses. The ordered three most important variables and the R^2 of each respective equation is given where available by investigator and species.

Investigator	Species	From literature Variables	R^2	Our analysis Variables	R^2	SEE
Ike & Huppuch [23]	pitch pine	age, slope position, slope steepness	79	slope position, elevation, slope steepness	84	4.3
Ike & Huppuch	shortleaf pine	age, elevation, slope position	62	age, slope position, elevation	71	6.6
Ike & Huppuch	Eastern white pine	age, basal area, elevation	66	age, % silt, elevation	76	6.7
Ike & Huppuch	yellow poplar	elevation, age, slope position	69	age, slope position, elevation	73	9.4
Ike & Huppuch	black oak	% clay A, age, slope position	73	slope position, % clay A, soil series	66	5.8
Ike & Huppuch	chestnut oak	slope position, age, slope steepness	85	slope position, % silt A2, % silt B2	84	5.6

Source	Species					
Ike & Huppuch	Northern red oak	elevation age aspect	84	elevation age aspect	90	8.1
Ike & Huppuch	scarlet oak	age slope position aspect	50	slope position aspect elevation	82	6.3
Ike & Huppuch	white oak	elevation slope position age	69	age slope position elevation	73	9.4
Ike & Huppuch	yellow poplar	age slope position elevation	55	% slope A depth aspect	64	7.0
Kormanik [24]	loblolly pine	age depth B_2 % clay	61	wilting point B_2 depth B_2 available water	49	5.5
Kormanik	shortleaf	age drainage class depth B_2	78	section piedmont drainage class available water B_2	46	4.6
Kormanik	Virginia pine	age available water A_3B_1 available water A_2	72	available water B_1 available water B_2 plot location	32	4.6

Table 5. Comparative results of studies where all possible regression analyses were conducted. The ordered three most important variables and the R2 of each respective equation is given where available by investigator and species.

Investigator	Species	From literature Variables	R^2	Our analysis Variables	R^2	SEE
Hebb [21]	mixed oaks	—	—	aspect soil drainage slope length	41	12.0
McClurkin [29]	white oak	basal area slope position % clay A	54	% slope A depth depth mottling	50	6.0
Della-Bianca & Olsen [10]	shortleaf pine	age ridge distance % organic matter	75	age pH A_1 ridge distance	79	7.0
Della-Bianca & Olsen	black oak	age % slope organic matter	53	age A depth available H_2O B_2	57	8.6
Della-Bianca & Olsen	scarlet oak	age % slope ridge distance	08	age	67	7.4
Della-Bianca & Olsen	white oak	age organic matter A_1 ridge distance	11	age % sand A_2 % clay A_1	49	9.0

Della-Bianca & Olsen	yellow poplar	ridge distance age latitude	50	—	—	—
McComb & Einspar [30]	white oak	—	—	total soil depth % slope aspect	50	9.2
Miller [31]	loblolly pine	—	—	texture A texture B_2 A depth	29	10.5
Grigsby [19]	loblolly pine	—	—	plasticity % clay A A depth	78	4.1
Copeland ([7], [8])	Western white pine	—	—	A depth depth permeability water retention	77	10.0
Walker & Reed [39]	catalpa	—	—	A depth % slope pH	57	6.1

report. The R^2 values were higher by a factor of three than we
obtained using the same variables. The reason for this was
difficult to ascertain because no summary information was given
and the manner in which the results were presented was difficult
to interpret. All of his data were combined in our analysis.
The equation developed (Table 5) had an R^2 of 41 and was low
chiefly because of the type of variables measured.

McClurkin [29] derived two equations from data collected on
white oaks in Mississippi and Tennessee. R^2 values of 73 and
75 were reported for these equations. Both equations, however,
contained basal area and this variable was twice as important
as all others in both equations. In both, the logarithm of
site index was the dependent variable. Duplicating the published analysis was difficult. The coefficients and the SEE
were accurate, but his R^2 values were either calculated incorrectly or their derivation was incorrectly stated. The latter
case is more probable for the two values were similar to the
R values we obtained (Table 5). It is interesting to note that
in both equations, when tree growth was expressed as site
index, slightly better results were obtained.

Della-Bianca and Olsen [10] conducted an extensive analysis
of four hardwoods and shortleaf pine in the Piedmont of Virginia
and the Carolinas (Table 5). Considerable field work and statistical analyses were conducted to gather and analyze the data.
However, due to the methods they used in manipulating this
data, much of the work may not have produced many meaningful
results. In their method as outlined below the 63 soil-site
factors initially measured were reduced to 15 by plotting
scatter diagrams and single-variable regressions. While data
reduction was important, these techniques may eliminate variable regressions. While data reduction was important, these
techniques may eliminate variables whose importance was masked
by other site variables. In a screening program, equations
with height as the dependent variable were then developed from
these 15 variables. Step-wise techniques were then used to
increase the accuracy of the equations and the results were run
through a general multiple regression program to determine
exact equations. This procedure is complex, providing many
sources for error. It is further complicated by the original
inclusion of so many independent variables. One should not

Table 6.   Summary results from data of Della-Bianca and Olsen [10] showing comparative R^2 values for each of three equations by species.

Species	Deviation (%) of mean tree height found here compared to that reported	R^2 of equation reported in literature	R^2 of our duplicate equation	R^2 of our best equation
black oak	0	56	52	57
scarlet oak	0	65	08	67
shortleaf pine	6	76	75	79
white oak	3	56	11	49
yellow-poplar	0	57	56	64

endeavor to measure all possible site variables unless they can be properly tested. Sixty-three exceeds by a factor of two the number that many early regression programs for digital computers could accurately handle.

The discrepancies which appear between our analyses and those of Della-Bianca and Olson [10] seem to be due to the complexities of the investigator's analysis (Table 6). For all cases, we were able to develop a better predictive equation than that in the literature. This improvement was accomplished by using the screening program of Furnival [17] as an initial rather than intermediate step in the analysis. A general multiple regression was then used to further reduce the data and examine the plots.

Table 5 also contains the results of analysis of studies by McComb and Einspar [30] in Iowa, Grigsby [19] in Louisiana, Miller [31] in Mississippi, Copeland ([7], [8]) in Idaho, and Walker and Reed [39] in Oklahoma. For these studies, no equations or quantitative results were given in the literature for direct comparison to our results. The studies that have been detailed here are representative data sets taken from a more complete analysis by Wright [40].

3.4   Conclusions from comparative analyses

In order to examine the relative importance of the different site variables included in our equations, the data were segmented into three major geographic regions based on divisions in species type and available data. The categories and singular variables which were most important in each geographic region are given in Table 7. The rank was determined as the mean relative importance of a variable in all equations weighted by the number of terms in each respective equation. The variables we have shown to be important fit well with most accepted opinions on this subject. On the poorly drained soils of the Coastal Plains, soil productivity is inversely related to surface drainage. Those factors which control soil drainage-- maintaining water in the profile and yet providing adequate growing space for roots--proved to be most important. On the physiographically different areas of the Piedmont-Appalachian region, topographic factors determine, to a large degree, the quality of a site. Overall, the depth of the A horizon is probably the most singularly important variable for a wide range of species and geographic areas.

Table 7. Comparison of important site parameters for various
regions of the United States.

Location	Rank
Southeast (pine species; 19 equations)	
Coastal Plain	
texture	3.0
impermeable layers	3.0
Piedmont	
topography	2.9
available water	2.7
South-central (hardwoods; 28 equations)	
soil depth A	1.9
topography	3.1
Northwest (conifers; 12 equations)	
soil moisture retention	3.1
total soil depth	3.1

In the models reviewed in the literature as well as the
models developed in this study, the multiple correlations generally are low. One should work toward developing models by
understanding those factors which have caused the models to be
inadequate.

Various errors can cause low correlations in prediction
equations. Some of these errors are inherent and impossible
to correct. Genetic difference among trees of the same species
will decrease correlations. Stands often do not fit the ideal
conditions required for accurate site index determinations.
For example, some stands may have an unknown history of logging, fire, or cultivation. Frequently, errors result from
sample sizes that are too small. One should strive to sample
a wide range of commonly occuring values for all the variables
used.

Another factor causing low correlations involves cases where
important measurements are omitted. This may be due to an

oversight, difficulties in measuring the parameter, or the investigator simply starting out with a predetermined notion of what he thought would be important variables and measuring only those.

About 50% of the equations from the literature which we tested appeared to be calculated incorrectly. Most of the discrepancies resulted in an inflation of the existing R^2 values. The basis for some of the inconsistencies may be due to a misinterpretation on our part of the results reported in the literature. The data that we used may have been altered in some way so as not to conform with that upon which the reported analyses were conducted. Inaccurate computational procedures provide a large area for explaining many of the discrepancies.

Those equations produced in the literature which used standard multiple regression procedures were, on the average, more accurate than those produced by step-wise or all possible regression approaches. From the sample investigated here, it appears that the all possible regression approach produced the highest number of discrepancies probably due to its inherent complexity and unfamiliarity of the researchers with this form of analysis.

4.   LINEAR PROGRAMMING IN DETERMINING THE OPTIMUM SITE REQUIREMENTS

4.1   Multiple species evaluation

More than half of the nation's hardwood timber output is a product of the south-central hardwood forests. Unlike many southern pines, the approximately 40 commercial hardwood species are not often found growing in pure stands. The problem facing the forest manager is knowing what species to grow on a given type of land. There are two main approaches to the problem of determining the productivity of different species growing on the same land area. Both methods use the results of comparative site index equations to find those species best suited to the site concerned. These traditional quantitative techniques are generally imprecise and often quite subjective. It is intended that the method presented here will provide a means for more objectively solving the above problem.

Current economics usually dictates which species are selected for soil-site studies. Market values, especially for south-central hardwoods will probably not remain constant in rank or amount. We wanted to know what site parameters would produce maximum growth of a given species and still assure better than average yields of other species. Land evaluation could then be based on its ability to produce several species.

4.2 A case study with oaks

Carmean ([4], [6]) collected data on four hardwood species on each of two different sites--white oak, chestnut oak, black oak, and scarlet oak. We utilized these data to develop multiple regression equations to predict site index for each species from the same site variables. We incorporate these as an objective function, for white oak on well-drained sites, and used as constraints the other species, i.e., chestnut, black, and scarlet oak, in a linear programming formulation.

The objective function is:

$$SI_{white\ oak} = .02X_1 + 1.4X_2 + .17X_3 + .01X_4 + 1.8X_5 + .16X_6$$

where: X_1 = % gravel B X_4 = distance to ridge top x A horizon depth

X_2 = sine of the aspect X_5 = A horizon depth

X_3 = length of slope X_6 = % silt B

To maintain a realistic solution, the constraints imposed on each of the variables in the objective function were the maximum and minimum values occurring in the field data following the procedure of Van Dyne [37] (Table 8). The maximum and minimum values for the site index of each of the seven species x site combinations in the constraining equations are also given in Table 8.

The optimum solution is subject to the constraints that at least average or better-than-average growth can be achieved by each of the other species that may be considered for the site. These constraints are formulated from the multiple regression equations based on the respective data for each species. Each equation is set equal to the mean site index value less the constant term for the respective species.

Table 8. Ranges of site factors and site indexes of oaks as used in linear programming analyses.

Site factors		Site indexes	
Range	Name	Range	Species-location group
$2.2 < X_1 \leq 66.0$	percent gravel in B horizon	$43 < Y_2 \leq 97$	black oak-well drained
$-1.0 \leq X_2 \leq 1.0$	exposure	$51 \leq Y_3 \leq 77$	black oak-imperfectly drained
$2.2 \leq X_3 < 30.0$	slope length	$58 \leq Y_4 \leq 89$	scarlet oak-well drained
$34.0 \leq X_4 < 760.0$	distance to ridge x	$46 \leq Y_5 < 73$	scarlet oak-imperfectly drained
$1.6 < X_5 \leq 9.5$	A horizon depth	$44 \leq Y_6 < 80$	white oak-imperfectly drained
$18 \leq X_6 < 76$	percent silt in B Horizon	$37 < Y_7 \leq 89$	chestnut oak-well drained
		$44 < Y_8 \leq 72$	chestnut oak-imperfectly drained

Black oak (well-drained site)

$$0.14X_1 + .97X_2 + .02X_3 + .02X_4 + 1.4X_5 + 1.6X_6 = 18.04$$

Black oak (imperfectly drained site)

$$.102X_1 - .24X_2 + .82X_3 - .004X_4 + 2.9X_5 + .03X_6 = 16.26$$

Scarlet oak (well-drained site)

$$.542X_1 + 1.2X_2 - .23X_3 + .01X_4 + .95X_5 - .002X_6 = 5.50$$

Scarlet oak (imperfectly drained site)

$$.610X_1 + 2.0X_2 - .02X_3 + .02X_4 + .50X_5 + .22X_6 = 13.04$$

White oak (imperfectly drained site)

$$.02X_1 + .66X_2 + .37X_3 + .006X_4 - .10X_5 + .22X_6 = 12.87$$

Chestnut oak (well-drained site)

$$.080X_1 + .22X_2 - .10X_3 - .01X_4 + 2.7X_5 + .06X_6 = 12.82$$

Chestnut oak (imperfectly drained site)

$$.03X_1 + .25X_2 + .47X_3 + .005X_4 + 1.1X_5 + .04X_6 = 9.13$$

The linear programming solution for the site index of white oak on well-drained sites was 44 ft. When added to the intercept of 49 ft. obtained from the regression producing the objective equation, this gave an optimum site index of 93 ft. The solution vector elements, i.e., the values for each of the site parameters that produced the above figure were as follows. X_1 was 61%. The figure is slightly below the maximum alloted, indicating a high percentage of gravel which contributes to good drainage and is beneficial. X_2 was 1.0, i.e., the maximum. This refers to a northern exposure, the most favorable site. X_3 was 30 ft. Generally, length of slope is directly proportional to site quality, thus a maximum was chosen. X_4 was 695. Since this is an interaction term, the results are not clear, but there appears to be some value less than the maximum which produces the best site quality. X_5 was 9.5 in. It is only when soil depth is uniformly deep that further increments in depth become negligible. A depth of 9.5 in. does not approach this level and, thus, a maximum was chosen. X_6 was 70%. At the optimum solution, a value slightly less than the maximum for this variable was found. The effect of this variable above

a certain percentage is curvilinear. The percentage obtained
here appears to be the point at which further increases in the
silt content are detrimental to site quality.

The six parameters above, provide the necessary information
to evaluate the site potential for white oak. In this study,
the optimum site index obtained of 93 ft. exceeded by 3 ft. the
maximum encountered in the data and suggests that no plots in
the study area occurred on such optimum sites. All factors in
the objective function contributed positively toward the solution. Three of the six variables in the objective function
were solved at a maximum value and the remaining three were
very close to maximal values. It thus appears that the upper
bounds on the independent variables played as important a role
as did the several constraint equations in obtaining an optimum
solution. This model was formulated to permit quantitative
land evaluation for any of the four species. The results show
that this intent has been satisfied in this study. However,
the lack of weight placed on the constraint equations may indicate a situation in which the growth of white oak alone is a
good indicator of site productivity for the other species
concerned.

5. INDIRECT ASSESSMENTS OF SITE PRODUCTIVITY

Those methods that will allow indirect assessment of productivity via gross physical features will play an ever-increasing
role in future forest site studies. In this section we show
how present soil mapping or classification units can be used
as a source of information to predict relative productivity
quantitatively for a given area.

5.1 Productivity prediction from survey data

Useful sources of soil and land form information are the soil
maps and interpretive material for woodlands produced by the
Soil Conservation Service, USDA. Grouped under the title of
Soil Survey Interpretations for Woodland Conservation, such
reports are available for many areas in the United States.
The soil survey interpretations contain varied edaphic and
physiographic data and are valuable because potential productivity is measured by site index. Most of the soil survey data
does not, being qualitative in design, lend itself to regression

analysis. Most investigators, e.g., Carmean [5], Dean and Case
[9], and Loftin et al. [28], who have attempted to relate site
quality to different soils have met with little success. In
preliminary analyses conducted here, we also found little relationship between soil survey data and site index. The analysis
of 15 soil survey studies produced an average R^2 of 35 with a
range of 15 to 50. In many instances these low R^2 values can
be justified they are derived from broad areas and qualitative
data. However, such results do not lend themselves to quantitative designs. The application of the variables directly from
the soil survey studies, and the type of measurements included
as opposed to those found for example in Table 7 appear to be
the chief reason for these low values. Most measurements are
highly stratified in form, e.g., erosion and slope class and
plot position, or directly measured with very little variation
when considered in view of their magnitude, e.g., precipitation
and frost-free days measurements.

5.2 A case study with loblolly pine

In the approach we used, the soil series, location and site
index values were taken directly from the soil survey interpretation data. Definitive information for a given soil series
and location are available from the USDA in standard soil survey
reports. Analysis revealed that seven measurements taken from
these reports for a given soil series were sufficient to develop
accurate prediction equations. Not all seven were found to be
important in any given equation, but all were used in initial
equations. The coding routine used to quantify the variables
taken from the soil survey reports is given in Table 9.

A complete analysis was run on three sets of data--all taken
on loblolly pine, the source of which was Ritchie, et al. [33]
for Georgia and Ellerbe and Smith [13] for South Carolina. The
resulting prediction equations developed are given in Table 10.
The results indicate that accurate prediction equations can be
derived on this basis for the data from soil survey interpretation manuals and the quantitative descriptions of the different
soil series. In the three equations, the most important variables were the soil texture measurements, with those of the
surface layer being slightly more important than those of the
subsoil. These studies were done in the Piedmont region of the
states listed above and these findings are consistent with the

Table 9.   Methods of coding variables used in developing site
            index equations from soil survey data.

Variable number		
1.	Drainage class	Divided into six classes with a 1 designating very poor drainage-- a 6 representing excessive drainage.
2. & 3.	Soil texture Surface and subsurface respectively	Both were arranged into 12 groups representing the 12 divisions of the "textural triangle." The numbers are arranged in increasing order of fine particles. A 1 indicates a low proportion, e.g. sand, a 12 a high proportion, e.g. clay.
4.	Depth of surface soil	Taken directly from soil-survey reports.
5.	Depth of subsurface soil.	Also taken directly from reports.
6.	pH	Average of range included in report.
7.	Available moisture content	The number of inches of water in an inch of soil; also directly from the report.

Table 10.   Site index (SI) equations developed from soil survey
             interpretation data on loblolly pine. All equations
             are in standard partial regression coefficients. The
             mean site index, coefficient of determination (R^2) in
             percent and the standard error of the estimate (SEE)
             is given.

Location	Equation	Mean SI	R^2	SEE
Georgia	$SI = 327 + .71X_4 + .07X_5 - .64X_2 + 1.0X_3$	90	86	7.0
South Carolina Piedmont	$SI = .28X_4 + .05 \ 1/X_5 + .62X_2 + .25X_3 - .13X_1 - .11X_6$	87	54	7.0
South Carolina Sandhills	$SI = 175 + 1.3X_2 - .27X_3 + .05X_5 - .59X_6 - .59X_6$	89	89	3.6

consensus of opinion regarding the important site factors in
this area. Both of the textural variables are easily obtained
and a part of every soil survey report. Available moisture
contnet was not included and drainage class was included only
once in the equations. In contrast, both of these variables
are not as easily obtained and not as consistent in their occurrence in soil survey reports.

6. DISCUSSION

The data utilized in this paper represent the culmination of
many thousands of man hours of effort and millions of dollars
over a period of the last two decades. Many of the data are
invaluable as many studies cannot be repeated due in part to
prohibitive costs and lack of suitable study areas.

6.1 Field and compilation methods

Many of the studies reflect a need for more representative
field data, with better techniques for recording the original
data. The need for proper sampling methods has been referred
to above. Good sampling designs which facilitate the comparison
of the results from different studies and allow the isolation
of important factors should be emplayed where possible. Such
designs have been used in several studies investigated throughout this paper and where they occurred, were emphasized in the
above discussion.

Often the published results of these analyses have been
fragmental, containing little comparative information other than
the final equations. Future use and comparison of the techniques
used requires a clear presentation of all the pertinent statistics and adequate description of the measurements used. An
excellent format to follow in both cases is that given in the
10 studies conducted by Ike and Huppuch [23]. For each species,
the regression equation, R^2, and SEE were given along with the
means and ranges of the variables included. All were presented
in a highly comprehensible format. This made comparison within
species and with our results quite simple. Precise units of
measurement were presented for each variable eliminating the
guesswork of qualitative descriptions. To be even more useful
for detailed comparisons, the standard errors of the regression
coefficients also should be reported.

6.2 Future considerations

In this discussion and throughout this paper, we have placed
considerable emphasis on proper data collection procedures.
As more and more information concerning our environment is accumulated, such techniques will become increasingly important.
Large regional data storage banks will have to be developed to
synthesize and collate this information making it readily available to all wishing its use. The predictive theories which
will form the foundation for many of the decisions in the management of our resources will be based to a large extent on
these pools of data which have been and are being collected.
Efficient use of data in the models of tomorrow will require
better recording and compilation of the original data today.
The criticisms we have given in this paper are directed toward
this goal. Our intent is not to degrade existing investigations
but rather to positively point out deficiencies in certain
methodologies in a desire to promote better analyses in resource
management.

6.3 Linear programming

The knowledge and techniques of the linear programming method
are being acquired by more and more foresters. Many of the uses
deal with different methods of production and scheduling of
cutting rotations and seek to determine the best choices from
a number of alternatives most of which have economic overtones
([15], [25]). In this paper, we have examined a basic problem
of forest management, i.e., that of determining which species
to grow on a given area of land. All land is not equally productive nor of equal quality. Therefore, we sought to determine
which site parameters would produce maximum growth of a given
species while still assuring better than average yields of all
other species.

Capital restrictions played a role only in the selection of
an objective function and this restriction was subjective. The
framework for employment of linear programming is the multiple
regression equations developed and examined in this paper which
show the relation between site quality and productivity. Because
site index measures the integrative effect of all environmental
conditions affecting survival and growth over several years,
the prediction equations we included in the linear programming
framework are conceptually deterministic. However, during the

lifespan of a tree, many site factors may influence and limit
growth for only a short time period. These agents are difficult
to determine and the deviations from expected growth patterns
produced essentially introduce a probabilistic quality into
both the objective and constraint equations.

The prediction equation developed for white oak on well-drained
soils was used as the objective function. On such sites, the
influence of limiting factors is important and had these been
accounted for, the constraints may have been altered considerably.
Our failure to do this is one reason why the variables describing
the optimum site were solved at maximum or nearly maximum values.
Because of this omission, our description of the optimum site may
not be entirely realistic. A more realistic solution might have
been attained by choosing as the objective function, one of the
equations for a species growing on the imperfectly drained sites.
On these sites, poor drainage overshadows all others in determining productivity. However, these sites, because of their
location, are sub-optimal.

Since the upper bounds on the independent variables played a
more important role than did the constrain equations in obtaining
an optimal solution, it appears that the growth of white oak alone
is a good indicator of the site quality for other species concerned.
Those sites that are best suited to optimum gorwth of white oak
and good for all species concerned would be a site with a north
exposure and a gentle slope with deep well-drained soils.

6.4 Analytical techniques and tools

In many instances in the past the regression analyses were conducted by laborious hand methods of computation. Evidently the
tedium of making the calculations prevented a check of the calculations. When we reexamined such data we often found the same
mean values and variances for the variables, but the more labor-
iously obtained regression coefficients were in error. Perhaps
with the widespread influx of use of digital computers such pit-
falls can be avoided. Hopefully, investigators will put more
time and thought into meaningful collection, presentation, and
interpretation of their data.

The modern high-speed digital computers greatly reduce the
tedium of data processing and statistical analyses, but computers
themselves are introducing new problems in analytical studies.

For many modern computers as many as 11 digits may be carried

in the computations and printed in the output. A plea has been
made to scientists to put into their publications, however, only
as many digits as are actually significant in their data. These
and related problems are discussed by Van Dyne [37]. Certainly
when considering sampling variability in the field and laboratory,
many of our natural resource data contain only two or three truly
significant digits.

A second major problem introduced by the use of computers is
that it appears most biologists are now having their data analyzed
by others, i.e., rather than their doing their own computations.
In fact, it appears there is a third person in the completion of
analyses in many instances, a programmer. The recently analyzed
data we have examined suggest there often appears to be poor communication between the biologist collecting the data, the
programmer processing the data and running the analyses, and the
biometrician advising on analyses.

A third kind of problem concerns future comparative studies in
which there is need to know the actual procedure used in an analysis. All too frequently the computer programs used in an analysis
are not specified. Even when a program is specified, it is difficult to locate a listing or deck of it for detailed checking of
procedures. The kind of computer used, especially as it may
influence accuracy of computational operations (e.g., matrix in-
version), may be of importance in some instances. For example,
in site factor equations many of the site variables may be inter-
correlated. If such correlations are high, the sums of squares
and cross-products matrices and the correlation matrices may
approach singularity. Then there may be a considerable difference
in results obtained using a computer with 7 significant digits as
compared to one with 11 significant digits, even though the same
computation method is used. Results obtained may be spurious in
fact.

For the above reasons, although computers greatly lighten the
tedium of computation for biologists in site factor analyses, the
computer results should be interpreted with care. The biologist
should be especially careful to transmit his problem to the analyst.
The analyst should be careful that the proper methods are used and
that the biologist clearly understands the value and limitations
of the analyses. With reasonable care in their use, computers
should allow better predictive relations to be developed in site
factor studies.

ACKNOWLEDGMENTS

Most of this work was supported by subcontract #2765 from Union
Carbide Corporation through the Radiation Ecology Section of the
Oak Ridge National Laboratory. Part of the analytical studies and
completion of the manuscript were supported by National Science
Foundation Grant GB-7824.

REFERENCES

[1]   Allen, J. C. 1960. Pine site index relationships in the
       Northwest Alabama and adjoingin portions of Mississippi
       and Tennessee. Tennessee Valley Authority Forest Rel.
       Tech. Note 32.

[2]   Avery, T. E. 1967. Forest measurements. McGraw Hill Corp.
       N. Y. 290 p.

[3]   Carmean, W. H. 1956. Suggested modifications of the stan-
       dard Douglas-fir site curves for certain soils in
       southwestern Washington. Forest Sci. 2:242-250.

[4]   ——————. 1959. Forest soil-site research in Southeast
       Ohio. Paper presented Ohio Chap. Soil Conserv. Soc.

[5]   ——————. 1961. Soil survey refinements needed for accu-
       rate classification of Black oak site quality in southeast
       Ohio. Soil Sci. Soc. Amer., Proc. 25:394-397.

[6]   ——————. 1965. Black oak site quality in relation to
       soil and topography in southeastern Ohio. Soil Sci. Soc.
       Amer., Proc. 29:308-312.

[7]   Copeland, O. L. 1956. Preliminary soil-site studies in the
       western white pine type. Internountain Exp.-Sta. Res.
       Note 33.

[8]   ——————. 1957. Soil-site index studies of western white
       pine in the northern Rocky Mountain region. Soil Sci. Soc.
       Amer., Proc. 22:268-269.

[9]   Dean, H. C. and Case, J. M. 1959. Forest Coastal Plain--
       Arkansas. IN: Soil Survey interpretations for woodland
       conservation. Progress report for period ending 1959.
       U. S. Dept. Agr. (Little Rock, Ark.).

[10]   Della-Bianca, L. and Olsen, D. F. 1961. Soil-site studies
       in Piedmont hardwood and pine-hardwood upland forests.
       Forest Sci. 7:320-329.

[11]   Dickenson, J. E. 1950. Height growth of planted slash pine
       (Pinus elliotti, Englm.) as influenced by some physical
       properties of the soil. MS Thesis. Univ. Fla. 21 p.

[12] Doolittle, W. T. 1957. Site index of scarlet and black oak
      in relation to southern Appalachian soil and topography.
      Forest Sci. 3:114-124.

[13] Ellerbe, C. M. and Smith, G. E. 1961. South Carolina Coastal
      Plains and sand hills. IN: Soil survey interpretations for
      woodland conservation. Progress report for period ending
      1961. U. S. Dept. Agr. (Columbia, S. C.).

[14] —————————. 1964. South Carolina southern Piedmont. IN:
      Soil survey interpretations for woodland conservation.
      Progress report for period ending 1964. U. S. Dept. Agr.
      (Columbia, S. C.).

[15] Fasick, C. A. and Sampson, G. R. 1966. Applying linear
      programming in forest industry. Southern Exp. Sta. Paper
      50-21.

[16] Furnival, G. M. 1961. An index for comparing equations used
      in constructing volume tables. Forest Sci. 7:337-341.

[17] —————————. 1964. More on the elusive formula of best fit.
      Paper presented Soc. Amer. Forest, Proc. 1964.

[18] Goggans, J.F. and Schultz, E. F. 1958. Growth of pine plan-
      tations in Alabama's Coastal Plain. Alabama Agr. Exp. Sta.
      Bull. 313.

[19] Grigsby, H. C. 1952. The relationship of some soil factors
      to the site index of loblolly pine in southern Louisiana.
      MS Thesis La. St. Univ. 44p.

[20] Grosenbaugh, L. R. 1958. The elusive formula of best fit:
      A comprehensive new machine program. Southern Exp. Sta.
      Occa. Paper 158.

[21] Hebb, E. A. 1962. Relation of tree growth to site factors.
      Tennessee Agr. Exp. Sta. Bull. 349.

[22] Hodgkins, E. J. 1959. Forest site classification in the
      southeast: an evaluation, p. 34-38. IN: Burns, P. Y.
      (ed) 1960. Southern Forest Soils. Annu. Forest. Symp.,
      Proc., 8.

[23] Ike, A. F. and Huppuch, C. D. 1968. Predicting tree height
      growth from soil and topographic site factors in the Georgia
      Blue Ridge Mountains. Georgia Forest Res. Paper No. 54.

[24] Kormanik, P. P. 1966. Predicting site index for Virginia,
      loblolly, and shortleaf pine in Virginia Piedmont. South-
      eastern Exp. Sta. Paper SE-20.

[25] Leak, W. B. 1964. Estimating maximum allowable timber yields
      by linear programming. Northeastern Exp. Sta. Paper 17.

[26] Linnartz, N. E. 1959. Relation of soil and topographic
      characteristics to site index of loblolly pine in the Flor-
      ida Parishes of Louisiana. MS Thesis La. St. Univ.

[27] Linnartz, N. E. 1963. Relation of soil and topographic characteristics to site quality for southern pines in the Florida Parishes of Louisiana. J. Forest. 61:434-437.

[28] Loftin, L. L., Clark, W. M., Holcombe, E. D., Chaffin, B. F., and Fallin, H. F., 1959. Forested Coastal Plain, Western Louisiana. IN: Soil conservation interpretations for woodland conservation. A progress report for period ending 1959. U. S. Dept. Agr. Louisiana.

[29] McClurkin, D. C. 1963. Soil-site index predictions for white oak in north Mississippi and west Tennessee. Forest Sci. 9:108-113.

[30] McComb, A. L. and Einspahr, D. 1951. Site index of oaks in relation to soil and topography in Northeastern Iowa. J. Forest. 49:719-723.

[31] Miller, W. F. 1966. Soil-site predictions for loblolly. Miss. Farm Res. 29:6-7.

[32] Myers, C. A. and Van Deusen, J. L. 1960. Site index of ponderosa pine in the Black Hills from soil and topography. J. Forest. 58:548-555.

[33] Ritchie, F. T., McFarland, T. A., Sands, N. E., and May, J. T. 1961. Georgia. IN: Soil survey interpretations for woodland conservation. Progress report for period ending 1961. U. S. Dept. Agr. (Georgia).

[34] Spurr, S. H. 1964. Forest ecology. Ronald Press Corp., N. Y. 352 p.

[35] Trimble, G. R. and Weitzman, S. 1956. Site index studies of upland oaks in the northern Appalachians. Forest Sci. 2: 162-173.

[36] Van Dyne, G. M. 1965. REGRESS A multiple regression and correlation program with graphic output for model evaluation. Oak Ridge National Lab. TM-1288. 56 p.

[37] —————. 1966. Application and integration of multiple linear regression and linear programming in renewable resource analyses. J. Range Mgmt. 19:356-362.

[38] —————, Wright, R. G., and Dollar, J. F. 1968. Influence of site factors on vegetation productivity: A review and summarization of models, data, and references. Oak Ridge National Lab. TM-1974.

[39] Walker, N. and Reed, R. M. 1960. Site evaluation for western catalpa in north central Oklahoma. Oklahoma Exp. Sta. Bull. 544.

[40] Wright, R. G. 1968. Analyses and development of models depicting the influence of site factors on forest tree productivity. MS Thesis Colorado St. U. 132 p.

[41] Zahner, R. 1958. Site quality relationships of pine forests in southern Arkansas and northern Louisiana. Forest Sci. 4:162-176.

RECORD OF PREPLANNED AND SPONTANEOUS DISCUSSIONS

O. L. LOUCKS (University of Wisconsin)

Critical re-examination of the multiple regression techniques used
in site evaluation has been long overdue. Perhaps the major reason
why we have not had a paper of this type before is that there has
not previously been an appropriate forum combining the areas of
statistics, environmental biology and resource management. One of
the achievements of this meeting is that it provided that forum,
and Dr. R. G. Wright and Dr. Van Dyne have responded to it with a
most timely paper.

Having worked for some years in yield prediction and the inventory of land for its potential yield, I can say from experience
that the area of site evaluation has been one of the weak links
in forest research and in our forest management programs. I view
the paper by Wright and Van Dyne as a major step in strengthening
and stimulating higher quality studies in forest productivity.

In spite of my enthusiasm for the general thrust of the paper,
however, I cannot let it pass without comment on two significant
deficiencies. I am a little disturbed by the over-simplification
of the potential use of soil survey information for prediction of
site index. There is a great volume of literature on this topic,
much of it lying outside the papers reviewed by the authors, and
there are very few examples where existing soil surveys can be
brought into use without years of additional study. The case study
cited by the authors is encouraging, of course, but it cannot be
a basis for much optimism on the general availability of suitable
data.

Much more important, is the failure of this paper to treat
adequately the role of irregular stand development in introducing
error to the plot data and therefore in the resulting analysis.

The equations involve variables with many sources of error in
both independent and dependent variables. The error sources in
the site factors are readily understood, but the effects of differences in conditions for height growth, particularly the presence
of a few overstory trees during the early years of stand development, induces a response that often cannot be detected, least of
all measured in the mature stand. The error introduced is not
random but is a bias toward an underestimation of the site index
for the given environments.

An ecologist familiar with the autecology of the species analyzed is able to see that the highest R^2 coincides with those species that consistently develop as pure, even-aged stands, for example, scarlet oak, northern red oak, and some of the southern pines. Species that only tend to develop as even-aged stands, with reproduction extending over a 20 year period, frequently mixed with other species, can be analyzed by these methods, but the sampling errors are larger and the results less satisfactory. White oak is a good example of such a species. Differences in the way stands develop are species related, and account for much of the residual error. Significantly, the authors have used none of the studies on site index in shade tolerant species such as the northern hardwoods where multiple-aged structure is almost a rule.

This point is important because the authors extend their conclusions to the use of these methods in resource management over large areas in the near future. But the real world is one in which most of our forest land in the eastern United States, and significant portions of the west, is occupied by multiple-aged stands and by species that rarely develop in an even-aged structure. The techniques of site index evaluation generally have not been successful here and multiple regression analysis offers little potential for prediction. The authors point out that the need for good estimates of the productivity of all our forest lands is an absolute requirement for the intensive resource management we face in the immediate future. While the evidence assembled by Wright and Van Dyne for even-aged species is impressive, it is, I believe, inappropriate of the authors to leave the impression that we now have techniques for making the necessary site evaluations and for utilizing the great potential of linear programming on the majority of our forest lands.

I do not want to take anything away from the potential of the methods reviewed in this paper for land occupied by forests that lend themselves to the technique. This is why I value the contribution of the paper so highly. But since the authors themselves have identified and discussed the national need for application of these techniques at this time, it is necessary for them to consider as well the biological bases of the error terms within the studies they reviewed and the extension of these error sources in the regions of all-aged stand structure. Development of statistical and analytical techniques suitable for these regions, and a re-examination of the biology and environmental base and variance of the

productivity of these forests must continue to be a priority concern of statistics, forestry, and ecology, as it has been in recent
years.

H. L. LUCAS, JR. (North Carolina State University)

At the outset I wish to commend the authors for making such a
comprehensive study of some important aspects of the management
of renewable natural resources. The idea of developing satisfactory site factor equations for predicting tree growth and then
using those equations in formal optimization procedures to arrive
at practical decisions is a good one.

The authors are to be complimented also for making several
cogent points of broader relevance. Among these are the need to
use existing data as effectively as possible, proper forethought
and care in the gathering of data, intelligent choice of statistical procedures and computer programs, adequate description of
data and methodology when publishing, and sufficiently complete
presentation of results. Further, they urge better communication
between biologists, biometricians and programmers to ensure that
relevent mathematical and statistical results are obtained, and
to guard against the distressing computational errors evidenced
in the literature. Such items cannot be overemphasized.

It is with hesitancy, therefore, that I venture some criticisms,
make a number of comments intended to be constructive, and raise
some questions. I hope that the authors will view my efforts in
the same spirit as they presented their paper. To paraphrase,
my intent is not to degrade but rather to contribute in a positive
way.

Despite the pleasant and easy literary style, some reasoning
was hard to follow. Difficulty was encountered mainly in connection with a point strongly emphasized in the introduction and
reiterated later. At the base were certain statistical matters
and related terminology which illustrate the problem of communication between biologists and biometricians. Since the trouble arose
in connection with a point deemed critical by the authors, the
associated statistical questions seem worthy of discussion even
though they are ostensibly picayune in the context of the paper.
Attention will be devoted to those questions before proceeding to
more general issues.

The authors argue that using site index (height adjusted to a
standard age using tabled adjustments) as the dependent variable

and site factors as the predictors is preferable to using height as the dependent variable and introducing age as an additional predictor. A reason given in the introduction is that the use of age as a predictor "inflates the accuracy" of the prediction equation. In discussing results the authors note that the "prediction" when using site index was low as compared to the prediction when using height with age as a predictor.

The above statements were based on the relative magnitudes of R^2. Clearly, if age is one of the predictors, R^2 mirrors age effect as well as site-factor effects, and, if using site index, the effect of age is essentially eliminated. Thus, the R^2 values are not comparable. The authors note, however, that there might be reason to use age as a predictor. The relation of height to age can be conditioned by the site factors; hence the tabled corrections for age might not be appropriate.

There are better ways of assessing the "goodness" of prediction equations than to examine values of R^2. Interest really lies in two things, the prediction error and the amounts of variability attributable to different groups of predictors. The analysis of variance is one approach that is effective. Here, for example, if height rather than site index is the independent variable, the analysis of variance would consist of subdividing the total sum of squares for height into a part due to age, a part due to site factors and a part not accounted for. Such can be done even when the correction for age is determined from other data as in the site-index approach. Such an attack yields both an estimate of prediction error (SEE), a relevant item, and, via the F-statistic, an assessment of the impact of the site variables. The method generalizes. The sum of squares for site factors can be subdivided into parts corresponding to biological and logical groupings of the site factors, even individual factors, and the impact of each group can then be assessed separately. Also, provision can be made to test whether the age correction in a site index formula is compatible with the data at hand.

As an aside, it should be noted that standard partial regression coefficients do not have standardized variances as stated by the authors. The variances are highly dependent on correlation patterns among the predictors and on numbers of observations. Further, since such coefficients are normalized, they do not measure in appropriate terms the change of a dependent variable with respect

to an independent variable. For comparative purposes it is best
to examine "non-standardized" regression coefficients along with
their variances.

Now to more general considerations. After looking over the
results on best predictors in the several tables, one can ask,
"Since the sets of factors decided as best vary greatly over the
range of data, do they reflect biological reality?" Reasoning
from plant physiology, the nature of the site factors and the
assumptions involved in the models compared, one might answer,
"Yes." In fact, the authors offered general biological explanations of the fluctuations obtained. At the same time, the tenuous
nature of decisions made when using step-wise and all-possible
regression techniques cause doubt.

Even though step-wise and all-regression approaches in various
modifications are currently the only statistical approaches available to decide between alternate models, the results are subject
to high uncertainty. This should some way be taken into account.
Unfortunately, statisticians have not yet devised methods for
assessing, from internal evidence of a set of data, the confidence
which can be placed in the model chosen. The best test known is
to see how well a chosen model predicts in other sample of data.
In practice this might often mean that models are decided from a
part of the data at hand and "tested" on the remainder.

In addition to the statistical testing of models, it is desirable to invoke biological and physical reasoning in every way
possible. I could cite a number of instances from other areas in
which such reasoning led to models which involved small numbers of
predictors, gave excellent fits to data and yielded good predictions for situations far outside the range of the data used in the
fitting. The predictors in these cases were derived functions of
the same or, in several cases, of a larger number of measured
variables which, when used empirically, gave poor predictions in
other data.

There were obviously some slips in presenting or in solving the
example of linear programming found in the preprint. These are
correctable, so I shall confine attention to questions of other
sorts.

One question is the following: Why did the restricting inequalities involving species encompass poorly drained sites when the
objective function was for a species on well-drained sites?
Another is: What would be done if the programming problem posed

had no solution; i.e., was "unfeasible?" Still another is: Why
was a rectangular bounding region defined for the site factors?
In view of the correlations which can exist in nature, bounding
space probably should be non-rectangular. One example restriction
on the space is % gravel + % silt $\leq$ 100.

Another problem is this: The regression coefficients in the
objective function and in the restricting inequalities involving
other species, and even the bounds stipulated for the site factors,
are really random variables, not constants. Thus, the programming
solution is a random vector. Faced with this circumstance, one
would prefer the solution to define, with specified confidence,
a region in the factor space which included the best point rather
than yielding simply a point. This problem is very difficult and
has not been solved. Nevertheless, "sensitivity" analyses provide
insight on the matter, and are easy to apply. It seems desirable
to include sensitivity studies because one would hardly wish to
"pin" practical actions too tightly if the uncertainty in recom-
mendation is considerable.

A number of more general questions can be raised. For example
"Why should one seek a solution which maximizes yield for one
species but guarantees only average yield for alternative species?"
Perhaps one should seek the region of factor space which ensures
yields some amount greater than average for the few most desirable
of the several species. Also one can ask, "How should costs and
profits be brought into the picture." Even more generally one can
ask, "When is the point of view adopted by the authors relevant?"
They ask the question, "Given a desired outcome, what resources
are needed?" An alternate is, "Given a set of resources, what is
the optimum way to use them?" Perhaps some of these matters best
be left to researchers and advisors in resource management, busi-
nessmen and politicians.

In closing, I must thank the authors for putting before us a
useful conceptual framework in which to attack important practical
problems. Progress is always limited without new and improved
overviews. It is proper, nevertheless, to be critical of individual facets of the total picutre, such as data, methodology and
specific ideas. If such are not sound enough, results can be misleading or of questionable value.

Part of my discussion merely emphasized the authors' simple,
cogent plea for better communication between biologists, biometricians and programmers. I hope that my efforts have amounted to

a little more. If I am fortunate, they will have contributed to overcoming the communication problem and to the overall goals of the authors.

D. S. ROBSON (Cornell University)

I should first like to commend the intestinal fortitude which underlies a work purporting to expose the errors and weaknesses in a number of specifically identified published papers--intestinal fortitude on the part of the critics who have the courage to assert that 50% of the published site factor equations examined are erroneous, and intestinal fortitude on the part of the victims who were willing to submit their data for this computational check of their published analyses.

Such a paper as this naturally presents a challenge to see whether any of the criticisms leveled at the victims can be turned back upon the critics themselves. One complaint of the authors was that the actual procedure used in computing site factor equations is often not described in sufficient detail in the published papers to permit the critics to ascertain whether disparities revealed by their recalculations are due to actual computational errors by the victim or due to more legitimate differences in analytic procedures. Without attempting a third recalculation of the equations I cannot turn this gun directly upon the critics, however I did note instances in their paper where their description of their analytic method was frustratingly unclear. To illustrate this, I point to Table 3, second to the last line, where--if I interpret the authors correctly--the victim found a multiple R^2 of .74 by regressing log height of white oak on the three ordered independent variables of elevation, slope position and age (Table 4) while the critics recalculated this to be $R^2 = .69$. However, the critics also performed a stepwise regression analysis to find the ordered three most important independent variables, and found the same set of three in reverse order, but now obtained a multiple $R^2 = .73$. Since the value of R^2 is independent of the ordering of the included dependent variables, it appears that the critics have erred in finding $R^2 = .69$ and $R^2 = .73$ by their two different computational techniques for fitting the same regression function. If this is not a computational error but simply a misinterpretation on my part then my criticism echoes their own in complaining of unclear and confusing presentations of analytic procedures.

On the constructive side I should like to point out that the
site index defined by foresters as the average tree height at the
age of maturity is directly comparable to the use of age factors
in dairy science to adjust the measured milk production of cows of
differing ages to milk production at the age of maturity. Biases
inherent in the site index due to environmental trends, differential selection pressures through time, and due to selection of
dominant and co-dominant sample trees have direct counterparts in
dairy cow age factors. The extensive literature on age factors
in dairy science might well prove useful to the field of forestry.
Entrance to this literature could be gained through the following
two references:

[1]   Miller, R. H., etc. 1966. "Maximum likelihood estimates of
      age effects", J. Dairy Sci. 49:65.

[2]   Hendersen, C. R. 1966. "Outline of sire evaluation which
      accounts for unknown genetic and environmental trends,
      herd differences, seasons, age effects, and differential
      culling", Symposium National Tech. Workshop on Dairy Sires
      and Cows, Washington, D. C.

W. C. SCHMIDT (Forestry Sciences Laboratory, U. S. Forest Service,
    Missoula, Mont.)

I realize that the authors used what data they could obtain from
other researchers for this paper. If they are planning to continue
this work into other forest types, I suggest that they take another
look at the type of data that they use. I suggest that site index,
as a measure of productivity, leaves much to be desired in many of
our forests. Instead, there are many other units of measure such
as basal area, bole area, cubic-foot volume, or in some cases even
sapwood basal area, that give a much better index of productivity.
Overstocking, particularly in many of our western conifer forests,
strongly affects height growth and as a result often precludes
good measurement of site index. In addition, using any of the
above measures also eliminates some of the problems in obtaining a
meaningful site index in stands composed of a large number of
species.

P. DAGNELIE (Faculté des Sciences Agronomiques de l'Etat, Gembloux,
    Belgium)

I would like to stress the fact that the differences between R^2
values might perhaps come from the use of different estimates.
R^2 values are generally calculated using either the total and the

residual sums of squares of deviates:

$$R_1^2 = \frac{SSD_t - SSD_r}{SSD_t} \ ,$$

or the total and the residual mean squares:

$$R_2^2 = \frac{MS_t - MS_r}{MS_t} \ .$$

The difference between these two estimates might be rather important in such situations (numerous explanatory variables and/or
small number of observations).

Moreover, both estimates are biased and some authors are therefore using even other formulas.

J. S. OLSON (Oak Ridge National Lab., Oak Ridge, Tenn.)

The Symposium Committee, Professor Sokal, and other Symposium
gadflies have rightly urged this distinguished group to use well,
the time and money which were spent to bring you together. In
particular, we have been urged to examine fundamental questions of
ecology, as well as the corollaries of practical application which
follow from basic principles.

There have been many discussions bearing on the static conditions of an ecosystem, mostly its plants. The preceding paper on
insect populations could be related to problems of animal (or
secondary) production. The first paper of this section on productivity will concern one practical aspect of plant (or primary)
production - specifically of the growth left after losses to con-
sumers and pathogens and attrition to the environment have
detracted from plant growth.

Aside from points of detail from discussants, these seem to
agree with the desirability of making the most of conventional
data on height-growth as a prelude for predictions of actual and
potential production, whether measured as biomass, dollars, or
energy units--as in the next paper.

In the next paper by R. V. O'Neill, attention is turned back
to secondary production, in this case by an invertebrate predator.

R. G. WRIGHT (Colorado State University)

We would like to thank all of the discussants for their helpful
and cogent criticisms of our paper. In particular, Dr. Lucas'

comment concerning the desirability of having a means of defining
a random vector for the linear programming problem is well taken.
We agree that this problem is very difficult and have also looked
at sensitivity analyses as a partial solution. We would also like
to thank him for pointing out the fact of a non-standardized variance which is held by the standard partial regression coefficients.
As to his comment on the linear programming problem we would like
to point out that the restriction he indicated concerning the gravel
plus silt content was incorporated although not clearly into the
problem.

It is true that there are many considerations that should be
applied to analyses designed to look at the optimum allocation of
our natural resources. The alternatives we chose to examine were
done so rather subjectively and his insight on this approach is very
helpful.

We did not mean to imply that our use of soil-survey data for the
prediction of site index was a solution concerning the utilization
of this type of information as indicated by Dr. Loucks. It was intended only as an example of one possible method whose incorporation
will have to await many subsequent tests. Dr. Loucks' comments
concerning the care needed when extending the results obtained in
this paper to multiple-aged stands are well taken. However, we
endeavored to incorporate all of the data we found available, and
did not purposely avoid the use of data from the shade-tolerant
northern species.

The computational error discussed by Dr. Robson is a misinterpretation on his part. The table to which he refers to (4), in
indicates that the variables given are the three most important for
each respective equation. The equation in which we obtain an
$R^2 = .73$, contains one additional variable other than the three given
in the table, which thus gives it an R^2 higher than the .69 obtained
from the equation presented in the literature. We would like to
thank Dr. Robson for his example illustrating the comparable use of
age factors in the field of dairy science.

When working on a problem such as this involving a large compli-
cational effort with many data sets, it is easy to become engrossed
with the details of the analyses and miss much of the philosophy
underlying analyses of forest productivity. In this regard, the
comments of the reviewers and the additional ones from the audience,
have been most helpful and refreshing in reintroducing us to some of
the important considerations of the big picture.

A STOCHASTIC MODEL OF ENERGY FLOW IN PREDATOR COMPARTMENTS OF AN ECOSYSTEM*

R. V. O'NEILL
Radiation Ecology Section, Health Physics Division
Oak Ridge National Laboratory,
Oak Ridge, Tennessee

SUMMARY

A Monte-Carlo simulation model has been devised to describe the
energy dynamics of a predator. The caloric pool of a randomly
selected individual is decreased by respiration until a hunger
threshold is reached. At this point, several porbability functions
are used to decide which species of prey will be consumed and
during which hours. The assimilated portion of that prey is added
to the predator compartment and respiration continues. The model
is illustrated with data on a centipede, Otocryptops sexspinosus,
and demonstrates that an entire energy budget can be generated
along with estimates of variance. Applications of the model to
analysis of sensitivity and components of variance are briefly
discussed.

*Research sponsored by the U. S. Atomic Energy Commission under
contract with the Union Carbide Corporation.

1. INTRODUCTION

The analysis of population energy dynamics has largely been restricted to studies of herbaceous or saprophagous organisms ([4], [7], [8]). A number of difficulties are involved in making comparable studies of predators. For example, the assumption of feeding as a continuous function of time is implicit in the gravimetric and radionuclide methods used in the above studies. Although the assumption is not limiting in long-term studies of predators, it is invalid for short, discrete sampling periods.

Frequently, in large ecosystem studies, an investigator wishes to evaluate an energy budget by a method which will maximize the available information, while restrictions of time and complexities of the system prevent him from gathering sufficient data to quantify a general and sophisticated predator model. Such a general approach has been taken by Holling ([2], [3]), and the present work is based on the excellent conceptual framework he has developed. However, the model as presented by Holling is not directly applicable to problems in which (1) data are available as probability distributions, (2) the predator feeds on several prey species, or (3) a complete quantification of the system is impractical.

The present paper presents a method for simulating predators and studying their energy dynamics based on data which include probability distribution. In order to circumvent the complexities of stochastic mathematics and to present a practical approach, Monte-Carlo simulation methods have been used. The motivation for the development of the model was a study of energy dynamics in a centipede, Otocryptops sexspinosus (Say), and data from this study are used as an illustrative example.

2. CENTIPEDE ENERGETICS

To place the discussion on an empirical basis, a summary of the information available from the centipede study is presented. Although details of method and discussion will be presented elsewhere, an understanding of the type of information used is important to a discussion of the structure and development of the model.

Otocryptops sexspinosus (Say) is a large scolopendromorph centipede which is common in deciduous forests of eastern United States. The animals have been found to feed on eight prey species [5] and the probability that the centipede will feed on a particular species can be calculated from the probability of encounter and the conditional probability of feeding following encounter [6]. Field sampling indicated the average caloric content of adult centipedes to be 647.1 cal (SD = 120.8, N = 20). It also was determined that this value did not change significantly in the field from May 1 to June 1, the period which is simulated.

Metabolic rates were measured in the laboratory and expressed as a log function of temperature. Continuous temperature recordings were made on the forest floor and analysis revealed that an average May day could be described adequately by a sine curve with a 24-hour period, fluctuating from 20 to 25°C, and with a standard error of estimate (Sxy) of 1 degree. Measurements of assimilated energy were made in the laboratory by gravimetric methods and showed a percentage assimilation of 94.8. This estimate seems high for arthropods but is in accord with estimates for other predators (e.g., [1]).

Laboratory observations of centipedes that were hungry and actively seeking prey revealed that feeding activity is largely restricted to two periods of the day. These periods occurred between 0600 to 1000 and 1800 to 2200 hours. By placing the centipedes in terraria containing leaf litter, soil and cricket nymphs, it was possible to observe feeding during particular activity periods. The probability that feeding would occur during any one period is described by the distribution in Table 1. Based on observations of successful feeding, it is possible to construct a distribution which describes the conditional probability that feeding will occur during any one of the four hours of each period. This information is shown in Table 2.

3. MONTE-CARLO SIMULATION MODELS

Using this limited information it is possible to construct a complete energy budget for the centipede. An individual animal, regarded as a homogeneous energy pool, is randomly selected from the normally distributed field population. This individual is simulated for a one-month period with respiration losses decreasing the energy pool at hourly intervals until a hunger threshold is reached. At this point, a complex process, described below, is used to decide when a prey will be successfully captured and which species it will be. The energy content of this prey is added to

Table 1. Probabilities for feeding by Otocryptops sexspinosus
          during successive 4-hr activity periods. Activity
          periods occur at dawn and dusk of each day.

Period	Probability of feeding during each period
1	0.50
2	0.25
3	0.13
4	0.07
5	0.03
6	0.01
> 6	0.01

Table 2. Probability for feeding by Otocryptops sexspinosus
          during a specific hour of an activity period.

Hour of Active Period	Probability for feeding during each hour
1	0.15
2	0.51
3	0.20
4	0.14

the centipede's energy pool and respiration continues until the hunger threshold is again reached. At the end of a month, another individual is chosen and the precess is repeated.

The Monte-Carlo method basically involves simulating the behavior of a random individual during a series of random steps. Repetition of the procedure for many individuals yields estimates of the expected results of the process. Stochastic elements enter the model in two ways. The first stochastic element is the random selection of an individual. In the centipede example, this choice is made by a machine-generated random number chosen from a normal distribution. A similar normally distributed variable is used to fluctuate the temperature about the sine curve model. The second stochastic aspect involves decision-making on the basis of a probability distribution. This is accomplished by comparing a machine-generated, uniformly distributed, random variable with a known cumulative probability distribution. For example, once it is determined that the centipede will feed during a particular activity period, a decision must be made on the hour of feeding. Transforming the data in Table 2 to a cumulative probability distribution (0.15, 0.66, 0.86, 1.0) shows the probability that an animal will have fed by the end of a given hour. A random number chosen between 0.0 and 1.0 is compared to this distribution. For example, the number 0.5 would cause a decision to be made for feeding during the second hour.

Because of the double aspects of random choice of individuals and distinct random decision process, the general model is most easily envisioned as composed of two parts. One part chooses a predator, generates a temperature set and subtracts respiration until a hunger threshold is reached. The other portion utilizes the probability distributions to decide on the time of capture and nature of the prey, adds the appropriate calories and transfers control back to the main program.

4. A MODEL FOR CENTIPEDE PREDATION

The application of this approach to the centipede predation problem is shown in Figure 1. The flow chart illustrates the program as functionally split into a main program and a decision-making subprogram. Double lines around a box indicate that the computation involves a random variable. An initial centipede is chosen

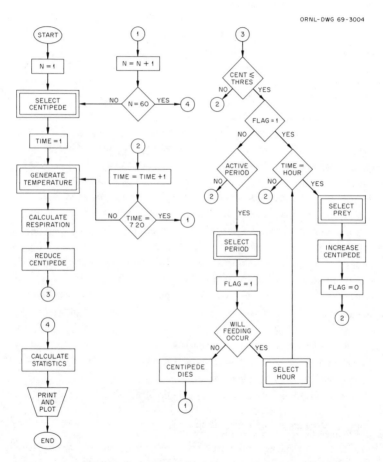

Figure 1.   The computer flow diagram for the centipede model.
Double lines indicate that a machine-generated random
variable is involved in the calculation.   The circled
numbers indicate the points of connection between segments of the program.

and followed for a total of 720 hours to approximate a monthly
interval. Each hour, a temperature is generated by a sine curve
function plus a random variable, N (0, 1). The respiration rate
appropriate to this temperature is calculated and the caloric pool
of the centipede reduced according to the simple iterative formula

$$C_t = C_{t-1} (1 - e^{-Rt}) \tag{1}$$

where C is the energy pool, t is time and R is the respiration rate
in cal cal^{-1} hr^{-1}.

Control is then transferred to the subprogram which tests to see
if the threshold has been reached. Field sampling of the centipedes
had revealed that the animals remained at an average energy content
of 647.1 cal during the month of interest. This was interpreted as
meaning that the animals were maintaining this level as an equilib-
rium. It was assumed, therefore, that when the caloric pool of an
individual was below this value, the animal would attempt to feed
in order to maintain a balance at or above this level. When the
caloric pool was above this level, hunger would not be a driving
motivation. Based on this argument, the threshold was chosen at
647.1 cal. Results of the simulations showed this to be a reasonable assumption, as discussed below.

When the animal falls below the threshold, the value of a variable FLAG determines whether the decision-making process has been
completed previously. This variable is initially set at 0, and the
program proceeds to test whether the current hour is in one of the
two daily activity periods. If it is not, the program continues
to decrease the caloric pool by respiration until the beginning of
the next period. This aspect of the program is consistent with the
observation that feeding occurs only during these active periods.
When such a period is reached, a decision is made for feeding to
occur during this period or some successive period. If the unlikely decision is made that no feeding will occur during the next
twenty periods, the centipede is considered as starved, and the
next centipede is generated and followed. If feeding does occur,
the variable FLAG is set equal to one and the hour of feeding is
chosen. The variable HOUR then indicates the number of hourly
intervals between the current time and the hour chosen for feeding.
Examination of the flow chart will then reveal that the variables
FLAG and HOUR revert control to the respiration-reduction program
until the feeding time is reached. Then a particular prey species

is chosen and its caloric value added to the centipede's energy
pool. The variable FLAG is again set equal to zero and respiration
continues until a new threshold is reached.

At the end of 720 simulated hours, information stored during the
program is used to calculate the total energy ingested, excreted,
assimilated and respired. The difference between energy assimilated and respired is used as an estimate of the productive energy.
After a total of 60 animals have been followed under independent
temperature regimes, the program prints all the relevant information along with standard deviations. The program can be modified
so that all simulations are made with the same temperature regime,
and various random variables can be set equal to constants.

5. RESULTS FROM THE CENTIPEDE SIMULATIONS

The results of the simulations illustrate the kind of information
the model will yield. Figure 2 shows results from two individuals.
Since the caloric content of the prey varies with the species,
large deviations are possible in the number of prey consumed. The
lower figure represents a centipede which encountered and fed on
small prey. Therefore, it had to consume nearly twice as many
individuals as the centipede in the upper figure. Variations in
the time needed for a successful capture can be seen in the upper
figure where 30 hours were needed to capture the first prey and
only 15 hours required to capture the second prey.

The slow respiration rate characteristic of the centipedes and
the relative success of the animals in capturing prey explains why
most of the month is spent above the threshold. This finding is
reflected in the reproductive energy calculation shown in Table 3.
The table shows predicted estimates for the energy budget along
with standard errors. The model predicts that the animals will be
successful predators, assimilating energy in excess of respiratory
losses and ending the month with a greater caloric pool size.
Since the field populations remained at the same caloric value,
this excess can serve as an estimate of the reproductive energy
expended by centipedes during the May breeding period. Examination
of the standard errors reveals that the size of the centipede
compartment can be predicted with some degree of accuracy, but the
items of the energy budget can be expected to show large fluctuations, particularly the energy available for reproduction. The

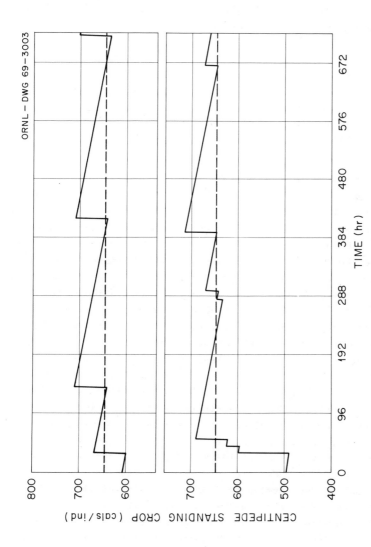

Figure 2.   Simulated feeding behavior of two centipedes. The dashed line indicates the hunger threshold and the continuous line follows the caloric pool of the centipede through time. Decreases are due to respiration and increases to feeding.

Table 3.   Predicted energy budget for Otocryptops sexspinosus
during May. Values for energy fluxes are in cal
individual^{-1} month^{-1}. Numbers in parentheses are
standard errors, N = 60.

Initial centipede caloric pool	628.8 (46.9)
Final centipede caloric pool	797.7 (64.0)
Energy ingested	398.2 (100.2)
Energy excreted	20.7 (5.2)
Energy assimilated	377.5 (95.0)
Energy respired	208.6 (13.0)
Energy for reproduction	168.9 (87.0)

ability to evaluate accuracy of predicted values by associated
error statements is one of the powerful advantages of the stochastic simulation approach.

Because the original problem was to utilize limited information
to estimate a total energy budget, it is difficult to evaluate the
validity of the model by comparison with extensive field data.   If
such data were available, the model would never have been constructed.   In this type of application, the aim is to produce "best
estimates" of parameters that cannot be measured directly due to
the complexity of the system or limitations of time.   The usefulness of the model must therefore be judged on its ability to
perform this task and not on its ability to interpret known data
by a process analagous to curve fitting.   It is possible, however,
to make some comparisons of predictions with observations made in
the laboratory.   Since the laboratory cultures were maintained on
cricket nymphs, simulations were made with these nymphs as the only
available prey.   The model predicts that 3.1 (SE = 0.34, N = 20)
prey will be consumed during the month.   This value is not significantly different from the 3.4 (SE = 0.56, N = 20) which was
measured in the laboratory.   Similarly, the predicted compartment
size after a month is not significantly different from that measured on the laboratory culture.   The model also predicts that
larger centipedes, beginning well above the threshold, may not eat
at all since slow metabolic rates would not reduce them to the

hunger threshold during the month. This was confirmed by large
centipedes in the laboratory which ate only once or not at all.
An additional check is the prediction that substantial reproductive
energy should be available to centipedes in the field. Since the
period simulated represents the normal reproductive period, such
production energy would be expected.

In addition to making preliminary estimates from limited data
and estimating variance, this type of model can be used for investigations into the nature of the system under study, e.g., an
analysis of the sensitivity of the centipede to various parameters
of the model. Figure 3 shows a plot of the final pool size as a
function of the hunger threshold. This plot shows that the partial
derivative of the compartment with respect to the threshold is a
constant equal to the slope of this line. Similar graphs can be
drawn which would vary other parameters and would yield information
on the nature of the predation relationship. For example, Figure 3
suggests that increased demand for productive energy could be
counteracted by a higher hunger threshold.

Another interesting by-product follows from the ability to hold
various random variables constant and examine the resultant reduction in variance. Table 4 shows the percentage reduction in
standard deviation associated with various estimates when the
parameters are held constant. The majority of the variance seems
to be due to the range of potential prey that can be eaten, with
variation in initial centipede compartment size ranking second.
Much smaller errors are due to the random temperature fluctuations
and variations in the time of feeding. The lack of importance of
feeding time was not anticipated but can be understood by an examination of Table 1, which shows that an expected 94% of the animals
will succeed in feeding during the first forty-eight hours after
reaching the threshold. This period is short in comparison to the
centipede's slow metabolic rate. It cannot be demonstrated by an
F-test ($p = 0.05$) that reductions in standard deivations of less
than about 25% in Table 4 are significant for 19 and 19 degrees of
freedom. Since feeding time and random temperature factors may not
contribute significantly to the variance, little information would
be lost in this case by simplifying the model and setting these
parameters equal to their expected mean values.

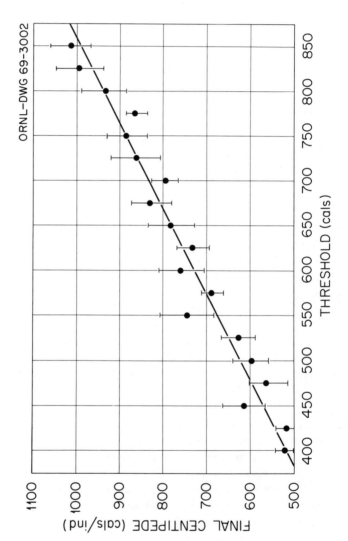

Figure 3.   Sensitivity of the final caloric pool of centipedes to changes in the hunger threshold. The data were generated by the simulation model and are indicated as means plus or minus a standard error (N = 20)

Table 4.   Percentage reduction in standard deviations of estimated energy parameters produced by setting various random variables at a constant value

Factor held constant	Energy Parameters				
	Ingested energy	Excreted energy	Assimilated energy	Respired energy	Reproductive energy
Species of prey consumed	60	60	60	83	54
Initial caloric pool	41	41	41	32	42
Time of feeding	16	16	16	35	14
Temperature	17	17	17	27	17

6. DISCUSSION

Methods for making first approximations of transfer dynamics are becoming increasingly important as more investigations are made into the complexity of ecosystems. The desire to make maximum use of limited data and to generate confidence limits for subsequent predictions should result in increased use of simulation techniques. In order to utilize discrete time functions and probability data, stochastic methods must play a part in these simulations. The complexity of the mathematics involved in deriving expected values for means and variances analytically in complex processes is formidable and is not likely to be available to the ecological researcher in the near future. The Monte-Carlo simulation technique, however, circumvents most of the complexities and provides a method simple in concept and implementation.

The ability of the stochastic simulation technique to predict energy parameters and confidence limits from minimal data demonstrates its usefulness in drawing information from experimental results which would otherwise be concealed by the complexity of the system. Sensitivity analysis and analysis of the components of variance introduce additional tools for the exploration of complex predator-prey interactions. The ability of Monte-Carlo simulation techniques to produce this type of analysis argues strongly for the increased use of such models following the general procedure outlined here.

REFERENCES

[1] Golley, F. B. 1960. Energy dynamics of an old-field community. Ecol. Monogr. 30:187-205.

[2] Holling, C. S. 1964. Analysis of complex population processes. Can. Entomol. 96:335-347.

[3] ——————. 1966. The functional responses of invertebrate predators to prey density. Memo. Entomol. Soc. Can. 48:1-86.

[4] O'Neill, R. V. 1968. Population energetics of a millipede, Narceus americanus (Beauvois). Ecology. 49:803-809.

[5] ——————. 1968. Prey preferences of a forest centipede. Radiation Ecology Progress Report for 1968, ORNL-4316, 99-101.

[6] ——————. 1969. Indirect estimation of energy fluxes in animal food webs. J. Theor. Biol. 22:284-290.

[7] Reichle, D. E. 1967. Radioisotope turnover and energy flow
     in terrestrial isopod populations. Ecology. 48:351-366.

[8] White, J. J. 1968. Bioenergetics of the woodlouse Tracheonis-
     cus rathkei Brandt in relation to litter decomposition in a
     deciduous forest. Ecology. 49:694-704.

ANALYSIS OF WILDERNESS ECOSYSTEMS

ROBERT R. REAM*
LEWIS F. OHMANN
U. S. Dept. of Agriculture, Forest Service, North Central
Forest Experiment Station, Folwell Avenue, St. Paul, Minn.

SUMMARY

Using the analysis of the natural communities of the Boundary
Waters Canoe Area, an accurate model of the natural wilderness
ecosystem is built. The model, in two parts, identifies and
describes the communities of the area and shows how they are
related to the environment and to each other, predicts the behavior of each species along environmental gradients and also
interspecific association patterns.

*present address:

    School of Forestry
    University of Montana
    Missoula, Montana

1. INTRODUCTION

Development of systems analysis in ecology has primarily been
related to productivity, energy flow, and nutrient cycling of
ecosystems. However, productivity is of little concern in Wilderness Areas and National Parks where nonconsumptive uses of
ecosystems prevail, except, perhaps, when comparing these reserves
to man-dominated landscapes. In such reserves the usual objective
of management is to maintain the ecosystems in, or where necessary,
restore them to a "natural" condition. However, this does not mean
that systems analysis cannot be applied to the problems of park or
wilderness management. An accurate model of the natural wilderness
ecosystem is needed before maintenance or restoration techniques
are applied. The nature of the model will be determined by the
stated objectives of management. In regard to such objectives, the
following questions might be asked: What is the composition and
structure of "natural" communities? How are these communities
related to the environment and to each other? How do individual
species relate to the environment and to each other? What conditions are necessary to maintain or restore natural communities?

We have undertaken a large-scale inventory and systematic
analysis of the natural communities of the Boundary Waters Canoe
Area to build a model that will help answer these and related
questions. Although we initially concentrated on the vegetation,
animal studies are now being initiated, and will be related to the
communities studied. This model consists of two separate but related parts, which are models in themselves. The first model
identifies and describes the communitites of the area and shows
how they are related to the environment and to each other. The
second predicts the behavior of each species along environmental
gradients and also interspecific association patterns.

Figure 1 is a flow chart illustrating our approach to systems
measurement, analysis, and description--the first three of five
steps involved in scientific management of a resource [7]. The
fourth step, simulation, is being approached through a simulation
of ecological processes using the present data. The fifth step,
optimization, will consist of determining the optimum strategy for
maintaining or restoring natural conditions in this wilderness.

We are using many of the statistical techniques being discussed
at this symposium in carrying out these steps. We have developed
computer programs for most of them, starting with summarizing plot

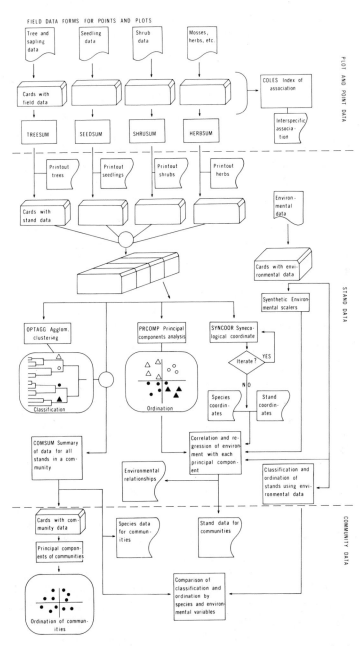

Figure 1.   Flow of information for ecological survey of the
boundary waters canoe area

data that has been key-punched directly from field data sheets. Each of the four summary programs provides a printout as well as punched output for each stand. After checking each of the stand printouts for possible errors, the punched output or stand data can be used for a number of different analyses. Instead of pursuing only one analysis technique and ignoring all others, we are using a number of techniques to gain a better understanding of the species and stand relationships.

Principal components analysis is used to provide an abstract spatial model (ordination) of the relationships of stands; the analysis also indicated the variance accounted for by each of the principal components used to construct it [4]. Using environmental data collected, we can determine the contribution of these variables to each principal component or geometric combination of several components (rotated axis). A polythetic agglomerative clustering technique, also developed by Orloci [5], is used to classify all stands into communities based on the same species data. Stands can also be classified and ordinated using only the environmental data collected. It is then possible to determine the degree of agreement between the classification based on species data and that based on environmental data [6]. We have also used a technique developed by Bakuzis [1] for assigning "synecological coordinate" values to species and stands. This technique provides an understanding of how species are related to moisture, nutrient, heat and light coordinates. The values for stands can then be related to the spatial ordinations to see how they fit the spatial model. To better assess interspecific relationships we can also determine indices of association for all species combinations. This has been done with stand data using a program that provides Cole's [2] index of association and chi-square significance for each species compared with all others.

After all stands are classified or clustered into communities we can describe the composition and structure of a community in a predictive fashion. Data from all stands included in a community are summarized. Thus species data for a community represent an average of the stands in that community. The prevalent species are listed for each community by taking the "n" species having the highest Index of Commonness [3], "n" being the average number of species found in all the stands of that community. Using the prevalent species data for each community a principal components

analysis ordination of communities illustrates the relationship of
the communities to each other and to the environment.

Using the above approach, we are building a model of this wilderness ecosystem that represents the composition and structure of
communities and also the behavior of individual species. Our research station has initiated a simulated wildfire study as a
possible technique for maintaining natural communities. A co-worker,
Dr. M. L. Heinselman has completed a study of the frequency and
extent of past wildfires and logging history. With information
derived from these two studies and a knowledge of the synecological
coordinates of species, we plan to develop "successional coordinates" or "successional functions" for each species. Then, using
the community data as a base, we will try to run a simulation of
succession for each community to predict what composition and
structure will be in 10, 50, or 100 years. With such information
and practical field experience with techniques, it will be possible
to determine the optimum strategies necessary to maintain or restore the ecological integrity of this wilderness area.

2. DISCUSSION

Now that we have described our study, we will briefly mention some
of the problems we have run into and some of the advantages we have
found using this approach. First, the systematic approach that we
have set up for collecting, summarizing and analyzing data has paid
off very well. By setting up our field data sheets for transfer to
punch cards, we eliminate the need for transcribing data and summarizing stand data manually, which is slower, more costly and
subject to more errors. Second, by having all our data on cards we
have been able to try a number of analytical techniques in examining and evaluating them. This helps in acquainting us with the
relative advantages of the various techniques and to gain more
knowledge about species and stand relationships.

One of the first problems we encountered in working with these
classification and ordination techniques was deciding which attribute or value of each species was most appropriate to use in
determining degree of similarity between stands. For example, in
our present study we have 106 stands of upland natural vegetation
that were sampled, with a possible 111 values for species occurring in three or more stands. A stand similarity matrix could be

computed based on presence/absence species scores or on quantitative data for species. For quantitative data we have the choice of frequency (of occurrence at sample points), dominance (basal area for tree and shrub species), or some combination of these. Orloci [6] has stated that "the information content of a species may be visualized as a composite of two independent components, arising from the factors of establishment and survival, and variable performance, respectively in the different sites." Establishment and survival then is represented by presence/absence species scores and variable performance is represented by one of the quantitative measures indicated above, or others. Although they are not strictly independent these two components are only loosely related. Orloci (see [6], p. 278) further states:

> It can be postulated that quantitative data are potentially more informative than the presence/absence species scores. This does not mean, however, that quantitative data necessarily represent an ideal type for the analysis of phytosociological relationships. Such data, being related to variable performance, may contain 'too much' information about details thereby obscuring important relationships at the level of establishment and survival in the different sites.

The contribution of the presence/absence component to the information content is at a maximum when a species is present in half the stands and when the non-zero values are constant. Conversely, if a species is present (or absent) in most of the stands and its quantitative values vary considerably, most of the information is found in the quantitative component.

Therefore, there is no simple answer to the question of the most appropriate attribute to use in comparing stands; it depends on the nature of the data. If we had distributed our 106 stands randomly over the entire state of Minnesota, we would have diverse data and most of the information would reside in the presence/absence component. As it was, we restricted these samples to random points within the half million acres of natural forest in the Boundary Waters Canoe Area, and further limited them to upland sites. However, this includes sites with a great variety of dominant species, such as Pinus banksiana, P. resinosa, P. strobus, Abies balsamea, Picea glauca, P. mariana, Thuja occidentalis, Populus tremuloides, Betual papyrifera, Cladonia spp. (lichens), and several shrub species. We mentioned above several kinds of quantitative data that could be used to represent the quantitative component. Some of these are "more quantitative" or contain more information about variable performance than others. Frequency is simply an average of presence/absence data for a number of sample plots or points in a stand, and as such it roughly indicates species abundance. The dominance data we collected contain the most information about variable performance of each species. For a given species, cover or basal area is directly related to biomass. However, when treating all 106 stands, we find that the dominance data contained "too much" information about details, so we have settled on frequency data for our initial ordination and classification of all stands. When working with a segment of the classification, for example, the Pinus banksiana - Picea mariana complex, we found that most of the stands are similar in species composition and much information would be lost if the dominance data were not used for more detailed analysis of this group of stands. Much more work needs to be done on determining the appropriate attribute to use at different levels of study. It appears that partitioning of information content could contribute a lot in such an effort. Such work could also contribute greatly to the design of ecological inventory studies to optimize the information gained per unit cost. We need to weigh the information lost by using more easily collected data against the reduction in cost in obtaining this data.

There are other interesting properties of these techniques that we have encountered in our analyses, but time will not permit full discussion of these. In determining relationships of environmental variables to principal component ordinations the level of generalization we are working with is important. Again, this is related to information content, not only of the species data but also the environmental data. Also, because the first principal component accounts for the greatest variance in the similarity matrix and represents the greatest distance in factor space, it covers the extremes in sites. In other words, instead of relating to moisture or temperature directly, this axis may go from a "hot-dry" stand at one end to a "cold-wet" stand at the opposite end. In working with synecological coordinates for species we have found it useful to look not only at the coordinate or mean value of a species but also at the variance about this mean. This gives us a much better picture of the range or amplitude of tolerances of individual species on a given coordinate.

We have tried to show how we are applying to our study a number

of statistical techniques recently put forth in ecological literature. Each of these techniques has contributed something to our
knowledge of the Boundary Waters Canoe Area Wilderness ecosystem.
There is much that remains to be done, but we are gradually
building the data base that will be necessary to ecologically
"manage" this large natural area.

REFERENCES

[1] Bakuzis, E. V. 1967. Some characteristics of forest ecosystem
     space in Minnesota. Pap. 14th. Congr. International Union
     of Forest Research Organizations, Munich. 2:107-125

[2] Cole, L. C. 1949. The measurement of interspecific associa-
     tion. Ecology. 30:411-424

[3] Curtis, J. T. 1959. The Vegetation of Wisconsin. University
     of Wisconsin Press, Madison.

[4] Orloci, L. 1966. Geometric models in ecology. I. The theory
     and application of some ordination methods. J. Ecol. 54:
     193-215.

[5] ————————. 1967. An agglomerative method for classification
     of plant communities. J. Ecol. 55:193-205.

[6] ————————. 1968. Information analysis in phytosociology:
     partition, classification and prediction. J. Theoret. Biol.
     20:271-284.

[7] Watt, K. E. F. 1968. Ecology and Resource Management. McGraw-
     Hill, New York.

RECORD OF PREPLANNED AND SPONTANEOUS DISCUSSIONS

F. C. HALL (U. S. FOREST SERVICE)

Questions:
1. Which "classification" system seemed most appropriate; the
classification or ordination approach?
2. Did you use regression analysis testing vegetation characteristics with environmental factors for both classification groups and
ordination gradients?
Answers:
1. Ordination is not really a "classification" system, but does
present a visual representation of the classification relationships.
In our work, both the classification used (5) and principal

components ordination [4] were found to be useful and Orloci [5] has shown that they are compatible. Therefore, neither is "most appropriate." The classification is useful for grouping stands into ecologically meaningful management units and the ordination is useful for determining stand-environmental relationships.
2. Yes. Samples selected for initial regression analysis were from classification groups. Ordination gradients were developed by combining similar groups for analysis. Example: groups A, B, C, and D were initially evaluated by regression. Then the two most similar groups were combined, i.e., A + B, and tested for regressions. Then group C was added and analyzed; and finally group D was added for the final regression analysis. This final analysis is essentially equivalent to ordination gradient evaluation.

We have tried using multiple regression with each stand's environmental variables as independent variables and stand location on each principal component as a dependent variable and this looks promising. The classification groups have been compared only on the basis of the means and deviations of the environmental variables measured for all stands of that group.

INFORMATION ANALYSIS OF QUANTITATIVE DATA

M. B. DALE
Department of Botany
University of Hull*

SUMMARY

This paper considers some of the problems which are encountered
when the classification method of Information analysis is applied
to quantitative data. Although there is a generally acceptable
procedure, much ecological data is in a form suitable for a special
modification.

This modification applies whenever the samples to be classified
can be treated as a priori groupings of subunits, and where the
recording is at a presence/absence level in the subunits. Examples
of data of this type include valence, percentage cover, cover repetition and density.

When data is collected using cover-abundance scales, the
resulting records can also be classified using Information analysis,
and this suggests one means of including data incorporating two
cross classifications, such as species and size class, in an Information analysis.

Two examples are presented of the use of the method. The
results suggest that polythetic methods may commonly produce
groupings which are very difficult to obtain by monothetic methods.
There is also some evidence that geographic separation is an important ecological feature, and rejecting the existence of plant
communities as entities.

*present address: Division of Plant Industry, CSIRO,
                   Canberra, Australia

1. INTRODUCTION

Studies of presence/absence data, using the heuristic classification
method known as Information analysis [17], have been shown to be
useful in ecology. Such studies may be criticized on the grounds
that qualitative data do not provide for the detailed analyses
which, some ecologists claim, are necessary in ecology. Such detail
is, in the opinion of these ecologists, only available from studies
of quantitative data. There is therefore some interest in the
possibility of extending Information analysis to encompass quantitative data. This paper considers the general problem briefly and
then outlines a simple extension which applies to much ecological
data and which permits Information analysis to be performed on these
data.

2. QUANTITATIVE DATA

The important problem with quantitative data is that of defining a
suitable measure of information for a continuous distribution. Provided the form of the distribution is known then such a measure can
be defined, but in most ecological work the distribution is not
known. Assumption of a normal distribution is almost always unten-
able due to the effects of presence/absence phenomena. In this
case the only alternative is to define a series of classes covering
the range of the quantitative variable and record only presence in
class, rather than the value. This is of course the method employed
by the various scales of cover-abundance which are often employed
as surrogates for quantitative data in ecology. The result is that
in place of the presence/absence dichotomy, the scales provide a
polychotomy. Any scales of this kind may be employed in Information
analysis so that the scales of vitality or sociability used by
European phytosociologists can also be analyzed if this seems desirable.

There is of course one major difficulty. The Information content
will depend on the choice of the class intervals. To define each
class to cover an equal range of variation provides one solution
but this can produce some difficulties. If, for example, a species
is rarely present, but happens to occur in large quantities when it
does appear, equal class intervals for coding into a multistate
variable leads to undue weighting of the species. Dale [2] has
suggested a means of defining the class intervals from the data

and this has been used with some success by Lance and Williams
([10],[11]) both in mixed data programs and in an extension of
Association analysis to quantitative data.

It should also be noted that Orloci's suggestion for evaluating
the relative importance of presence and quantitative information
[12] is also subject to the same criticism that the results will
depend on the particular classes chosen. Orloci also chooses to
ignore the partition of covariance which formed part of the original
procedure of Williams and Dale [13].

While there is little difficulty in employing this general method
of coding quantitative data, recent experience in numerical classification studies suggests that the closer the mathematical model
is to ecological reality the more useful the result. If the assumption is made that ecological considerations have determined the
necessity or desirability of employing quantitative data, it is then
possible to ask if the kinds of measures commonly employed in ecology have any special properties which would permit more efficient
analysis. There is little to be gained by inefficiently processing
data which requires greater expenditure of effort to obtain, and
presence/absence data is certainly very easy to record. However
there are two kinds of data commonly employed in ecology which do
have special features which permit the use of Information analysis
without having the difficulty of defining class intervals.

3. THE TWO TYPES OF DATA

In ecology it is common practice to employ measures which are percentages either of some fixed number of samples, or relative to the
total vegetation. These two form the two types of data which can
be efficiently incorporated into Information analysis.

For percentage data calculated from fixed numbers of samples the
extension to Information analysis rests on a conceptual distinction
between the samples recorded and the operational units classified.
For example consider a sample taken with a one meter square quadrat,
subdivided into 100 10 x 10 cm subunits. Recording the presence of
rooted individuals in the subunits provides a measure of the rooted
frequency or valence for the sample. At the level of the subunits
the recording is presence/absence only, and the sample is obtained
by an a priori grouping of the subunits, based in this case on spatial contiguity. This is, of course, exactly the process of

grouping carried out by Information analysis, so that analysis of
this data is exactly the same as classifying presence/absence data
by starting part way through the analysis after some initial classifications have been made. The modifications required to a
computer program are trivial.

This extension is not restricted to valence, since if the
presence of any part of a plant is recorded for the subunits of
the sample, the measure for the sample becomes shoot frequency,
and in the limit of very small subunits, this is percentage cover
(Greig-Smith [5], seems to expect a logarithmic relation of shoot
frequency and cover, not the simple relationship which actually
exists). It is clear that many of the measures used in ecology
derived from quadrats, point quadrats, and line transects do in
fact have the same form as valence or shoot frequency; line transect data for percentage cover use a very large number of subunits
in each sample (theoretically the number should be infinite, but
any measurement can only be made with finite precision), point
quadrats use frames as samples and points as subunits, and so on.
Not all measures fall into this class; for example density, yield
and cover repetition. Density could be converted to expected frequency and would then fit the analysis, although there are other
possibilities which will be considered in the next section.
However if density data are so converted, some ecologically interesting possibilities are apparent. The difference in information
content between a sample described by density data and one described
by valence data must be due to pattern, and by using this difference
it would be possible to classify, or ordinate, samples in terms of
their 'patternedness' at a fixed scale. Similarly the difference
between valence and cover information is due to size of plants and
it would therefore be possible to analyze this alone. Extension
beyond cover of this sequence of measures would seem to lie in the
direction of specifying more accurately the orientation of plant
parts, and their distribution in height, by using three-dimensional
sampling schemes. Such detailed description may prove to be essential for some purposes in ecology but for most purposes, cover
will be sufficiently informative.

Besides ecological data, some taxonomic data might also be
analyzed by a similar extension. If for example, the samples were
genera and the subunits species, then the recording of the number
of species with a given character in each genus is in the required
form. Here of course the number of subunits in each sample is not

fixed, and the sensitivity of information measures to sample size
might suggest caution in interpreting the results. There is however no theoretical problem in having varying numbers of subunits
in each sample. This situation might also occur in ecology if a
synthesis of samples collected by a variety of workers was attempted
since there could well be variation in the details of the sampling.

The possibility of having varying numbers of subunits in each
sample leads to a second extension which permits relative values to
be analyzed. In this case each sample is described by a single
property, for example number of plants, and this total is divided
between mutually exclusive types, for example the species. The
number of subsamples is now the total number of plants and there is
a single multistate variable with as many states as there are species. The analysis here is of the relative amounts of the species
and takes no account of the total amount of vegetation at all.
Whether this is desirable depends on the purposes of the user (see
[1] p. 840 for a discussion of the components of variation in vegetation).

This second type of analysis has proved useful in several non-
ecological fields. Relative data of this type are the standard
records of pollen analysts and an application to the zoning of
pollen diagrams is under study. Dale, Williams etc. [3] have
discussed the application of this type of analysis to the comparison
of transition matrices; these can be derived from temporal or spatial sequences such as DNA base sequences, amino-acid residues in
proteins, growth curves, leaf shapes and ecosystem models, all of
which are currently under study using this method.

4. METHOD

The method of analysis follows the method of Information analysis
with very minor modifications. For the first type of data, the
percentages, the modification consists only in the calculation of
an initial information content for each sample. For presence/
absence data this initial value is zero necessarily. Consider a
sample composed of n subunits and for each of p species, the number
of subunits which possess the appropriate property of the jth
species is a_j. Then the information content of the samples is
given by

$$I = p \times n \times \log(n) - \sum_{j=1}^{p} [a_j \times \log(a_j) + (n-a_j) \times \log(n-a_j)]$$

If any value is zero then $0 \times \log(0)$ is taken as 0, and the base of the logarithms is arbitrary, although computer operation will usually provide natural logarithms. As an example consider two samples, the first with twenty subunits, ten with species A, twenty with species B and one with species C, while the second sample has fifty subunits, fifty with A, five with B and none with C. The information contents of these samples using natural logarithms equals 17.834 and 16.254 respectively. Now combine these two samples into a composite of seventy subunits, sixty with A, twenty-five with B and one with C and the information content is now 79.573. The change in information caused by fusing is given by 79.573 - 17.834 - 16.254 = 45.485. This change in information is the measure of likeness between the two samples.

In the case of relative values the information content of a single sample is given by

$$I = n \times \log(n) - \sum_{j=1}^{p} [a_j \times \log(a_j)]$$

but the process is otherwise identical. In this case of course $n = \Sigma a_j$.

In the general case, where some classes have been defined, each species forms a multistate variable similar to the single variable of the relative data type. In this case the information content must be summed over all species. It is possible to use a similar method if, for example, the samples were described by species in size classes, as is common in forest survey. However if there is a possibility of interaction between the two groupings (of species and size class) then some care is necessary in the choice of the appropriate information statistic since five are available (see [6] for details of information statistics in multidimensional contingency tables).

5. APPLICATION

The methods outlined above have been programmed for a KDF 9 computer, in ALGOL, and for the CDC 3600 computer in FORTRAN IV. These programs only cover the percentage and relative data types, as

other programs exist [11] for general numeric data and others are
under development for the multidimensional (species x size class)
problem. The results are presented here of two analyses of percentage data. These data form part of a wider survey of grasslands
near Sheffield, Yorkshire. The samples were positioned by a random
walk method and each sample was divided into 100 subunits each of
10 x 10 cms. The measure used was valence, and a limited amount
of environmental information was simultaneously recorded. The low
sampling density of the broad survey means that the interpretation
of results from small subsets cannot be very detailed.

The first set of data consists of twenty-five samples from acid
grassland in Edale, Derbyshire. The data are shown in Table 1 and
the hierarchy obtained by noraml analysis shown in Figure 1. Following Lambert and Williams [8] suggestion, all hierarchies will
be read as if divisive although the program is in fact agglomerative.

The initial separation distinguishes areas with high and low
frequencies of Vaccinium myrtillus. This floristic division is
associated with a significant difference in pH at 10 cm depth.
(t = 3.52, p < .002). This does not decide the causal sequence
but does illustrate one means of testing the distinctions produced
by the analysis. The separation distinguishes between two samples
each of which has a 79% frequency of Vaccinium. (This division is
impossible to reproduce exactly by monothetic division and a quantitative analogue of Association analysis which also divides on
Vaccinium, distinguishes those samples with less than 79% from those
with at least 79%.)

The subsequent divisions seem to be related to the frequency of
Festuca ovina on the low Vaccinium side and to Nardus stricta on
the high Vaccinium side. These divisions are interpretable on
general ecological grounds but none of the environmental information
obtained showed marked relationship with these distinctions.

The second set of data is presented in Tables 2 and 3 and the
hierarchy is shown in Figure 2. The twenty-nine samples fall into
two groups, twenty-one from Lathkilldale, eight from Wardlow, both
areas being limestone grassland in Derbyshire. Both floristically
and in terms of site factors, these two areas overlap, and no simple
monothetic division can separate them. The Information analysis,
however, makes the geographical distinction its primary division.

Within the Lathkill sites, numbers 3, 4, 6, 7, 8, 9, 10 and 11

Table 1. Edale data: % frequency rooted & environmental data

Species		1	2	17	22	24	3	18	4	20	9	13	23	11	12	14	10	7	8	15	16	5	21	6	25	19
Agrostis canina.	1																	7	36							
A. tenuis.	2									13	11	8		23				9	20			38				28
Anthoxanthum odoratum.	3									2								8								
Blechnum spicant.	4															1										
Calluna vulgaris.	5																	2	3							
Carex binervis.	6																	64								
Deschampsia flexuosa.	7	94	71	95	95	81	99	70	99	100	48	86	66	86	94	98	70	17	74	94	98	84	93	12	24	22
Festuca ovina.	8	95	92	59	66	53	47	18	23	54	30	26	51					12	57	26	30	67		24	100	
Galium saxatile.	9									13	13	11	24		1		17	4	34					6	6	
Holcus lanatus.	10										2															
Juncus squarrosus.	11											3							2							
Luzula campestris.	12								1												1	4				
Nardus stricta.	13		5	1		8		31	33			27	36	13	3	16	59	80	81	59	53	46		4	2	
Potentilla erecta.	14																	2								
Pteridium aquilinum.	15			2	17	17			7	1					2	2		13	1					14	17	
Rumex acetosa.	16																								28	
Vaccinium myrtillus.	17	100	100	100	100	99	100	100	100	99	79	99	98	97	100	98	100	1	45	10	39	31		27	79	
V. vitis-idaea.	18					9																				
Slope : minimum (in °'s).		27	25	5	22	20	26	7	25	22	21	25	27	6	35	28	7	13	21	30	6	20	25	18	24	13
Slope : maximum (in °'s).		33	27	10	25	20	32	7	29	22	58	25	28	26	35	35	25	16	32	30	8	27	25	21	34	13
Aspect : (in °'s).		20	⟵——— 225 ———⟶		⟵—— 20 ——⟶		⟵—— 225 ——⟶		⟵—— 20 ——⟶																	
Soil Depth : (inches to C horizon).		11	11	15	20	21	11	23	11	26	10	13	16	16	19	22	21	19	16	15	30	15	24	16	20	15
pH : surface.		3·7	5·3	3·8	3·7	3·9	3·5	3·7	3·8	3·9	3·7	3·6	3·7	3·7	3·8	3·7	3·8	3·7	4·1	3·7	3·8	3·7	3·8	3·5	3·8	5·5
pH : 4 inches depth.		3·7	3·6	3·8	3·7	3·6	3·7	3·9	3·6	3·8	3·6	3·7	3·6	3·4	4·4	3·9	3·8	3·7	3·9	3·8	3·7	3·9		3·9	3·7	3·9

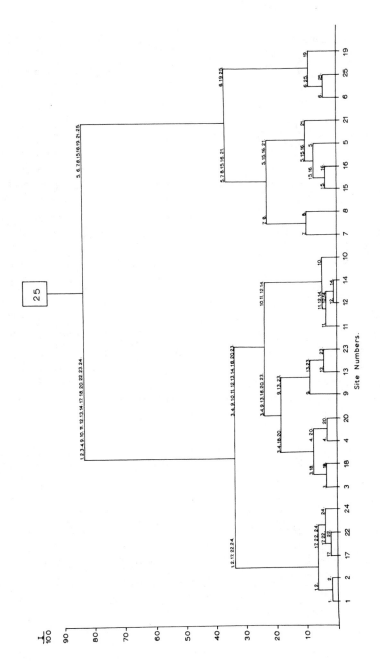

Figure 1.   Edale data-normal information analysis: % frequency rooted

Table 2. Lathkill & Wardlow data: % frequency rooted

Species		12	20	14	21	13	15	16	17	18	19	1	5	2	8	10	4	3	6	7	9	11	28	29	24	25	26	22	23	27
Achillea millefolium.	1					3									19	1											15	27	19	
Agrostis canina.	2								11		18				4	25	13	24				27	70	62	62	2	93			31
Agrostis tenuis.	3								9	79	82					62	38	59	36	5			59	72	97	100	19	91	100	98
Anemone nemorosa.	4																1						7	1	19	10	3	18	3	21
Angelica sylvestris.	5														3	1				10	2									
Anthoxanthum odoratum.	6													8	33	33	26	11	20	34	33	53	15	32	16	21	44	38	70	50
Aphanes arvensis.	7			41	6			5																						
Arabidopsis thaliana.	8			20				5																						
Arabis hirsuta.	9	14			2	2		16	10																					
Arenaria serpylifolia	10			30	8	4																								
Arrhenatherum elatius.	11	13		60	17	92	68	92	73	45	13				1		4		59	32										
Betonica officinalis.	12																													8
Briza media.	13	1													6	3											5	30	13	
Campanula rotundifolia.	14	3		19	20	3	1	1	1													1	1	15	10	35	56	41	15	
Cardamine pratensis.	15																	1												
Carduus crispus.	16			10																										
C. nutans.	17	6				4																								
Carex caryophyllea.	18			37	2										6	13			8	5			8	1			47	26	7	
C. flacca.	19	3				3		2								64			8	45	1						5			
C. panicea.	20														16	5				5					8	12		4		14
C. pulicaris.	21														18															
Carlina vulgaris.	22	4																												
Centaurea nigra.	23														6	10														
Cerastium holosteoides.	24	1				2	6								5		2	3										1	1	
Chamaenerion angustifolium	25		20																											
Circium arvense.	26													7																
C. vulgare.	27																											1	2	
Crataegus monogyna.	28						3												1				1							
Cruciata chersonensis.	29						21												11											
Dactylis glomerata.	30		2				1	1												6	2									
Deschampsia caespitosa	31																						32			2		18	48	
D. flexuosa.	32									100	41						6						97	100	35	72	82	13		
Epilobium montanum.	33																					5								
Festuca ovina.	34	67	54	22	89	18								10	18	12	58	85	50		26	11								
F. rubra.	35		95	76	81	92	87							80	35	55	7	96	98	98	100	80			5		77	68	44	
Fragaria vesca.	36						3								2	1														
Fraxinus excelsior.	37						1								1															
Galiopsis angustifolia.	38							8																						
Galium aparine.	39			3																										
G. saxatile.	40								14	28	41				5								84	99	90	98	67	92	72	78
G. sterneri.	41	9	11			5									11	14				6										
G. verum.	42																	49	79	4	10	14								
Geranium molle.	43	3		7	17	23	7																							
G. robertianum.	44						10	2	11	11																				
Geum rivale.	45														6	2				32	7	7								
Helianthemum chamaecistus	46	39	74	57																										
Helictotrichon pratense.	47	36	42												14	5						5	24	28	2	47	4	48	8	23
H. pubescens.	48														70	22	16	3	22								5			
Heracleum sphondylium.	49						1													2										
Hieracium pilosella.	50	6					23																							2
H. sps.1.	51																			1										
H. sps.2.	52		29				3																							
Holcus lanatus.	53			8		22						21			16	21	33	3	14	24										
Hypericum perforatum	54														1															
Koeleria cristata.	55	27	33	26	23	4									26	32	10	32	12	4	14	15	5	9	37	48	41	92	51	

Table 2 (continued).

Species	No.	12	20	14	21	13	15	16	17	18	19	1	5	2	8	10	4	3	6	7	9	11	28	29	24	25	26	22	23	27
Lathyrus montanus.	56															90							2	11	3					
Leontodon hispidus.	57														4															
Linum catharticum.	58	26	21	2	2																									
Lotus corniculatus.	59	23	12	6	4	66					7	5	3	1				10	4									23	6	
Luzula campestris.	60										1		2	2								2	99	97	100	91	89	89	72	81
Medicago lupulina.	61	29	34	83	22	26	1	12																						
Mercurialis perennis.	62														3		28		35											
Myosotis ramosissina.	63	1		2		3	4	7																						
Nardus stricta.	64										12															46				
Orchis mascula.	65														2															
Pimpinella saxifrage.	66		8												3	3														
Plantago lanceolata.	67	12	5			3																						2		
Poa pratensis.	68	3		96	93	91	87	94				13	94	55		2												2		
Potentilla erecta.	69														3	30	56	4	13	19	48	62	2	11		3	2	10		
P. sterilis.	70																	4												
Poterum sanguisorba.	71											48	63	9	2				23	31	50									
Prunella vulgaris.	72	1																										1		
Ranunculus acris.	73	1																										1		
R. repens.	74															1														
Rubus idaeus.	75									3																				
Rumex acetosa.	76											1	3	7	1		1	6	1											
Saxifraga granulata.	77																		3											
Sedum acre.	78	1		4	2																									
Senecio jacobea.	79	2		1	2				6	6																	1			8
Sieglingia decumbens.	80																										1			8
Sonchus sps.	81	2	12			1				5	10	1																		
Stellaria holostea.	82																16													
Succisa pratensis.	83											27	54						20		59							1	1	
Taraxacum laevigatum.	84	3	5																											
T. officinale.	85					1															1									
Teucrium scorodonia.	86	31	47	2	23	32	15	26	43	34																				
Thymus drucei.	87		66			5																								3
Trifolium repens.	88					47																							10	34
Trisetum flavescens.	89	13		7	2	27	19																							
Urtica dioica.	90					1	3								1								17		100					
Vaccinium myrtillus.	91																													
Valeriana officinalis.	92															1				1	4									
Veronica chamaedrys.	93		9						63							2										14				
V. officinalis.	94																									4				
Viola hirta.	95	2					14																							
V. lutea.	96																								5			7	9	3
V. riviniana.	97						2					29	17	7	15				33	13	11							37	32	1
Zerna ramosa.	98																				1									

Table 3. Lathkill & Wardlow: environmental data

Properties	Site Nos.																												
	12	20	14	21	13	15	16	17	18	19	1	5	2	8	10	4	3	6	7	9	11	28	29	24	25	26	22	23	27
National Grid Reference.				←176656→										←175653→											←178740→				
Slope : minimum (in °'s).	48	42	18	46	35	20	43	30	23	28	5	0	10	28	35	12	15	20	30	28	35	20	20	20	20	15	20	10	20
Slope : maximum (in °'s).	48	42	18	50	35	20	43	32	23	32	5	5	10	28	35	17	15	33	30	28	35	25	25	25	20	20	25	20	20
Aspect : (in °'s).	180	180	195	240	210	180	180	180	180	180	320	20	280	0	0	10	65	0	0	0	0	10	0	0	0	20	20	350	10
Soil Depth : (inches to C horizon).	9	9	8	9	13	8	9	12	12	12	14	20	13	12	10	9	10	9	9	14	8	8	8	12	12	10	12	7	10
pH : surface.	7.5	7.7	6.4	7.5	7.2	5.8	7.7	7.4	7.9	7.9	4.1	4.9	3.9	7.1	6.5	4.9	4.7	5.3	5.8	6.9	6.8	4.1	4.0	4.7	4.7	3.9	5.2	5.7	4.9
pH : 4 inches depth.	7.7	7.8	7.4	7.3	7.5	7.1	7.8	7.6	8.0	7.9	3.5	6.9	3.6	7.3	7.3	7.1	4.5	6.6	6.6	7.5	7.4	4.6	5.1	4.9	6.9	4.3	7.5	7.5	5.9

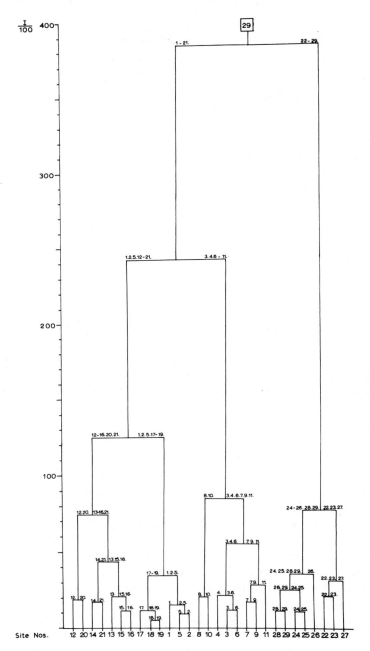

Figure 2.   Lathkill & Wardlow data information analysis:
            % frequency rooted

are north- or east-facing on slopes of at least 12 degrees. Sites
12, 13, 14, 15, 16, 20 and 21 are on shallower soils (using Tukey's
counting-in test as a rapid approximation the difference in depth
is significant, p < 5%). Sites 1, 2 and 5 and 17, 18, 19, are all
species poor sites. Festuca rubra is the only species common to
both parts of this group, which parts are separated at a lower
level. It would appear that this group is formed primarily because
of absences. Such a ragbag group has been noted previously [8] and
techniques are known for distinguishing them.

The Wardlow samples are much more homogeneous, but the separation
of sites 22, 23 and 27 appears to be related to pH differences both
at the surface and at 10 cms.

6. CONCLUSIONS

While the examples presented hardly provide a sufficient test of
the power and applicability of the method, there are some points
which merit discussion. Both the examples suggest that groupings
difficult to obtain by monothetic methods are common and may be
ecologically important. If a monothetic result is used as an
approximation to a polythetic result this must certainly be considered, although the use of a monothetic method to produce a key
is unexceptional. On testing various modifications, it was found
that the original version of Association analysis ([15], [16])
came closest, on a subjective rating, to the polythetic result
obtained here.

Ecologically, perhaps the most interesting result is the suggestion that geographic separation may be an important factor.
This is suggested by the Lathkill-Wardlow result and by unpublished
results from chalk grassland in East Yorkshire. This, if substan-
tiated, would provide evidence against the existence of plant
communities as entities, although this does not imply unqualified
support for the continuum hypothesis, as alternative models can be
suggested. Goodall [4] has previously considered the importance
of spatial separation on a small scale. It is also of interest that
a classification method could supply evidence against the existence
of plant communities and this reinforces the separation of method
and model made earlier [7].

ACKNOWLEDGMENTS

I am grateful to Drs. Rorison, Lloyd, Anderson and Grime, of the Department of Botany, University of Sheffield, for permission to use portions of the survey data. Dr. Lance, of the Computing Research Division, CSIRO, for the FORTRAN computer program.

REFERENCES

[1]   Austin, M. P. and Greig-Smith, P. 1968. Quantitative methods in vegetation survey. II. Some methodological problems of data from rain-forest. J. Ecol. 56:827-844.

[2]   Dale, M. B. 1964. Application of multivariate methods to heterogeneous data. Ph.D. Thesis, University of Southampton.

[3]   —————, Macnaughton-Smith, P., Williams, W. T., and Lance, G. N. 1970. Numerical Classification of Sequences. Austral. Comput. J. 2:9-13.

[4]   Goodall, D. W. 1961. Objective methods in the classification of vegetation. IV. Pattern and minimal area. Aust. J. Bot. 9:162-196.

[5]   Greig-Smith, P. 1964. Quantitative plant ecology. Butterworth, London.

[6]   Kullback, S., Kupperman, M., and Ku, H. H. 1962. Tests for contingency tables and Markov chains. Technometrics 4: 573-608.

[7]   Lambert, J. M. and Dale, M. B. 1965. The use of statistics in phytosociology. Adv. ecol. Res. 2:59-99.

[8]   —————. and Williams, W. T. 1966. Multivariate methods in plant ecology. VI. Comparison of information analysis and association analysis. J. ecol. 54:635-664.

[9]   Lance, G. N. and Williams, W. T. 1966. Computer programs for hierarchical polythetic classification ('Similarity analyses'). Comput. J. 9:60-66.

[10]   —————. and —————. 1967. Mixed data classificatory programs I. Austral. Comput. J. 1:15.

[11]   —————. and —————. 1968. Mixed data classificatory programs II. Austral. Comput. J. 1:82.

[12]   Orloci, L. 1968. Information analysis in phytosociology: partition classification and prediction. J. Theoret. Biol. 20:271-284.

[13]   Williams, W. T. and Dale, M. B. 1964. Partition correlation matrices for heterogeneous quantitative data. Nature (Lond.) 196:202.

[14]   Williams, W. T. and Dale, M. B. 1965.   Fundamental problems
       in numerical taxonomy.   Adv. bot. Res. 2:35-68.

[15]   ————————. and Lambert, J. M. 1959.   Multivariate methods in
       plant ecology.   I.   Association analysis in plant communi-
       ties.   J. Ecol. 47:83-101.

[16]   ————————. and ————————. 1960.   Multivariate methods in
       plant ecology.   II.   The use of an electronic digital
       computer for association analysis.   J. Ecol. 48:689-710.

[17]   ————————, ————————, and Lance, G. N. 1966.   Multivariate
       methods in plant ecology.   V.   Similarity analysis and
       association analysis.   J. Ecol. 54:427-445.

ANALYSIS OF VEGETATION DATA:
THE USER VIEWPOINT

P. GREIG-SMITH
School of Plant Biology
University College of North Wales
Bangor, U. K.

SUMMARY

A systematic discussion of the analysis of vegetation data from
the user's viewpoint. The main objectives in collecting such data
are pointed out. The author strongly emphasizes the vital importance of a full appreciation of the biological situation in deciding
techniques of analysis of vegetation data.

1. INTRODUCTION

The crucial test of a technique is essentially an empirical one.
Does it prove practical in use? Are the results capable of readier,
or more exact, or more certain interpretation than those obtained
by other means? In most branches of biology the development of
technique and its assessment by users have gone hand in hand. The
situation is straightforward when an advance in technique is essentially one of instrumentation, as in the devising of an improved
microscope or a more accurate pH meter, though even then the final
test must be a user one. The use of statistical analysis represents the opposite extreme, where advance in technique normally
requires considerable understanding of the biological questions
being asked, and interpretation requires a full understanding of
what statistical analysis is doing, though not necessarily of the
precise mechanics of how it is done.

When statistical analysis was first applied to biological data
few biologists had the mathematical grasp, and few statisticians
the appreciation of the biological situation, to achieve an overall
understanding. Today the position is considerably improved though
it is still often true that the biologist either hands over his
data 'for analysis' or, more industriously, himself applies a
technique of analysis culled from a textbook and only half-under-
stood. Equally, the statistician too often accepts the data without
understanding the explanation, given in biological terms, of the
questions to be asked of them. Neither side is blameless--there
is a need, as Cox [6] has put it, for textbooks of "Biology for
Statisticians" as well as "Statistics for Biologists."

In relation to biology in general this is probably an unduly
pessimistic picture, but I believe that it is not unfair in relation
to analytical (as opposed to experimental) ecology. The field
situation is so highly multivariate that any reasonably efficient
technique of analysis is inevitably complex and beyond the scope
of the elementary textbook to which the biologist normally turns
and which often represents the limit of his statistical understanding. The advent of the computer, permitting more complex analyses,
has accentuated the difficulties. Experience of advising ecologists
on the application of numerical methods to vegetation data indicates
that many of them, faced by a multiplicity of available techniques,
are only too willing to accept a program that is at hand and then
treat the computer, like a pH meter, as an instrument that 'gives
the answer'.

The aim of this paper is to consider what the field plant ecologist asks of numerical methods and the limitations that the field
situation sets. Much of what I have to say will appear commonplace
to one or other side in the dialogue between ecologists and statisticians, but it is offered in the hope that it may contribute to a
greater mutual understanding.

I am concerned here with the analysis of vegetation data, but I
suspect that the problems presented by animals are not as dissimilar
as is sometimes implied. There are five main objectives in collecting such data, any or all of which may be important in a particular
investigation.

(i) Comparison of composition between stands of vegetation, in
terms either of single species or of overall composition.

(ii) Deduction of generalizations about the range of variation of
composition present. Such generalizations are normally expressed
as classifications or ordinations.

(iii) Correlation of vegetation with environment.

(iv) Assessment of potential (in terms of management or land use).

(v) Prediction of future composition.

2. COMPARISON OF STANDS

The comparison of representation of a single species between stands
presents, in principle, little difficulty, being readily handled
by t-test or analysis of variance, after appropriate preliminary
transformation of the data if necessary, or by contingency X^2,
according to the measure of representation used [10]. Comparison
of overall composition raises similar problems to those of classification and ordination, discussed below. What is often not realized is the much greater scope for observer bias than is general
in biological data. This has two sources, difference between
observers in accuracy of observation and the tendency of observers
to be either 'includers' or 'excluders'. The former can be reduced
by experience, though always likely to be important in species-rich
communities of uniform growth form e.g., grassland, epiphyllous
bryophyte communities. The second source of bias arises from the
tendency for some observers to include any individual plant lying
more or less on the boundary of the sampling unit, and for others
to exclude such plants, thus sampling either a slightly greater
or a slightly smaller area than that supposed. There is even

some evidence that one observer may be an includer in relation to one species and an excluder in relation to another [4].

The practical implication of observer bias is that we cannot accept the results of even simple tests of significance of difference at their face value without taking into account the procedure used in the field. A corollary is that when planning field observations with more than one observer, it is important that between-observer variance is combined with error variance and not with any comparison in which we are interested.

3. THE SIMPLIFICATION OF DATA

The central problem facing the ecologist concerned with vegetation data is to reduce the large body of information before him to more manageable proportions--to simplify and to make generalizations.

The data are typically in the form of records of the occurrence of species (either as presence or as some measure of amount) in stands. The amount of information (in the nontechnical sense) available or obtainable about each stand is so great that each stand is unique; to produce any useful simplification we must ignore some information. The first step is therefore to decide what information is to be rejected. The occurrence of some species in some stands is the result of circumstances peculiar to that stand and should clearly be rejected, e.g. a species with low establishment rate may have succeeded in establishing in one stand, but have failed to do so in another comparable one because precise conditions at the times propagules arrived--and conditions are never completely static--were less favorable. The occurrence of some species is thus, at the level of detail at which it is useful to examine vegetation, a matter of chance. Put another way, the data include background noise. The amount of background noise varies with the type of vegetation and sometimes becomes an important factor in the choice of technique [11]. The existence of background noise in the initial data is comparable with the situation in data on variation in individual plants. There is, however, an important difference; experience has shown that variation in certain attributes of individual plants can generally be considered as background noise, and variation in other attributes as generally of 'taxonomic value', whereas with vegetation no such confident judgements can be made. The attributes are all of the same kind, occurrence of species, and there is no general means of assessing which, of the species that are relatively rare in the data, are likely to be ecologically informative and which represent background noise.

Accepting for the moment that the elimination of background noise is possible [11], more information will still commonly remain in the data than is relevant to the questions that the ecologist is asking. An analysis that extracts all the remaining information will generally itself be too complex for interpretation. Thus an essential prerequisite is clear understanding of the aims of the investigation.

Ecologists have long been aware of these difficulties, though they have rarely been explicitly formulated, and have attempted to overcome them by selecting as criteria in classification particular categories of species, e.g. dominant species, faithful species, constant species. Without concerning ourselves here with the exact meaning of these categories (not always clear in any case) we may note that the choice of criteria is based on unproved theories about the nature of the plant community and control of its composition. The greatest contribution that numerical methods have made is that they permit the data themselves to determine the species to be used as criteria. Their present limitation is that any one method appears to emphasize certain aspects of the initial information, but we do not yet know enough about them to assess with any certainty which will give the most interpretable result for a particular set of the data, e.g. the interpretability of agglomerative classifications using different standardizations appears to vary in a complex way with the structure of the data matrix in terms of species richness, standing crop and species predominance [3]. It is clear that a technique must be judged in terms not only of efficiency in retaining the maximum amount of the original information, but also of which information is retained and which is rejected. Understanding of this has depended so far on trial in a variety of field situations.

4.   CLASSIFICATION AND ORDINATION

Two approaches to a set of stand data are available, classification and ordination. Discussion of their relative merits has been confused by considerations of the nature of variation in vegetation, which are largely, if not quite, irrelevant. Concepts of the plant community vary from a closely integrated system, analogous to an organism, to the view that it represents the result of a sifting by the environment of species having negligible interactions with one another. Likewise, variation in composition of vegetation has been regarded as stepwise, as continuous and as sometimes stepwise, sometimes continuous. Not unnaturally, techniques of ordination appealed at first especially to those who viewed vegetation as a continuum, and the mistaken idea arose that ordination is necessarily more satisfactory if variation is continuous and classification more satisfactory if discrete communities exist. This unfortunate juxtaposition of ideas has hindered the use and practical test of numerical techniques, though more recently various authors have discussed the problem more objectively in terms of efficiency and interpretability, e.g., Anderson [1], Lambert and Dale [17] and various contributors to a symposium on ordination and classification of vegetation in Abstracts Xth Int. Bot. Congr. (1964) pp. 283-287.

Recently Yarranton [23] has proposed abandoning classification and ordination as tools, at least in correlating vegetation with environment, on the grounds that either involves a view of vegetation as something more than the sum of its component species and that any influence of species A on species B can be accommodated by regarding A as one of the characteristics of the environment of B. He goes on to suggest analysis of the occurrence of an individual species by multiple regression on the factors of its environment. This suggestion is logically defensible but, I believe, misguided. The fact that classification and ordination are possible indicates that species do show parallelism in their response to environment. It is thus more economical to work in the first place in terms of vegetation, considering species together; this may well lead on to correlation of individual species with those environmental factors suggested by vegetational analysis as likely to be most important. The multiple regression approach itself is a well-tried one in autecological studies ([5], [20]).

Theoretical considerations suggest that choice between classification and ordination should rest on the range of difference in the set of stands being analyzed and this is confirmed in practice [12]. Data for n stands containing p species can perhaps be most usefully considered as n points defined by their positions in relation to p species axes. If the range of habitats (and hence of species composition) is great there will be some degree of clustering of the

points because of the non-random distribution of values of habitat
factors; some site-types occur more frequently than others. Such
clustering will occur whether the vegetation shows continuous or
discontinuous variation in relation to environmental gradients.
Consider the simplest case in which there are only two clusters.
The first axis of an efficient ordination will join the centers of
the two clusters. The second axis must be perpendicular to the
first, but the major axis of variation within either or both of the
clusters will commonly not be perpendicular to the line joining
them. Only if it is perpendicular in both clusters will an efficient ordination result; at worst the result will be an ordination
whose first axis is an expression of a difference obvious on inspection of the data, with the remainder of the relevant information
so distributed among subsequent axes as to be virtually uninterpretable. In the more likely situation of several clusters the
difficulty will be accentuated. Thus with a wide range of
composition it is clearly desirable to break the data down into
relatively homogeneous subsets by an appropriate classificatory
procedure. Within clusters ordination, with its greater flexibility
as a framework on which to examine correlation with environment,
is likely to be more rewarding. At what level in a classification
is the switch to ordination optimally made? The level is likely
to vary according to whether the principal controlling factors of
the vegetation are the same throughout the data, but the problem
needs further examination and can only be clarified by practical
tests.

In addition to making the choice of classification or ordination,
or a combination of both, the ecologist must decide the kind of
data to be collected. This decision is frequently made with little
prior consideration of the procedure of analysis to be used and
with little regard to the ultimate objectives of the investigation.
It has become increasingly clear, and here I modify my earlier
opinion ([10], p. 160), that in many cases the bulk of the interpretable information lies in qualitative differences--the presence
and absence of species ([17], [19]); only within groups of closely
similar stands do variations in amount of a species contribute
significantly. Thus, at least with survey data, presence and absence data are likely to be sufficient for classification and
ordination per se. Many practical ecologists, accustomed to the
undoubted importance of relative amounts of different species to

the understanding of the functioning of communities, find this difficult to accept. The reason for the success of qualitative data is presumably that, while variations in amount of species are undoubtedly important and meaningful ecologically, differences in the amount of one species are correlated with the presence and absence of other species; this concept is not unlike that of some of the older, subjective, systems of vegetation classification, especially that of the Scandinavian School, with its emphasis on 'constant species'. That qualitative data may be sufficient to produce satisfactory classifications and ordination does not mean that they will necessarily be adequate in the subsequent use made of the analyses. Particularly when the objective is assessment of potential, or prediction of future change, quantitative description will be needed before the results of analysis can be applied to the objective [11]; the decision on type of data to be collected cannot be based solely on the technique of analysis to be used.

Numerical techniques of classification and ordination have al-ready proved themselves valuable tools in dealing with vegetation data. What limitations do the field situation and the inherent limitations of the techniques themselves impose on their use?

It is customary to talk of species as having ranges of tolerance in relation to features of the environment, ranges that are, in principle, definable and which are commonly assumed to be contin-uous. This does not imply that tolerance is definable in relation to an individual environmental factor, but rather to the complete complex of environment. In other words, if we regard the environ-ment as specified by a large number of independent axes, the range of tolerance of a species forms a hypersolid, often of irregular shape, in the space so described.* Unfortunately, the situation may not always be as straightforward. That the 'hypersolid of tolerance' is irregular in shape, with concavities in its surface, may itself introduce difficulties into analysis. The limited volume of environmental space represented by a given set of samples may include discrete portions of the total hypersolid of tolerance. This means that the indication given by presence of the species in a sample is ambiguous, and the ecological information conveyed, which may be important, is lost in the background noise.

If the volume of environmental space is increased, this

---

* The 'hypersolid of tolerance' is equivalent to the hypervolume, defining the niche, of Hutchinson [16]

difficulty will disappear and the indicator value of the species will at least be retained in relation to larger differences in vegetation. It is likely, however, that in some cases there may be two or more hypersolids of tolerance, and then the ambiguity will remain; obvious examples are those species which occur in both maritime and montane habitats but not in habitats apparently intermediate in their physical characteristics, but most ecologists are familiar with less obvious cases.* Whether the cause of this behavior is the interaction of different environmental factors in controlling the species or the existence of ecotypes not recognized in the field is ecologically interesting but does not affect the immediate problem.

It may, optimistically, be supposed that there will generally be sufficient indication from other species to compensate for ambiguities caused in this way. There is, however, another difficulty, the effect of competition between species. Ellenberg's [7] well known experiments have demonstrated that at least in a simple experimental situation the optimum of a species in relation to an environmental factor may shift in the presence of another species and the field behavior of some species is plausibly explained in this way e.g., the occurrence of Pinus sylvestris in southern Britain only on acid peat and poorer mineral soils. The effect, in terms of our model, will be either two or more hypersolids of tolerance, or a 'hole' in the hypersolid. Though the effect on interpretability of an anlaysis may not be severe, it is particularly unfortunate that the effects of competition between species are likely to be excluded from the analysis. Though there is a considerable amount of information about interspecific competition in simplified experimental situations ([22], [15]) and it is possible to make suggestions about its effect at small scales in more complex vegetation [9], the role of competition in determining the overall composition of vegetation is still obscure.

5. CORRELATION WITH ENVIRONMENT

Perhaps the commonest objective of the ecologist setting out to survey vegetation in the field is to detect correlations between

---

* That the maritime and montane habitats are geographically separated is irrelevant; the important fact is the discontinuity in range of habitat conditions.

composition of vegetation and environmental factors. Such correlations are then used to formulate hypotheses about the factors determining composition. (The testing of such hypotheses and the elucidation of the mechanisms effecting control are outside this discussion.) It is important therefore to consider the effectiveness of the results of numerical analyses in the search for correlations. We may note first that correlations of environmental factors with classes is usually less easily detected than with position in ordinations, except in very simple situations. Where the initial data are very heterogeneous it may be possible to relate the major divisions in a hierarchy to corresponding difference in environment, as Grunow [14] has shown for a range of South African bushveld types. At a lower level of heterogeneity, but still one that is more effectively dealt with by classification, this is often not possible; this need not cause difficulty, as it is possible to ordinate the classes themselves [12].

Having prepared an ordination (and in this context a principal components or a factor analysis is an ordination), we may use it as a display of stand relationships (in terms of floristic composition) against which to examine the behavior of environmental variables. There are two approaches to this and here we can distinguish the characteristic habits of thought of the statistician and the ecologist. The ecologist is accustomed to look directly for differences in habitat between contrasting stands. The corresponding approach to an ordination is to plot values of any environmental variable on the ordination and examine the resulting pattern; if the values are ordered there is evidence of correlation. The assessment is normally subjective and is justifiably criticized for its crudity.

The statistician's approach, having extracted axes of variability, is to attempt to interpret these axes by examining the relation between position on an axis and environmental variables by, for example, regression techniques. Superficially this is a much more satisfactory approach, but may, I believe, obscure important features of the biological situation. In a set of vegetation samples (stands) with corresponding environmental (site) data there will normally be strong correlation between environmental variables, both because such environmental variables are, in part, determined by common causes, and because the set of samples represents a non-random selection of possible habitats (even if the selection is

only of samples within a defined geographical area). Thus the main
axes of site variability will oblique to the individual environmental axes. If we accept that vegetation responds in some degree
to all differences in environment--and this to most ecologists is
axiomatic--it follows that the main axes of vegetational variation
are unlikely to relate to individual environmental axes in any
clear-cut way.

From a statistical viewpoint the solution seems obvious. Let us
ordinate the environmental data as well and compare the axes of the
two ordinations. As far as I am aware this has not been attempted
by any of the more efficient techniques from any extensive body of
data, though Austin [2] has made a critical comparison of vegetational and environmental ordinations within one relatively
homogeneous vegetation type. If it were, there would probably
be a satisfactory correspondence between the axes of the two
ordinations, though perhaps complicated by the effect of historical
factors on the vegetation, but the only conclusion that could be
drawn would be that vegetation is determined by environment. The
analysis of vegetation data is essentially exploratory and what
the ecologist seeks from it are indications about the effect of
particular environmental factors or the control of individual
species, indications that can lead to experiment, if necessary
after collection of further field data.

6. APPLICATIONS

Relatively little attention has been paid to the application of
numerical methods of vegetation analysis to applied problems of
management, land use, etc., but they are of considerable potential
value in these fields [11]. If intensive study or experience of
a limited number of samples of a range of vegetation allows us to
make statements about, for example, the optimum sylvicultural or
grazing treatment for a particular stand or the amount of usable
timber or other resource in it, and we have means of grouping
further stands with those already known, there is a sound basis for
decision about such stands. It is desirable that allocation of
additional stands to groups previously established should be possible on a minimum of data, and by a simple technique, so that an
agricultural or forestry officer in the field can assign a stand
to a group from his own resources.

In principle either classification or ordination could be used to summarize the relationships of the known stands. Preference has been expressed above for classification where data are markedly heterogeneous, as they are likely to be in this context. Moreover, any procedure of placing a stand in relation to an existing ordination, e.g. by using the previously calculated species loadings, is likely to require a complete enumeration of the new stand. Lastly, there are practical reasons for preferring classification as the basis of management.

The relative advantages of divisive and agglomerative strategies of classification have been argued largely in relation to the analysis of one set of data and subsequent interpretation ([21], [18]), but no clear cut conclusion has been reached. If, however, the objective of an analysis is to use it as the basis of action for a much wider set of stands within the same range, there are strong reasons for preferring a monothetic divisive strategy. At least in principle only the occurrence of the limited number of species which serve as division criteria need be recorded, and the position of a new stand in the classification can then be 'keyed out'. These advantages are so great that they override possible lesser efficiency.* Again considerations of the field situation are crucial in choice of technique.

The final objective of analysis listed at the beginning of this paper is the prediction of future composition. In situations where vegetation changes slowly over a long period, e.g. in the recovery of degraded tropical forest, it is difficult to determine the course of change. If, however, stands are available at various stages and they can be re-examined after a relatively short interval, an ordination treating each enumeration of a stand as a separate individual can be made. If the positions of a stand on the ordination at successive intervals of time are joined by a line this indicates the changing composition of the stand. If the lines for all stands follow the same trend, prediction of future composition, in so far as this is expressed by position in the ordination, becomes possible. This approach has evident applied value and we are greatly interested in it at Bangor. Initial investigations are promising but conclusions must await the completion of work in progress.

---

* R. Sokal (pers. comm.) has since pointed out that a classification need not necessarily be monothetic divisive to provide 'key' characters.

In this paper I have aimed to show the vital importance of a full
appreciation of the biological situation in deciding techniques of
analysis of vegetation data. I have deliberately avoided discussion
of rival techniques in an attempt to concentrate attention on the
broader issues. I have also chosen to direct attention to variation
between stands and have not mentioned within-stand variation--the
structure of plant communities. The latter is equally interesting
to the ecologist but its analysis presents fewer statistical, though
perhaps more ecological, problems ([8], [13]).

REFERENCES

[1]   Anderson, D. J. 1965. Classification and ordination in
      vegetation science: controversy over a non-existent
      problem? J. Ecol. 53:521-526.

[2]   Austin, M. P. 1968. An ordination study of a chalk grassland
      community. J. Ecol. 56:739-757.

[3]   ————, and Greig-Smith, P. 1968. The application of
      quantitative methods to vegetation survey. II. Some
      methodological problems of data from rain forest. J. Ecol.
      56:827-844.

[4]   Beshir, M. E. 1968. A study of the accuracy of estimates of
      the specific composition of vegetation. M.Sc. Thesis,
      University of Wales.

[5]   Blackman, G. E. and Rutter, A. J. 1946. Physiological and
      ecological studies in the analysis of plant environment.
      I. The light factor and the distribution of the bluebell
      (Scilla non-scripta) in woodland communities. Ann. Bot.
      10:361-390.

[6]   Cox, C. P. 1968. Some observations on the teaching of
      statistical consulting. Biometrics 24:789-801.

[7]   Ellenberg, H. 1953. Physiologisches und ökologisches Ver-
      hatten derselben Pflanzenarten. Ber. dt. bot. Ges. 65:
      350-361.

[8]   Greig-Smith, P. 1961. Data on pattern within plant communi-
      ties. I. The analysis of pattern. J. Ecol. 49:695-702.

[9]   ————. 1961. The use of pattern analysis in ecological
      investigations. Recent Advances in Botany, vol. 2:
      1354-1358. Toronto.

[10] ————. 1964. Quantitative Plant Ecology, 2nd ed.
      London.

[11] ————. 1969. Application of numerical methods to tropi-
      cal forests. International Symposium on Statistical
      Ecology, New Haven, vol. 3:195-206.

[12]   Greig-Smith, P., Austin, M. P. and Whitmore, T. C. 1967.
       The application of quantitative methods to vegetation sur-
       vey. I. Association-analysis and principal component
       ordination of rain forest. J. Ecol. 55:483-503.

[13]   ——————, and Chadwick, M. J. 1965. Data on pattern within
       plant communities. III. Acacia-Capparis semi-desert scrub
       in the Sudan. J. Ecol. 53:465-474.

[14]   Grunow, J. O. 1967. Objective classification of plant com-
       munities: a synecological study in the sourish mixed
       bushveld of Transvaal. J. Ecol. 55:691-710.

[15]   Harper, J. L. 1967. A Darwinian approach to plant ecology.
       J. Ecol. 55:247-270.

[16]   Hutchinson, G. E. 1965. The Ecological Theater and the
       Evolutionary Play. New Haven and London.

[17]   Lambert, J. C. and Dale, M. B. 1964. The use of statistics
       in phytosociology. Adv. ecol. Res. 2:59-99.

[18]   ——————. and Williams, W. T. 1966. Multivariate methods
       in plant ecology. VI. Comparison of information-analysis
       and association-analysis. J. Ecol. 54:635-664.

[19]   Orloci, L. 1966. Geometric models in ecology. I. The theory
       and application of some ordination methods. J. Ecol. 54:
       193-215.

[20]   Rutter, A. J. 1955. The composition of wet-heath vegetation
       in relation to the water table. J. Ecol. 43:507-543.

[21]   Williams, W. T., Lambert, J. M. and Lance, G. N. 1966.
       Multivariate methods in plant ecology. V. Similarity
       analyses and information analysis. J. Ecol. 54:427-445.

[22]   Wit, C. T. de. 1960. On competition. Versl. landbouwk.
       Onderz. Ned. 66:8.

[23]   Yarranton, G. A. 1967. Organismal and individualistic con-
       cepts and the choice of methods of vegetation analysis.
       Vegetatio. 15:113-116.

RECORD OF PREPLANNED AND SPONTANEOUS DISCUSSIONS

R. C. CHAPMAN (U. S. Forest Service)

The analysis of vegetational data as Mr. Greig-Smith indicates can
be a major problem for the statistician and the ecologist. The
increased availability of the computer, although it allows the
ecologist an opportunity to look at the multivariate complex that
exists in his biological sampling universe has emphasized many
areas that have previously received relatively little attention.

The ability of the computer to process mountains of data and produce reams of output it sometimes seems to me to have encouraged the collection of as much data as time and funds allow. In many cases insufficient prior attention appears to be devoted to the formation of a model.

The emphasis on hypothesis testing seems to be somewhat overdone when one considers the lesser emphasis on the details of the model. In some cases estimation would seem to be more appropriate.

Some of the problems such as observer bias that Mr. Greig-Smith mentioned can be eliminated by the use of well thought out sampling designs with clear objectives. The lack of statistical significance of important biological variables is sometimes an indication that the sampling design has not received sufficient attention.

The suggestion by Mr. Greig-Smith to combine the between-observer variance and the error variance has the same objection that pooling interaction variances and the error variance in the usual analysis of variance situations, namely that pooling the two variances tends to swell the expected value of mean squares and decrease the sensitivity of the tests. There are a number of sample designs that have been developed to isolate the differences between observers.

To some extent I believe that Mr. Greig-Smith has used the term classification when he is really talking about clustering. In many instances it appears that the question is not what group to assign an individual to but do groups exist and how many are there. Cluster analysis may bridge the gap between classification and ordination. The question however arises which clustering technique to use.

As Mr. Greig-Smith suggests ecologists often need more statistics and statisticians probably need more biology. The irritation that occasionally arises between the two species is due in part to communications--jargon--and in part to lack of understanding of the other assumptions and objectives.

P. DAGNELIE (Faculté des Sciences Agronomiques de l'Etat, Gembloux, Belgium)

We have to be grateful to Professor Greig-Smith for having given us such a good paper on the viewpoint of the ecological use of statistical methods. Generally speaking, I completely agree with these views and I will only comment on or stress a few points.

The first point is the absolute necessity of having in mind a well defined aim for every investigation. A clear understanding of this aim is really a prerequisite at every stage of the investigation, especially when defining the data to be collected (qualitative or quantitative for instance), especially also when selecting the best way of collecting these data and the most adequate methods of analysis (ordination or classification for instance). It seems to me that many research workers do not pay enough attention to this problem.

Although Professor Greig-Smith did not talk much of what he has written on this subject, I would also comment on the need for a reciprocal training of ecologists and statisticians, the ecologists having to learn a lot of things in statistics and computing, and the statistician needing a good understanding of biological and ecological problems. I would probably go further than Professor Greig-Smith in saying that ecological research is becoming more and more a job for teams of research workers and not for individual research workers. As a matter of fact, most ecological research projects imply at least a good understanding of ecological problems, a well defined objective, the ability to collect data, a good knowledge of statistical and computing methods and their limitations, and the ability to attain a final interpretation of the results. This is frequently far beyond the capability of a single man and is the work of a team of scientists.

By a team, I mean at least two research workers acting in close cooperation right from the beginning to the end of the investigation. I do not mean an ecologist doing the research by himself plus a statistician or a computer man whose help is asked only at the time of the analysis of data!

JON T. SCOTT (State University of New York, Albany)

I would like to comment on two statements made by Dr. Greig-Smith. First, he said to the effect that the vegetation ordination is a framework upon which to examine the "behavior" of the environment. I think I know that he means to see how the environmental variables plot on the ordination, but I think it is conceptually better to think of vegetation as varying in an environmental framework because this is much more like what happens in nature. In other words it is better to think of the vegetation as the dependent variable. This is of course a trivial comment but the reverse thinking could perhaps lead to misconceptions about the nature of the vegetation-environment relation.

The second statement I should like to comment on is one in which he said that it is "axiomatic" that vegetation is correlated with environment. This needs some clarification because it depends upon how you define environment and also on what you mean by "correlated". If you define environment as the set of physical phenomena which directly impinge upon an organism to affect its mode of life sometime during its life cycle then the statement is circular. You have defined environment based upon the effect it produces upon vegetation. The two are not independent and I must therefore agree with the statement. But from the users' viewpoint the best we can do about environment, at least with present methods, is to make a site analysis for each stand. Then if we want to determine if (and how) vegetation is correlated with environment we can consider the analytic analog of a vegetation-environment plot.

For the ordinate use the mean value for a restricted range of some type of ordination (say principal component analysis). I call this the vegetation gradient (or compositional gradient). On the abcissa plot the mean value for the same restricted ranges of an independently measured value of a site factor (which we will call environment). If the result is a straight line then you must say that there is no valid reason for classification because you cannot tell where one "type" leaves off and another begins. You therefore classify only for utilitarian purposes. It is the same as classifying red, orange, yellow, etc. There is no physically realistic border between red and orange and so the classification is arbitrary. In the case of vegetation we would say that the "continuum" hypothesis of Curtis and McIntosh or the "individualistic" hypothesis of Gleason are correct.

However, if you find that there is a region of gradual change in the slope of the plot of the vegetation ordination with change in the environment followed by a region of steep change and again by a gradual change, then we must conclude that vegetation can be validly classified. We have two "vegetation types" with an "ecotone" between them. There must therefore be an interesting biologic reason causing this particular kind of plot which merits investigation. We could then perhaps say that there is some reason for species to "associate" into classifiable types.

H. LIETH (University of North Carolina)

Dr. Greig-Smith made the comment that qualitative informations are much more important in vegetation analysis than quantitative data. I agree that one usually starts with the analysis of qualitative

changes, but I think that in many cases the utilization of quantitative changes remains the only one or the better means for
classification. This becomes increasingly more important as we go
into "harsher" environments, where the number of species is greatly
decreased and therefore the chances for qualitative changes decreases as well.

T. F. ALLEN (University of Ife, Nigeria)

Concerning the points of quantitative and qualitative data, I report
that in an ordination of microscopic terrestrial algae on rock surfaces, quantitative data gave a long thin cluster of stands. The
long axis of the sausage represented the variation of the dominant
Schizothrix calcicola, while the variation of the other species was
obscured. A conversion of the data to presence and absence gave a
much more satisfactory ordination. It is encouraging to find microscopic plants behaving statistically like higher plants, for there
are striking differences ecologically. For example the tropical
and temperate species are for the most part the same, and if anything my tropical forest stands are species-poor.

Inadvertently, species polymorphisms not species were used as
ordination taxa: this, however, did not obscure environmental
interpretation since, presumably, the polymorphisms were a product
of the environment. Thus taxonomic problems may help ordination.
A full report of this work is in final preparation.

Greig-Smith's anticipation of ordinations spaced in time to
predict the future of vegetation may be particularly relevant to
microscopic vegetation, since the rapid changes that occur may give
us a chance, firstly to observe change, secondly to use it to prophesy and lastly to observe prophesy fulfillment or otherwise.

P. GREIG-SMITH (University College of North Wales, Bangor, U.K.)

I agree that "behavior of environmental variables" is a rather
unfortunate phrase and should perhaps have said "pattern of environmental variables."

Classification is certainly "utilitarian" if applied to a uniform
gradient of composition. I would go further and say that all classification is "utilitarian" in this context.

That quantitative features are important in harsher conditions
where there are fewer species is one aspect of their importance
where differences are less extreme.

SOME IDEAS ON THE USE OF MULTIVARIATE STATISTICAL METHODS IN ECOLOGY

PIERRE DAGNELIE
Faculté des Sciences Agronomiques
Gembloux, Belgium

SUMMARY

This review paper is mainly based on personal experience and is intended to be an introduction to a general discussion.

In order to avoid all misunderstanding during this discussion, the paper begins with a sketch of multivariate statistical methods (section 1). Then come several brief sections related to these different methods: factor analysis (section 2), component analysis (section 3), classification methods (section 4), discriminant analysis and analysis of dispersion (section 5), regression and correlation analysis (section 6). General comments will also be given as conclusions (section 7).

More detailed informations and numerous references concerning these different points may be found among others in several papers by the same author, some of which are listed at the end.

1. A SKETCH OF MULTIVARIATE STATISTICAL METHODS

To define in a few words the main aspects of multivariate analysis,
I will refer to a diagram (Fig. 1) already published elsewhere [4].

  I will first suppose that we always start from an observed p x n
matrix, the p variables being either vegetational (frequency, cover,
etc. of different species) or environmental, and the n individuals
corresponding to different stands.* On this basis, we may consider
that most multivariate statistical methods belong to five classes:
regression analysis, correlation analysis, component and factor
analysis, tests of homogeneity, classification and discrimination
analysis. For the first two classes, we assume that the p variables
are initially divided into two or three subsets, without any subdivision of the n individuals; for the last two classes, we assume
that the n individuals are divided or are to be divided in the same
way into two or more subsets, without any subdivision of the p
variables; on the other hand, in component and factor analysis, we
do not assume the existence of any division at all.

  The first technique I include in multivariate analysis, even if
it is not in accord with the English terminology, is multiple regression. At present, this technique ought to be known by almost
every ecologist. A less commonly known generalization of this
method is the theory of simultaneous equations, mainly developed
in the field of econometry. The purpose of these equations is to
relate not only one, but a set of dependent variables with a set
of "independent" or explanatory variables (for instance q dependent
variables and p-q explanatory variables).

  Multivariate correlation analysis has also to be considered in
its different aspects, namely multiple correlation, canonical correlation and partial correlation. In connection with multiple
regression, multiple correlation may be used to measure the quality
of the relation linking the dependent variable and the p-1 explanatory variables. In the same way, canonical correlation may be used
to measure the quality of the relations existing between two sets of
variables, for instance in simultaneous equations. Partial correlation too is a quite commonly known concept, used in the measure
of correlation between two variables, when one wants to leave out
of account the influence of one or some (say p-2) other variables.

---

* As a result of my own preoccupations, this paper is principally
  oriented to plant ecology: I nevertheless think that, broadly
  speaking, it applies as well to every biological community.

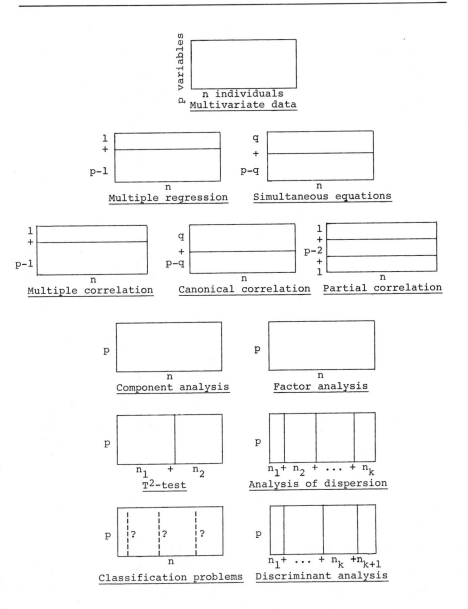

Figure 1. A sketch of multivariate statistical methods

The principal aim of component analysis is to reduce a set of variables to a smaller orthogonal subset containing the same or almost the same amount of information. Factor analysis, a group of techniques mainly developed in psychology, has a somewhat similar end, with a different mathematical model.

Tests of homogeneity include among others multivariate comparisons of means of two or several populations, on the basis of sampled individuals from these populations. The main statistical methods of this group are the T^2-test and the multivariate analysis of variance or analysis of dispersion.

Finally, classification problems must be considered as twofold. On the one hand, we have to define or to "make" classes, i.e., to subdivide a given set of individuals into a few subsets, using then classification or cluster analysis methods. On the other hand, we have to assign one or several individuals to one or the other of k predetermined classes, using discriminant analysis.

2. FACTOR ANALYSIS

As we have seen above, factor analysis starts from an unstructured data matrix, i.e., a data matrix with no distinction between variables as well as between individuals. In these circumstances, factor analysis is designed to express the observed variables in terms of a smaller number of fundamental non-observed variables or factors, generally on the basis of a linear model. The main problems and the main possibilities are then the estimation of loadings (the coefficients of this model), the estimation of factor variances and communalities (the sums of squares of loadings, measuring the importance of each fundamental factor and the importance of these factors for each observed variable), and the performance of rotation (to get more meaningful fundamental factors).

It seems to me that factor analysis is one of the most important multivariate methods in ecology. This opinion is based on the great number of problems that factor analysis enables us to tackle in an efficient way [1]. We may first analyze the correlations between species, getting then for instance:

(i) a sociological or an ecological interpretation of these correlations, by the way of the fundamental factors;

(ii) a measure of the importance of each of these sociological or ecological factors for each species, which is a measure of the

"indicator value" of each species for each factor, by the way of
the loadings;

  (iii) a measure of the general importance of each of these factors,
by the way of the factor variances;

  (iv) a measure of the general importance of these factors for each
species, which is a measure of the "differential value" of each
species, by the way of the communalities;

  (v) a basis of definition of sociological or ecological groups
of species, as well as of groups of stands;

  (vi) more generally, a basis of study of the homogeneity of stands
and structure of vegetation.

  In a similar way, factor analysis may be used to analyze correlations between environmental variables, these variables being then
expressed in terms of fundamental ecological factors.

  Moreover, factor analysis has been used to analyze matrices of
correlation or similarity between stands. Nevertheless, this is
probably neither the best processing of this kind of matrix, nor
the best ecological use of factor analysis. In the same way, although I myself did that kind of work some ten years ago, I would
now question the blind use of factor analysis in connection with
association coefficients other than the true correlation coefficient,
as well as the use of factorial methods to analyze mixed vegetational and environmental data. Regression and correlation analysis
would probably be more helpful in this latter case.

3. COMPONENT ANALYSIS

Component analysis has been applied by D. W. Goodall since 1953, in
conditions similar to those in which I have used factor analysis a
few years later. From a theoretical point of view, these attitudes
might seem more or less conflicting. Fortunately, from a practical
point of view, we may observe that both methods give very similar
results, provided that the number of variables is large enough [1].

  This is the reason why I will not spend more time on this topic,
considering that everything I have said about factor analysis applies to a large extent to component analysis.

  I must nevertheless add that the use of component analysis
would certainly be preferable when trying to get a reduced number
of variables with a view to some subsequent analysis (classification
for instance), instead of trying to get a final interpretation of
the observed variables and their relationships.

4. CLASSIFICATION METHODS

Classification has always been one of the ecologists' most important
concerns, and several modern methods of numerical classification
have even been set up in the field of ecology. But I must say that
I am quite afraid concerning these methods.

My apprehension comes from the ever growing number of new methods
appearing in many journals, not only in statistics and ecology, but
also in every branch of biology, in psychology, medicine, geology,
etc. ([5], [6]). Some of these methods are based on correlation or
association coefficients, others on similarity coefficients or even
on distances; some of them start from correlations, similarities or
distances between stands, and others from correlations, similarities
or distances between species or environmental variables; some work
by successive division and others by successive grouping, etc.
Moreover, in many cases, these new methods are published without
sufficient reference to previous similar works, with the consequence
that comparison of them becomes more and more difficult.

I would easily believe that nobody knows at present as much as
half of all existing methods. This would mean that we still need
the first scientist to be able to make a valid choice between these
methods or to give valuable advices about them.

Fortunately, many papers comparing these different methods are
now being published, but most of them are relevant only to a small
number of methods or to a small number of applications. We need in
fact one or a few methods having some optimal or suboptimal properties or, as far as possible, broad comparisons of all existing
methods.

5. DISCRIMINANT ANALYSIS AND ANALYSIS OF DISPERSION

The second aspect of classification, i.e., the allocation of individuals (species or stands) to well-defined groups, is now a
standard problem since R. A. Fisher's works on discriminant
functions. This concept too is now included in many advanced
courses and books of statistics and biometry.

Unfortunately, in ecology, as in many other fields, the classical
assumptions of multivariate normal distributions and equal variance-
covariance matrices are frequently doubtful. But much work has
been done to allow the use of discriminant functions in other cases:
discrete and qualitative variables, unequal variance-covariance
matrices, etc. [6].

In connection with discriminant analysis, the use of generalized distances and tests of homogeneity may also be helpful to compare well-defined groups of stands or species.

6. REGRESSION AND CORRELATION ANALYSIS

Simple and multiple regression and correlation have been used for a long time to express the development of species (frequency, cover, etc.) in terms of environmental variables.

Partial and canonical correlations may be used in the same context. Partial correlations between species, eliminating the influence of one or several ecological variables or factors, may be very helpful for instance when measuring the influence of these variables or factors on the relations between species; canonical correlation is also useful to relate a set or a group of species simultaneously to a set of ecological variables or factors ([2], [3]). Concerning this point, it might be interesting to stress the possibility of applying regression and correlation methods to results of factor or component analysis: canonical correlation may be used for instance to relate the results of a component analysis of correlation between species to the results of a component analysis of correlation between environmental variables.

As far as I know, the theory of simultaneous equations has never been applied in the field of ecology. Nevertheless, it seems to me that this might be worth trying.

7. CONCLUSIONS

At the end of this brief review, it becomes evident that multivariate analysis gives the ecologists a set of powerful tools to analyze their data. But some limitations do still exist, namely concerning the methods, the data, the users and the means.

With regards to the methods, one must always remember the basic assumptions, of normality and linearity for instance. The first is fundamental in most multivariate tests of hypotheses; the second is important even in purely descriptive factor or component analysis. The assumption of linearity is evidently seldom true, but transformations of variables may be very helpful when considering monotonic functions. Unfortunately, in many cases, the relations between species and ecological factors are not monotonic, and we certainly need more progress in non-linear multivariate methods.

Concerning the second point, we must acknowledge that only valuable <u>data</u> are worth using multivariate methods. This means that enough attention must be paid to sampling designs and observation methods. We certainly also have to search for more meaningful environmental observations: this is one of the main conclusions drawn from multivariate analysis of correlations between species and environmental variables.

Another limitation originates from the relatively low level of the mathematical and statistical knowledge of numerous <u>ecologists</u>. I am talking here about French-speaking countries, but I am sure that it will be true also in many other countries, and even, although to a lesser extent, in the United States. We certainly need more training programs in statistical ecology.

Finally, concerning the <u>means</u>, it is well-known that most multivariate methods require large amounts of computation. Even in the era of fast electronic computers, this problem is not completely solved: large computers are not available in each country or in each university or research center, and even when available, they are still very expensive. Once again, only valuable data are worth processing on these high speed machines. Moreover, we have to be careful and thrifty when choosing one method or another: this is especially true in classification problems, where a lot of very different methods are available.

REFERENCES

[1] Dagnelie, P. 1960. Contribution à l'étude des communautés végétales par l'analyse factorielle. <u>Bull. Serv. Carte Phytogéogr.</u>, <u>Sér. B.</u> 5:7-71 et 93-195.

[2] ─────────. 1965. L'étude des communautés végétales par l'analyse statistique des liaisons entre les espèces et les variables écologiques: principes fondamentaux. <u>Biometrics.</u> 21:345-361.

[3] ─────────. 1965. L'étude des communautés végétales par l'analyse statistique des liaisons entre les espèces et les variables écologiques: un exemple. <u>Biometrics</u>. 21: 890-907.

[4] ─────────. 1966. Introduction à l'analyse statistique à plusieurs variables. <u>Biom. Praxim.</u> 7:43-66.

[5] ─────────. 1966. A propos des différentes méthodes de classification numérique. <u>Rev. Stat. Appl.</u> 14(3):55-75.

[6] ─────────. 1968. Introduction aux problèmes et aux méthodes de classification numérique. <u>Biom. Praxim.</u> 9:87-111.

RECORD OF PREPLANNED AND SPONTANEOUS DISCUSSIONS

L. ORLOCI (University of Western Ontario)

I have enjoyed Dr. Dagnelie's presentation. I would like to make three points regarding the nature of ecological problems, the mode of implementation of models, and the testing of hypothesis.

Speaking in the broadest terms ecological problems tend to fall into two separate categories. They are either structure or function oriented. Structure may mean different things to different ecologists. It is often conceived as a set of similarity or dissimilarity relations, essentially in line with the requirements of the multivariate models reviewed by Dr. Dagnelie. According to an alternative definition, structure may imply the manner in which matter and energy are distributed between different parts of an ecosystem. Function, on the other hand, is a dynamic concept implying the changing structure in time. While the structure oriented problems can be handled by static models, such as cluster analysis, discriminant analysis, ordinations, etc., the function oriented problems require dynamic models which can sufficiently reflect the continuously changing state of the system.

Dr. Dagnelie's paper is principally concerned with multivariate models which are suitable for the analysis of structure as a static property. I am certain that most of us practicing ecologists would have liked him to discuss briefly the analysis of multivariate systems with continuously changing properties.

The models should preferably be probabilistic. It must not be forgotten, however, that a probabilistic model may not be required because the purpose may be only description and hypothesis generation [1]. Further, the use of a probabilistic model may not be appropriate because the conditions at hand do not satisfy the assumptions implicit in the model.

Hypothesis testing should logically follow hypothesis generation. Several alternative procedures are used by ecologists in this connection; e.g., consistency of results with known facts, consistency of results in repeated experiments or surveys, and reference to known probability distributions. The first of these alternatives may not seem too rigorous, it can nevertheless prevent the ecologist from attributing too much significance to results which according to previous experience could arise frequently by chance. Repeated experiments or surveys may provide the most reliable test but at

the same time they may also be the most tedious. The use of
standard variables of given probability distribution is probably
the most hazardous test of those mentioned since the ecologist may
be required to accept specific properties whose existence he may
not be able to verify experimentally or otherwise.

Finally, speaking about the field of plant ecology in general,
we were witness to many developments of great importance in the
past few decades. The most remarkable development appears to have
taken place in technique, particularly in data analysis. The trend
in part seems to point to the wide spread use of computer-based
systems of analysis and inference. Such systems, as Shubik [2]
points out in connection with political behavior, offer one fruitful
way or rendering complex problems manageable, and also providing
precision and speed in the process.

References

[1]   Lambert, J. M. and Dale, M. B. 1964. The use of statistics
       in phytosociology. In "Advances in Ecological Research".
       2:59-99.

[2]   Shubik, M. 1954. Readings in Game Theory and Political
       Behavior. Garden City, New York, Doubleday.

R. C. CHAPMAN (U. S. Forest Service)

Dr. Dagnelie has presented an interesting review of some multivariate methods in use in ecology. One of the oldest multivariate
methods that should be included is contingency analysis. Multiple
contingency tables are often a very useful way of looking at
ecological data.

Factor analysis has received wide attention in the last decade
with the increasing availability of digital computers. One of the
major problems still encountered by the factor analyst is the interpretation of factors. In many respects factor analysis is not
as objective as many of its users would appear to believe. The
intuition and knowledge of the researcher in the choice of variables, in the number of factors deemed appropriate, and the
interpretation of the factors seem to be the main criterion for
a successful factor analyst. All too often I have the feeling that
many users of factor analysis have little information about:

   (i) what are the most useful communality estimates;

   (ii) what set of variables should be considered;

  (iii) the type of relationships that exist between variables;

(iv) how many factors are expected;

(v) what type of factors are expected;

(vi) what relationships are expected to exist between the factors.

The problems associated with the mix of data that often occur in the analysis of the massed ecological data needs more attention. In particular I am referring to the mixture of discrete and continuous variables that often are found in ecological data.

Principal components have as Dr. Dagnelie suggests been used by ecologists in much the same way as factor analysis. In fact they have often been used interchangeably. The use of principal components in ecology suffers from the same sort of problems and interpretation that are associated with factor analysis.

The difference in philosophy of the two procedures needs, I think, to be emphasized. The purpose of principal components is the reduction in the dimensionality of the sample space. The components are not implicitly assumed to have any underlying importance. Factor analysis on the other hand attempts to determine "the" underlying structure of the population.

It is important to note that the principal component solution is not independent of scale. The problem of the scale and metric does not appear to be as widely known or acknowledged as it should.

I am in complete agreement with Dr. Dagnelie on the need for more information about the various methods of cluster analysis. Cluster analysis becomes sticky when one is dealing with data containing discrete and continuous variables. The selection of a cluster method I cannot help believing often depends on what clustering algorithm exists at the local computer facility. It would be interesting and useful to see more literature comparing the algorithms and similarity measures that have been developed.

Classical multivariate methods have provided the statistician and ecologists with reasonable first tools for analysis of associations among variables and among individuals. The problems that are currently being encountered need new and original methods of analysis. Too much of the data and methods are like the round peg and square hole.

P. J. LEE (Fisheries Research Board of Canada, Canada)

I would like to make two points: (1) Principal component analysis is used to create a hypothesis from a set of multivariate data. It is a hypothesis-generating technique. On the other hand, when we

use factor analysis, we will encounter the problem of determination
of communality or number of factors. Thus, maximum likelihood
factor analysis should be used to test our hypothesis whether our
data fit to the hypothesis. If the hypothesis is accepted, then we
estimate the factor loadings for the factor model. (2) Canonical
correlation analysis can be used to correlate two sets of random
variables or to predict a set of random variables from a set of
variables. It is powerful to use canonical correlation analysis as
a regression model.

FRANK RAYMOND (Ontario Dept. of Lands and Forests)

I would like to make two comments; one of which partly echoes Dr.
Chapman's remarks. The first is that, from a user's point of view:
Factor Analysis appears to be extremely subjective. Principal components require all of the variates to be measured on the same
scale. The two multivariate procedures which seem rigorously
objective are Discriminant Analysis and Canonical Correlation.
The second is that, since Professor Dagnelie has included Regression Analysis in his multivariate list, I am surprised that he did
not include Multivariate Analyses of Variance, although these are
difficult to interpret when they consist of more than two simple
analyses.

SOKAL (State University of New York, Stony Brook)

Dr. Dagnelie has outlined in an exemplary manner the complex subject
of multivariate statistical methods in ecology. I would, however,
like to raise several issues with him which occurred to me on reading his manuscript.

    First, and somewhat facetiously, I trust that the shape of his
rectangular matrices does not imply that he would always have more
OTU's than characters.

    Second, it seems to me that there is some danger that readers of
his paper will encounter terminological confusion over the word
"classification." Both taxonomists and ecologists have by now convinced even some of the statisticians that "classification" should
be the arrangement or partition into subsets of an initially unstructured set of objects based on whatever criteria are chosen on
which to erect a classification. The process of "identification" or
"allocation" is the assignment of a new object to one of several
previously defined sets or classes. This latter process to which

discriminant functions are admirably suited is often termed "classification" by mathematicians and philosophers, but much confusion
in discussion can be avoided by restricting ourselves to a uniform
terminology.

Third, I note that Dr. Dagnelie deplores the development of many
similarity coefficients without adequate comparison and evaluation
of their properties. I agree that much study in this field remains
to be done, but welcome the healthy outpouring of these methods in
fields as diverse as ecology and economics as a sign of the fundamental importance of these techniques. The Classification Society,
of which Dr. L. Orloci, a member of this Symposium, is the treasur-
er, is dedicated to the comparative study of the philosophy and
methodology of classification and I would hope that all of you who
share Dr. Dagnelie's and my interest in these problems would join
in our efforts by becoming members. Dr. Orloci will gladly furnish
you with membership applications.

Lastly, I cannot quite agree with Dr. Dagnelie that is is un-
desirable to "play" with computers. Only by a certain amount of
free experimentation can we begin to gain the experience necessary
to evaluate some of the methods he espouses. Until such work is
done it is often not possible to know which data are "good."

P. DAGNELIE (Faculté des Sciences Agronomiques de l'Etat, Gembloux,
            Belgium)

Dr. Orloci's first point is related to the dynamic properties of
ecological systems. Although I have no experience in it, I think
that these properties might be studied either by multiple time
series analysis or by factor analysis, considering the same stand
at different stages instead of different stands at the same moment.

Replying to Dr. Orloci's second point, I would like to stress
again the difficulties of doing valid multivariate tests of ecological hypotheses, because of the rather restrictive underlying
assumptions.

To Dr. Chapman, I must say that I agree with his views on the
importance of multiple contingency-table analysis. I also agree
with him and with a speaker from the floor concerning the fundamental distinction, from a theoretical point of view, between factor
analysis and component analysis.

Replying to Dr. Sokal's remark about the rapidly increasing
number of classification methods, I would also state precisely that
I am not deploring this number, but the fact that new methods are

often published without any reference to previous works, even by
the same author!

Lastly, I have to justify my inclusion of multiple regression
in multivariate analysis. The main reason is that we do not have
any distinction in French between "variable" and "variate" (i.e.,
random variable). I am thus making a distinction between one-
variable statistical methods, two-variable statistical methods,
and "multivariable," or, maybe loosely, "multivariate" statistical
methods.

NOTES ON THE MARCZEWSKI - STEINHAUS COEFFICIENT OF SIMILARITY

P. HOLGATE
Birkbeck College, London

SUMMARY

Some results are obtained for the distribution of the Marczewski-Steinhaus coefficient of similarity, in several particular cases.

1. INTRODUCTION

Many coefficients have been proposed as measures of the similarity
between a pair of ecological communities, or of the complementary
notion of the distance between them. Let the communities be numbered 1, 2, and let the finite or enumerably infinite set of species
which may appear in either of the communities be labeled 1, 2, ...
Suppose that a sample is taken from each of the communities according to some prescribed method, and let $F_{i\alpha}$ be a positive function
of the sample content of species α, from community i, taking the
value 0 when species α is absent from the sample. The following
functions have all been used in practical investigations:

(i) $F_{i\alpha} = X_{i\alpha}$, the number of specimens of species α present in
the sample from community i.

(ii) $F_{i\alpha} = p_{i\alpha} = X_{i\alpha} / \sum_{\alpha} X_{i\alpha}$, the corresponding proportions in
each sample.

(iii) $F_{i\alpha} = W_{i\alpha}$, the total weight of the specimens of species α
in the sample from community i.

(iv) $F_{i\alpha} = X_{i\alpha} = 1$ if species α occurs in the sample from community i, otherwise equal to 0.

(v) $F_{i\alpha} = \Delta_{i\alpha}$, defined for a plant community sampled by visual
inspection as taking the values 0, 1, 2, 3, 4, 5 according as
species α covers 0, 0-20, 20-40, 40-60, 60-80 or 80-100 per cent
of an area sampled in community i.

An important class of coefficients of similarity is obtained by
defining for a suitable function $F_{i\alpha}$,

$$L = \sum_{\alpha} \min_{i} F_{i\alpha} / \sum_{\alpha} \max_{i} F_{i\alpha}.$$

This clearly lies between 0 and 1, taking the extreme values respectively when the samples contain no species in common, and when
$F_{i\alpha}$ takes the same values for each sample, for every species. The
corresponding distance between the communities is

$$E = 1 - L = \sum_{\alpha} |F_{1\alpha} - F_{2\alpha}| / \sum_{\alpha} \max_{i} F_{i\alpha}$$

Other ratios of linear forms in the three quantities

$$M = \sum_{\alpha} \max_{i} F_{i\alpha}$$

$$m = \sum_{\alpha} \min_{i} F_{i\alpha} \tag{1}$$

$$d = \sum_{\alpha} |F_{1\alpha} - F_{2\alpha}|$$

contain the same information as L and E, and many ecologists have used the coefficients of similarity and distance

$$L' = 2m \ / \ (M + m) = 2L \ / \ (1 + L)$$

$$E' = d \ / \ (M + m) = E \ / \ (2 - E).$$

If alternative (iv) is chosen for $F_{i\alpha}$, L becomes Jaccard's well known formula, the ratio of the number of species occurring in both communities to the total number of species observed. If alternative (i) is chosen a coefficient is obtained which has been variously attributed. Gleason [3] seemed to be moving towards its definition, but he suggested

$$L^* = \sum_{\alpha}^{*} (X_{i\alpha} + X_{2\alpha}) \ / \ \sum (X_{1\alpha} + X_{2\alpha})$$

where the summation $\sum^{*}$ extends only over those species for which min $(X_{1\alpha}, X_{2\alpha}) > 0$. It is of historical interest to note that E with alternative (ii) was mentioned by Kulczyński [4] in his frequently quoted work, but he rejected it in favor of

$$E'' = 2d \ / \ (\ \frac{1}{\sum_{\alpha} F_{1\alpha}} + \frac{1}{\sum_{\alpha} F_{2\alpha}} \ ),$$

apparently on the grounds that E' was negatively biased under certain relevant conditions. Unfortunately E'' has come to be associated with Kulczyński as 'his' coefficient. Its drawbacks have been illustrated by several writers preparatory to introducing their own coefficients ([7], [6]). Discussions on the appropriateness of E for classification purposes as compared to, say, the Euclidean distance

$$E''' = \{\sum_{\alpha} (F_1 - F_2)^2\}^{\frac{1}{2}}$$

can be found in the papers of Austen and Orloci [1] and Bannister [2]. It is relevant to their discussion that Marczewski and Steinhaus [6] had already proved that E satisfies the triangle law. In another paper [5] these authors showed that the class of coefficients obtained by taking different functions $F_{i\alpha}$ all have this property, and can be regarded as generalizations of the Fréchet-Nikodym-Aronszajn distance between sets, with Jaccard's coefficient corresponding to the original distance. I attach these authors' names to the coefficient on the grounds that they were the first to study its mathematical properties, instead of merely trying it out on data.

The sampling distributions of E, L and their transforms are very difficult to study, because they involve the max and min functions, they involve ratios, and under many relevant sets of conditions it is not clear that they are asymptotically normal. The following three sections indicate directions in which some progress is possible, and present some elementary results.

2. AN APPLICATION OF NORMAL THEORY

If the structure of habitats 1 and 2 is identical, E will have its null distribution. Let the abundance of species α in each of the communities be defined as the mean number of specimens obtained per unit sampling effort and be denoted by λ_α, and suppose that the random number of specimens obtained has a Poisson distribution. A unit of sampling effort might be a quadrat which is exhaustively examined, or a period of time for which a trap is left open. Let $G(x)$ be the distribution function of species abundances, that is, the probability that a species chosen at random from the list will have abundance at most x. Let the moments of the abundance distribution be denoted by

$$\nu_s' = \int_0^\infty x^\nu \, dG(x).$$ (2)

The following assumptions will be made in this section:

(i) that the set of species in each of the communities is large but finite,

(ii) that the sampling effort, which will be denoted by t, is large.

It follows that $X_{1\alpha}$ which has the Poisson distribution with parameter $\lambda_\alpha t$ is approximately normal with mean $\lambda_\alpha t$ and standard deviation $\sqrt{(\lambda_\alpha t)}$. The variables $\max_i X_{i\alpha}$, $\min_i X_{i\alpha}$ and $|X_{i\alpha} - X_{2\alpha}|$ will not be normal, but by the central limit effect the quantities M, m and d defined by (1) will be approximately normal.

For the maximum and minimum of two observations on a normal variable with mean μ and standard deviation σ, the following results are well known

$$E[\min(X_1, X_2)] = \mu - 0.5642 \, \sigma$$
$$E[\max(X_1, X_2)] = \mu + 0.5642 \, \sigma$$

$$\text{var}[\min(X_1, X_2)] = \text{var}[\max(X_1, X_2)] = 0.6817 \; \sigma^2$$

$$\text{cov}[\min(X_1, X_2), \max(X_1, X_2)] = 0.3183 \; \sigma^2$$

Taking expectations with respect to the abundance distribution, for the finite number N of independent species, and using (2), the moments of m and M are found to be

$$E(m) = N(\nu_1 t - 0.5642 \; \nu_{\frac{1}{2}} t^{\frac{1}{2}})$$

$$E(M) = N(\nu_1 t + 0.5642 \; \nu_{\frac{1}{2}} t^{\frac{1}{2}})$$

$$\text{var}(m) = \text{var}(M) = 0.6817 \; \nu_1 t$$

$$\text{cov}(m, M) = 0.3183 \; \nu_1 t \; .$$

The familiar approximation to the expectation of a ratio is

$$E\left(\frac{m}{M}\right) = \left(\frac{E(m)}{E(M)}\right)[1 + \frac{\text{var}(M)}{[E(M)]^2} - \frac{\text{cov}(m, M)}{E(m) \; E(M)}]$$

and in the present case this includes terms of order $t^{-\frac{1}{2}}$. On substituting in this formula the expression in square brackets can be seen to be $1 + O(t^{-1})$, and hence

$$E\left(\frac{m}{M}\right) = 1 - (1.1284)\nu_{\frac{1}{2}}\nu_1^{-1}t^{-\frac{1}{2}} + O(t^{-1}).$$

Similarly the approximation

$$\text{var}\left(\frac{m}{M}\right) = \frac{\text{var}(m)}{[E(M)]^2} + \frac{\text{var}(M)}{[E(m)]^2} - \frac{2\text{cov}(m, M)}{E(m) \; E(M)}$$

leads to

$$\text{var}\left(\frac{m}{M}\right) = 0.7268 \, (N\nu_1 t)^{-1}.$$

It is easy to see that in the non-null case as well as in the one discussed here, that if the number of species in the population is finite, the bias in the sample expectation of the coefficient being studied will decrease with the amount of effort, at the order of $t^{-\frac{1}{2}}$. However, in models where the number of species is countable and new species are continually appearing in the samples with low representations, the bias may not disappear.

3.   AN APPLICATION OF GEOMETRIC THEORY

A situation in which some progress can be made with the distribution of d/m arises when $d_\alpha = |X_{1\alpha} - X_{2\alpha}|$ and $m_\alpha = \min(X_{1\alpha}, X_{2\alpha})$ are distributed independently which occurs when $X_{1\alpha}$ and $X_{2\alpha}$ have geometric distributions.

The conditions under which this is true are rather restrictive. It will be assumed that for each species the density varies throughout the habitat, so that at a particular point the probability of obtaining j specimens with an effort t is

$$p'_j = \exp(-\lambda_{i\alpha}t) \; (\lambda_{i\alpha}t)^j \; / \; j!$$

if for every species, the local abundance parameter $\lambda_{i\alpha}$ varies throughout the habitat according to an exponential law with probability density function

$$\frac{\exp\left(-\dfrac{\lambda_{i\alpha}}{w_{i\alpha}}\right)}{w_{i\alpha}}$$

the probability of obtaining j members of species α by sampling for time t at a random point in habitat i is

$$p_j = \int_0^\infty \{\exp[-\lambda_{i\alpha}(t + w_{i\alpha}^{-1})] \; (\lambda_{i\alpha}t)^j \; / \; (j! \; w_{i\alpha})\} \; d\lambda_{i\alpha}$$

$$= (1 + w_{i\alpha}t)^{-1}(w_{i\alpha}t)^j(1 + w_{i\alpha}t)^{-j}$$

$$= (1 - u_{i\alpha})u_{i\alpha}^j$$

where

$$u_{i\alpha} = w_{i\alpha}t \; / \; (1 + w_{i\alpha}t).$$

The joint distribution of m_α and d_α is easily found to be

$$\Pr[m_\alpha = j, d_\alpha = k] = (1 - u_{1\alpha})(1 - u_{2\alpha})(u_{1\alpha}u_{2\alpha})^j \qquad \text{if } k=0$$

$$= (1 - u_{1\alpha})(1 - u_{2\alpha})(u_{1\alpha}u_{2\alpha})^j (u_{1\alpha}^k + u_{2\alpha}^k) \quad \text{if } k \geq 1.$$

This shows that the distributions of m_α and d_α are independent, and their marginal distributions are

$$\Pr(m_\alpha = j) = (1 - u_{1\alpha}u_{2\alpha})(u_{1\alpha}u_{2\alpha})^j$$

$$\Pr(d_\alpha = k) = (1 - u_{1\alpha})(1 - u_{2\alpha})/(1 - u_{1\alpha}u_{2\alpha}) \qquad \text{if } k=0$$

$$= (1 - u_{1\alpha})(1 - u_{2\alpha})(u_{1\alpha}^k + u_{2\alpha}^k)/(1 - u_{1\alpha}u_{2\alpha}) \quad \text{if } k \geq 1.$$

The characteristic functions of these variables are thus found to be

$$(1 - u_{1\alpha}u_{2\alpha})/(1 - u_{1\alpha}u_{2\alpha}e^{it})$$

and

$$\frac{(1 - u_{1\alpha})(1 - u_{2\alpha})(1 - u_{1\alpha}u_{2\alpha}e^{2it})}{(1 - u_{1\alpha}u_{2\alpha})(1 - u_{1\alpha}e^{it})(1 - u_{2\alpha}e^{it})}.$$

Since $0 < u_i < 1$, each of these quantities may be expanded in a Taylor series about the origin, giving for the second characteristic functions of m_α and d_α the respective expansions

$$\sum_{j=1}^{\infty} \frac{(u_{1\alpha}u_{2\alpha}e^{it})^j}{j} - \sum_{j=1}^{\infty} \frac{(u_{1\alpha}u_{2\alpha})^j}{j}$$

and

$$\sum_{j=1}^{\infty} \frac{1}{j} [(u_{1\alpha}e^{it})^j + (u_{2\alpha}e^{it})^j - (u_{1\alpha}u_{2\alpha}e^{2it})^j - u_{1\alpha}^j - u_{2\alpha}^j$$

$$+ (u_{1\alpha}u_{2\alpha})^j].$$

If the further strong assumption is introduced that the exponential

distributions of local density are independent, then the second
characteristic functions of m and d may be obtained from those of
m_α and d_α by summation with respect to α.

If

$$\mu_{jk} = \sum_\alpha u_{1\alpha}^j u_{2\alpha}^k$$

they are

$$\Psi_1(t) = \sum_{j=1}^\infty \frac{\mu_{jj}}{j} (e^{itj}-1)$$

$$\Psi_2(t) = \sum_{j=1}^\infty \frac{1}{j} (\mu_{jo}e^{itj} + \mu_{oj}e^{itj} - \mu_{jj}e^{2itj} - \mu_{jo}$$
$$- \mu_{oj} + \mu_{jj}).$$

Now by the use of a saddle point approximation, the 'density'
function obtained by smoothing the probability function of m/d is

$$h(x) = \frac{\Psi_2'(-xt_o) \, \exp[\Psi_1(t_o) + \Psi_2(-xt_o)]}{\sqrt{\{2\pi[\Psi_1''(t_o) + x^2\Psi_2''(-xt_o)]\}}}$$

where t_o is the solution of the equation

$$\Psi_1'(t) = x\Psi_2'(-xt)$$

On differentiation, it can be seen that this takes the form

$$\sum_j [\mu_{jj} - x(\mu_{jo} + \mu_{oj} - 2\mu_{jj})]z^j = 0 \; , \; \text{with } z=e^{it}.$$

4.   A NOTE ON JACCARD'S COEFFICIENT

Finally, let $\chi_{i\alpha}$ be a characteristic random variable taking the
value 1 if species α is represented in the sample from habitat i,
and 0 otherwise. On the assumption of Poisson distribution made
in section 2,

$$\Pr[\chi_{i\alpha}=0] = \exp(-\lambda_{i\alpha}t) = x_{i\alpha} \text{ (say)} \tag{3}$$

Then for the variables m_α, M_α and d_α, the probabilities that they will take the value 1, and hence their expectations are easily obtained.

$$E(m_\alpha) = \Pr[m_\alpha=1] = 1 - x_{1\alpha} - x_{2\alpha} + x_{1\alpha}x_{2\alpha}$$

$$E(M_\alpha) = \Pr[M_\alpha=1] = 1 - x_{1\alpha}x_{2\alpha} \tag{4}$$

$$E(d_\alpha) = \Pr[d_\alpha=1] = x_{1\alpha} + x_{2\alpha} - 2x_{1\alpha}x_{2\alpha}$$

Further,

$$E(m_\alpha M_\alpha) = \Pr[m_\alpha=M_\alpha=1] = 1 - x_{1\alpha} - x_{2\alpha} + x_{1\alpha}x_{2\alpha}$$

$$E(d_\alpha M_\alpha) = \Pr[d_\alpha=M_\alpha=1] = x_{1\alpha} + x_{2\alpha} - 2x_{1\alpha}x_{2\alpha} \; .$$

Now let $\phi(t_1,t_2)$ denote the bivariate Laplace transform of the joint probability function of the abundances in the two habitats. Taking expectations with respect to the abundance distribution, and using (3)

$$E(x_{1\alpha}^{u_1} x_{2\alpha}^{u_2}) = E[\exp(-u_1\lambda_{1\alpha}t - u_2\lambda_{2\alpha}t)] = \phi(u_1t, u_2t) \tag{5}$$

Taking expectations over α for relations (4),

$$E(m) = N[1 - \phi(t,0) - \phi(o,t) + \phi(t,t)]$$

$$E(M) = N[1 - \phi(t,t)]$$

$$E(d) = N[\phi(t,o) + \phi(o,t) - 2\phi(t,t)].$$

Similarly, since the numbers of the various species caught are by hypothesis independent, the relations below (4) lead to

$$E(mM) = N[1 - \phi(t,0) - \phi(o,t) + \phi(t,t)]$$

$$E(dM) = N[\phi(t,o) + \phi(o,t) - 2\phi(t,t)] \ . \tag{6}$$

The first terms in the asumptotic approximation to $E(m/M)$ and $E(d/M)$ can be obtained from (5), and further terms can be derived from the product moments in (6), but they are complicated.

REFERENCES

[1] Austin, M. P. and Orloci, L. 1966. Geometric models in
     ecology II. J. Ecol. 54:217-227.

[2] Bannister, P. 1968. An evaluation of some procedures used
     in ordination. J. Ecol. 56:27-34.

[3] Gleason, H. A. 1920. Some applications of the quadrat method.
     Bull. Torrey bot. club. 47:21-33.

[4] Kulczyński, S. 1927. Die Pflanzenassoziationen der Pieninen.
     Acad. Pol. Sci. Math. Nat. Bull. Internat. Ser. B (suppl. 2):
     57-203.

[5] Marczewski, E. and Steinhaus, H. 1958. On a certain distance
     of sets and the corresponding distance of functions. Coll.
     Math. 6:319-327.

[6] ──────────. and ──────────. 1959. O odległości systematcznej
     biotopow. Zast. Mat. 4:195-203.

[7] Mountford, M. 1961. An index of similarity and its application
     to classificatory problems. Prog. Soil Zool. 1:43-50.

RECORD OF PREPLANNED AND SPONTANEOUS DISCUSSIONS

P. SWITZER (Stanford University)

We are discussing the measurement of similarity between two communities on the basis of the relative prevalences of various species.
It is supposed that the two communities have been sampled in some
way; Mr. Holgate has examined the sampling distribution of assorted
similarity coefficients. Knowing something about these distributions supposedly enables us to examine hypotheses about the two
communities based on the sample data. If this is one of the
objectives of this paper then we could be critical about his
selection of coefficients.

In particular, he assumes that the counts $x_{1\alpha}$ and $X_{2\alpha}$ are 2N independent Poisson variables (see section 2); personally I would be much happier with the more general multinomial model which does not require the set of species to be large. But in either case we are led to the same likelihood ratio statistic L which is known to have various large-sample properties of optimality, which has a well-known large-sample distribution, and which corresponds to none of the coefficients of this paper. In particular

$$-2\ln L = - \sum_{\alpha=1}^{N} 2X_{1\alpha} \ln \frac{2X_{1\alpha}}{X_{1\alpha} + X_{2\alpha}} + 2X_{2\alpha} \ln \frac{2X_{2\alpha}}{X_{1\alpha} + x_{2\alpha}} \qquad (1)$$

has an asymptotic χ^2-distribution with N-1 degrees of freedom when the communities being sampled are the same, and a non-central χ^2-distribution otherwise. The distribution is asymptotic to t, the sampling effort, rather than N, the number of species, as required for Mr. Holgate's exposition in section 2.

Alternatively the coefficient

$$\sum_{\alpha=1}^{N} (X_{1\alpha} - X_{2\alpha})^2 / (x_{1\alpha} + X_{2\alpha})$$

is also known to have a large sample χ^2-distribution with N-1 degrees of freedom. Finally, there may be some merit in the use of

$$\frac{1}{2} + \frac{1}{2t} \sum |X_{1\alpha} - X_{2\alpha}| \qquad (2)$$

as a similarity coefficient. If the quantities $X_{1\alpha}/t$ and $X_{2\alpha}/t$ are replaced in (2) by their population values, then the resulting expression has a direct interpretation: it is the probability that we will correctly identify the community affiliation of a randomly chosen individual or point on the basis of its species and using the best possible identification procedure. The sampling distribution of $|X_{1\alpha} - X_{2\alpha}|$ can be obtained from Mr. Holgate's paper by noting that it is equal to $\max(X_1, X_2) - \min(X_1, X_2)$.

L. ORLOCI (University of Western Ontario)

I think very highly about these attempts of discovering the sampling distribution of the different coefficients used by ecologists. My

comments relate to a specific metric property of the function E in
Mr. Holgate's paper and other closely related functions. For use
with presence data this function is defined by

$$E = \mu(A \doteq B)/\mu(A + B) = (b + c)/(a + b + c).$$

where $\mu(A \doteq B)$ and $\mu(A + B)$ express in terms of counts the symmetric
difference of set A and B and the size of the union set A + B
respectively. The other symbols, a, b and c, are those regularly
used in connection with 2 x 2 contingency tables. Note that
$(A \doteq B) = b + c$ is the squared Euclidean distance of set A and B.
   The prospective user may be interested to learn that E, as I have
defined it, is locally Euclidean. This implies that $\sqrt{}(E)$ represents
an Euclidean definition of distance in specific situations, as for
instance a set of stands with no species absent in more than one
stand. For the purpose of illustration consider the data in Table 1.
The function E is Euclidean in set II, non-Euclidean in set I and
also non-Euclidean in the union set I + II. The loss of Euclidean
properties in set I and I + II is a consequence of the changing
dimensions (a + b + c) of the model as it has already been pointed
out by Williams and Dale [5]. A non-Euclidean E is, of course,
greatly disadvantaged by the fact that there are very few techniques
that could efficiently handle dissimilarity measure of this kind.
   When (b + c) is divided by (a + b + c + d) then E represents the
complement of the matching coefficient of Sokal and Michener [4]
whose sampling distribution was derived by Goodall [3]. This coefficient is an Euclidean spatial parameter when used in its usual
context in connection with 2 x 2 contingency tables.
   We have considered with M. P. Austin [1] the properties of a
related function,

$$D = L_{max} - L, \quad L = 2a/(2a + b + c).$$

Function D, or more precisely a version of this function applied to
quantitative data, has been used in ordinations (e.g., Bray and
Curtis [2]). We have shown that when L_{max}, the maximum value of L
obtained in a repeated sampling of the same community, is less than
one the function D is not a metric. The use of D with the ordination technique for which it was devised, therefore, is quite
inappropriate.

Table 1. Data for illustration

Species	Set I			Set II		
	Stands					
	1	2	3	4	5	6
1	1	0	0	1	1	1
2	0	0	1	0	1	1
3	1	1	0	1	1	1
4	0	1	1	1	0	1
5	0	1	1	0	1	1

References

[1] Austin, M. P. and Orloci, L. 1966. Geometric models in
     ecology. II. An evaluation of some ordination techniques.
     J. Ecol. 54:217-227.

[2] Bray, J. R. and Curtis, J. T. 1957. An ordination of the
     upland forest communities of southern Wisconsin. Ecol.
     Monogr. 27:325-349.

[3] Goodall, D. W. 1967. The distribution of the matching coef-
     ficient. Biometrics. 23:647-656.

[4] Sokal, R. R. and Michener, C. D. 1958. A statistical method
     for evaluating systematic relationships. University of
     Kansas Sci. Bull. 38:1409-1438.

[5] Williams, W. T. and Dale, M. B. 1965. Fundamental problems
     in numerical taxonomy. In "Advances in Botanical Research,"
     Vol. 2, Academic Press, London.

APPLICATION OF NUMERICAL METHODS TO TROPICAL FORESTS

P. GREIG-SMITH
School of Plant Biology, University College of North Wales
Bangor, U. K.

SUMMARY

A review of the special problems of numerical methods in tropical forests and an assessment of the results so far obtained by their use.

1. INTRODUCTION

Tropical rain forest presents ecological problems which are among
the most urgent and the most intractable facing ecologists. Large
areas of rain forest have already disappeared and its destruction
continues at an increasing rate. The ecologist's curiosity about a
fascinating community would in itself be enough to stimulate investigation while adequate areas still remain for study. Much more
important, however, is the need to increase our understanding of the
most complex of all ecosystems as a basis of planning land use in
the moist tropics, where increasing population relies on natural
resources which are by no means unlimited, as once superficially
appeared, and are all too easily destroyed. The object of this
paper is to review the special problems of numerical methods in
tropical forest and to attempt an assessment of the results so far
obtained by their use.

2. CHARACTERISTICS OF RAIN FOREST

In the present context the most important feature of rain forest
is its species-richness, in terms not only of the large number of
species present in any limited geographical area, but also of the
large number in any one plot, with a correspondingly low representation of any one species. Types of tropical forest do exist which
contain a predominance of a few or even of one species, but these
are usually limited to less favorable habitats such as waterlogged
areas and bleached sands [13] and the greater part of rain forest
lacks species which are even moderately abundant. Richards [14]
has suggested that the uniqueness of rain forest in this respect may
have been exaggerated and that other, non-forest, communities, such
as some sclerophyll communities of the Cape Peninsula and of Australia and New Caledonia, may indeed be more species-rich than most
rain forest. Only in the rain forest, however, is great species-
richness combined with a physiognomy that dwarfs the observer. This
may seem a trivial point, but it does place serious field limitations on the type of data that can economically be collected. Not
only is the vegetation rich in species, but it commonly includes
numbers of congeneric species often not easily distinguished by the
ecologist working at ground level, and the state of taxonomic knowledge of most rain forest areas is still such that a proportion of
individuals is likely to be listed as 'unknown.' These practical

difficulties will clearly add to the volume of background noise, which is inherently high (see below).

Species-richness raises the problem of the minimum size of plot that can satisfactorily be used as a stand, a problem whose importance is emphasized by the practical difficulties and time-consuming nature of enumeration in rain forest. Classification and ordination of survey data can usually be satisfactorily based on presence and absence only of species [9], and experience indicates that a surprisingly small size of plot can be analyzed to give meaningful results. Ashton [1], in his pioneer application of ordination to rain forest, used plots of 0.4 ha, Greig-Smith, Austin and Whitmore [10] plots of 0.12 ha and both Austin and Greig-Smith [3] and Brunig [4] plots of only 0.04 ha (20 x 20 m). Though satisfactory analyses may be made of such small plots if the objective is correlation with environment, when the objectives of an investigation require the assessment of species density, as is likely in applied work [9], the situation is more difficult. The density of individual species is so low that to obtain even a reasonably satisfactory estimate of density (10% coefficient of variation) for even the more abundant species is likely to require plots of the order of size of 10 ha [8]. To enumerate fully such a plot is quite impractical and in any case it is unusual for an area of 10 ha to be even apparently homogeneous. There is no answer to this difficulty; we must be satisfied with a lower level of accuracy for density estimates than would be acceptable in most temperate forests. The necessary information is lacking for assessment of desirable plot sizes for estimates of other measures, such as basal area, of potential practical importance.

Species-richness has important implications for the techniques of analysis. Some aspects of the question have been examined by Webb etc. ([15], [16]) and by Austin and Greig-Smith [3], using very different sets of test data. Webb etc. examined complete enumerations, including all vascular plants,* of 18, 0.1 ha plots, spread over 7° of latitude in tropical Australia and including 818 species. The plots were deliberately chosen to include a range of physiognomic types, some of them only doubtfully rain forest as generally understood. Austin and Greig-Smith used enumerations from

---

* Webb etc. [15] state that 'all woody growth' was recorded but later [16] refer to the subset of 'herbs,' and make incidental references to species of fern.

Sabah by D. I. Nicholson of all trees of girth $\geq$ 12 in. (30 cm)
(198 species) in a block of 50, 0.04 ha plots in lowland dipterocarp forest, an exceptionally uniform type of rain forest growing
under optimum conditions.

With a large number of species in relation to number of stands,
the stands are necessarily over-defined [15]. Thus much of the
information is redundant. This redundant information may consist
largely of background noise, in which case its elimination is
desirable and may be essential. It may also include some meaningful
information but for practical reasons it is still desirable to
eliminate it.

Consider first the incidence of background noise. Of the 198
species in the Sabah data, 60 occur only as single individuals.*
Thus in 50 plots there are bound to be at least ten joint occurrences of such species, giving perfect correlation between them,
and 35 joint occurrences may be expected by chance. Species
represented by more than one individual, but still rare, will also
give rise to chance correlation. There is clearly a large amount
of background noise present. Although Webb etc. clearly demonstrated redundancy in their data, they did not consider directly the
amount of background noise, but in such very heterogeneous data it
is likely to be proportionally less. The two sets of data probably
represent the extreme conditions in relation to background noise.
The importance of background noise is illustrated by the fact that
an ordination of the Sabah data based on 50 species randomly
selected from the total of 198 was uninterpretable; R-analysis of
the correlation matrix between species showed that the first axis
of the ordination was primarily determined by three species all of
which occurred once only but in the same stand.

The need to eliminate meaningful but redundant information arises
from the desirability of reducing the extent of the data to be
handled. If such information can be identified in advance it may
be possible to avoid its collection in the field. It may also be
desirable to eliminate other information which is neither redundant
nor mere background noise. There may be groups of species which
are controlled in part by environmental factors with little effect

---

* This proportion of rare species is not unusual. Poore [12], in
  an enumeration of all canopy trees on a sample area of 23 ha
  selected for uniformity, found that 157 out of total of 381
  species occurred once only.

on the main assemblage of species, and, on the other hand, show
little response to environmental factors which are otherwise
important. The information on such species, which is clearly
ecologically important, can only confuse an initial analysis of
the vegetation. Thus Webb etc. [16] obtained quite different,
though interpretable, classifications on the basis of herbs only,
or of epiphytes only, from those based on all species or on 'big
trees' (which were closely similar). This, incidentally, validates
the general practice among rain forest ecologists of concentrating
attention on the trees of the canopy layer.

3. TECHNIQUES OF ANALYSIS

What, in practical terms, are the implications of these considerations for techniques of analysis? Truly redundant information will
not affect the results obtained, though it may in some circumstances
result in a data matrix of such a size as to limit the practical
choice of techniques available. Background noise presents a more
serious problem. In classification it might be expected to have
serious effects at the lower levels of the hierarchy, perhaps more
so with divisive than with agglomerative techniques, but to be
unimportant at the higher levels. With ordination, which it has
been suggested [9] is more appropriate within more homogeneous sets
of data, background noise is likely to be more important. Most
workers have, for, I suspect, largely intuitive reasons, ignored
the least frequent species in sets of data. It is precisely these
species which are the major source of background noise--random
absence of abundant species from a very few stands scarcely, if
ever, occurs in rain forest data--and the common practice is thus
a sound one.

The main question is then how far can the number of species used
be reduced without reducing the efficiency of the analysis to an
unacceptable extent. This can only be answered empirically. Webb
etc. [16] have examined the efficiency of various physiognomically
defined species-groups in association analysis of their data and
shown that the 269 'big tree' species gave a satisfactory analysis.
On the basis of an inverse analysis of the species they eliminated
204 of these big trees and finally obtained a fully satisfactory
analysis on the basis of 65 tree species (c. 16%) only out of the
total of approximately 400. Austin and Greig-Smith [3] prepared

ordinations of the Sabah data using the 10, 15, ..., 50, 75, 100
most abundant species and found little improvement in interpretability beyond the 25 (c. $12\frac{1}{2}$%) most abundant species. Ashton [1],
selecting the most abundant species that were certainly identifiable,
used 50 (c. $10\frac{1}{2}$%) out of 472 tree species of above 12 in. (30 cm)
girth for his Andalau ordination and 79 (c. 19%) out of 420 species
for his Kuala Belalong ordination. The heterogeneity of these sets
of data was intermediate between that of the Sabah and the Australian data. It may tentatively be suggested that the 10-20% most
abundant tree species provide an adequate basis for analysis; Greig-
Smith, Austin and Whitmore [10] probably used an unnecessarily large
number of species (91 out of 172, c. 53%) in their classification
and ordination of data from the admittedly less species-rich forest
of Kolombangara in the Solomon Islands.

One other consideration of technique calls for comment. Except
for relatively very homogeneous data, presence and absence is a
satisfactory measure of species representation [9]. In most temperate vegetation quantitative measures of species give ordinations
comparable to those based on presence and absence; in tropical
forest with no species having very high representation, however,
the species with the greatest ranges of representation will dominate
the ordination, unless a suitable standardization is applied to the
data first. Thus ordination based on unstandardized quantitative
data may only provide indications about two or three particular
species [3].

4. RESULTS

Tropical rain forest has proved less amenable to traditional approaches to classification than any other major vegetation type.
Atypical forest in less favorable habitats has been readily recognized but the greater proportion has generally been regarded
as 'mixed rain forest' and any subdivision has been based rather
on features of habitat than on the vegetation itself. Such a site
classification may be adequate as a basis of resource survey and
management, although it is not necessarily so; it is clearly not
satisfactory in studies of the relation between vegetation and
environment. It is the physical scale and, especially, the species-
richness of rain forest that make the use of traditional methods
difficult. (Significantly, only in the African rain forest, which

is markedly less species-rich than either the American or the
Asiatic rain forest, have such methods had even moderate success.)
This has led to the idea that within rain forest there is little
control by the environment, which, in turn, has strongly influenced
thinking on speciation in rain forest [5].

The few investigations in which numerical methods have been used
have already demonstrated not only that there is ordered variation
in rain forest composition, but that this variation can be correlated with environmental differences. Ashton [1], using a
relatively inefficient technique of ordination, found clear
correlation between composition and differences in physiography,
soils and drainage within each of two sets of 50, 0.4 ha plots in
Brunei. Re-examination of the same data by more satisfactory
techniques has confirmed and extended the conclusions (Austin,
Ashton and Greig-Smith, unpublished). Greig-Smith, Austin and
Whitmore [10], using classification and ordination, found comparable
relationships with physiography and drainage in a set of samples
from Kolombangara and were able to distinguish these from a geographical effect between the northern and western parts of the
island. In both these investigations there was also a clear
distinction in the behavior of related species in relation to
environment. Ashton [2] has discussed the relevance of such correlations to concepts of speciation. Williams etc. [17] have used
a novel multiple-nearest-neighbor method to investigate variation
in composition within a single 0.4 ha plot and have demonstrated
environmental correlations even at this scale. While further
evidence is desirable, it is clear that rain forest composition,
in spite of a considerable content of background noise, is essentially ordered, and that numerical methods have great potential for
the elucidation of its control. A possible development is to
analyze in terms of physiognomic and morphological characteristics
of the individuals present, rather than specific identity. This
would not only simplify the collection of data but might clarify
the adaptive significance of the morphological characteristics of
rain forest.* Such an approach has been tried, with some success,
in temperate vegetation [11]. It would be particularly appropriate

---

* Ashton [1] has suggested from examination of the proportions of
  different leaf-size classes in relation to the axes of his Brunei
  ordinations, that leaf-size is closely related to environmental
  differences.

to markedly heterogeneous data, such as those of Webb etc. ([15], [16]) and its use is being examined by Williams et al.*

The pattern of individual species in rain forest is of particular interest, not only, as elsewhere, because of the indications it may provide of the controlling factors most important for particular species, and of the direct relationships between species, but because of the supposedly intense competition between individuals in rain forest. Pattern, in this sense of departure from randomness of spatial arrangement of individuals, is concerned essentially with the same problem as in the ordering of vegetation composition discussed above, but the emphasis is on the individual species and its relationships and interest lies mainly in small-scale variation. The distinction between the fields of the two approaches is not a clear-cut one; the scale on which Williams etc. worked is one commonly looked at in terms of pattern, but their approach is a multispecies one.

If there is intense competition this may be expected to result in a tendency towards regularity of spatial distribution. The straightforward technique of analysis of variance of species representation in nesting blocks of increasing size [7] allows regularity to be detected. What little evidence is available suggests that intraspecific competition is not of great importance ([6]; [12], Fig. 9; [2], Fig. 5). Significantly the only case, as far as I know, in which there is clear indication of regularity is for the largest girth classes of Shorea albida in lowland peat forest in Sarawak, an atypical forest in which this species constitutes a very large proportion of the canopy (Greig-Smith and J. A. R. Anderson, unpublished). Whether this is due solely to readier detection of competition in populations of higher density, or partly also to the greater sensitivity of an individual to competition from other individuals of the same species, is uncertain. If density alone is responsible, then regularity might be expected in data for all trees; neither Greig-Smith [6] nor Poore [12] found evidence of this.

An important potential use of numerical methods is to provide a framework of classification for the practical forester. For this purpose the advantages of a monothetic divisive classification are considerable [9]. An extensive series of enumerations made as part of routine forest surveys in the Solomon Islands and totalling over

---

* Personal communication.

1800 plots, each either 0.4 or 0.8 ha, have been subjected to association-analysis (Greig-Smith, Austin and Swaine, unpublished). These formed six principal sets, from different areas. The sets were analyzed both separately and as a combination of four sets totalling 1103 plots. The analyses have only recently been concluded and a detailed assessment of the results has not yet been made. Preliminary conclusions are, however, encouraging. The same species recur as division parameters in different analyses and many of them are large, readily recognized species. It is probable that many of the enumeration data stored in the forestry departments of tropical territories could profitably be subjected to numerical analysis.

ACKNOWLEDGMENTS

Much of the work referred to formed part of a program financed by the Science Research Council. I am indebted to various workers, especially Mr. K. W. Trenaman, Conservator of Forests, British Solomon Islands Protectorate, for making data available. I am grateful to Dr. W. T. Williams for allowing me to see papers before publication.

REFERENCES

[1] Ashton, P. S. 1964. Ecological studies in the mixed Dipterocarp forest of Brunei State. Oxf. For. Mem. 25.

[2] ————. 1969. Speciation among tropical forest trees: some deductions in the light of recent evidence. Biol. J. Linn. Soc. 1:155-196.

[3] Austin, M. P. and Greig-Smith, P. 1968. The application of quantitative methods to vegetation survey. II. Some methodological problems of data from rain forest. J. Ecol. 56:827-844.

[4] Brunig, E. 1968. Der Heidewald von Sarawak und Brunei. Mitt. BundForschAnst. Forst-u.Holzw. 68.

[5] Fedorov, An. A. 1966. The structure of the Tropical Rain Forest and speciation in the humid tropics. J. Ecol. 54: 1-11.

[6] Greig-Smith, P. 1952. Ecological observations on degraded and secondary forest in Trinidad, British West Indies. II. Structure of the communities. J. Ecol. 40:316-330.

[7] ————. 1961. Data on pattern within plant communities. I. The analysis of pattern. J. Ecol. 49:695-702.

[8]   Greig-Smith, P. 1965. Notes on the quantitative description
      of humid tropical forest. Symposium on Ecological Research
      in Humid Tropics Vegetation, pp. 227-234. Government of
      Sarawak and UNESCO Science Cooperation Office for Southeast
      Asia.

[9]   ——————. 1969. Analysis of vegetation data: the user
      viewpoint. International Symposium on Statistical Ecology,
      New Haven, 3:149-166.

[10] ——————, Austin, M. P., and Whitmore, T. C. 1967. The
      application of quantitative methods to vegetation survey.
      I. Association-analysis and principal component ordination
      of rain forest. J. Ecol. 55:483-503.

[11] Knight, D. H. 1965. A gradient analysis of Wisconsin prairie
      vegetation on the basis of plant structure and function.
      Ecology 46:744-747.

[12] Poore, M. E. D. 1968. Studies in Malaysian rain forest. I.
      The forest on Triassic sediments in Jengka Forest Reserve.
      J. Ecol. 56:143-196.

[13] Richards, P. W. 1952. The Tropical Rain Forest. Cambridge.

[14] ——————. 1969. Speciation in the tropical rain forest and
      the concept of the niche. Biol. J. Linn. Soc. 1:149-153.

[15] Webb, L. J., Tracey, J. G., Williams, W. T. and Lance, G. N.
      1967. Studies in the numerical analysis of complex rain-
      forest communities. I. A comparison of methods applicable
      to site/species data. J. Ecol. 55:171-191.

[16] ——————, ——————, ——————, and ——————. 1967.
      Studies in the numerical analysis of complex rain-forest
      communities. II. The problem of species sampling. J. Ecol.
      55:525-538.

[17] Williams, W. T., Lance, G. N., Webb, L. J., Tracey, J. G. and
      Connell, J. H. 1969. Studies in the numerical analysis of
      complex rain forest communities. IV. A method for the elu-
      cidation of small-scale forest pattern. J. Ecol. 635-654.

RECORD OF PREPLANNED AND SPONTANEOUS DISCUSSIONS

D. W. GOODALL (Utah State University)

Species occurring uncommonly in a set of quadrat data may be compan-
ions in the Braun-Blanquet sense, or may be peculiar to a particular
community type or set of environmental conditions occurring rarely
among these particular stands. In the latter case, ignoring them
may involve a regrettable loss of information. It may then be preferable to use less common species as grounds for recognition and
elimination of special types of stand before proceeding to analyze
the more homogeneous subset formed by the residuum of the data remaining after stands containing them have been removed.

E. C. PIELOU (Queen's University, Kingston, Canada)

Professor Greig-Smith has said that he failed to find evidence of
regularity in the spacing of rain forest trees although there is
presumably intense competition among them. Are his conclusions
based on plot data only? Regular spacing might be revealed by
examining the distribution of ω (the squared distance from a tree
to its nearest neighbor). In the absence of competition the frequency distribution of ω will be monotonically decreasing. When
the trees are regularly spaced in the sense that short inter-tree
distances are impossible, the distribution of ω will have a mode
at some ω>0. To detect such a mode it is desirable to examine the
distribution of only the small values of ω. If all values are
considered the inevitable grouping of the data may obscure the
existence of a mode.

H. LIETH (University of North Carolina, Chapel Hill, North Carolina)

From visiting several tropical rain forest types I gained the
impression that the distribution pattern of plants is largely
influenced by leaf-cutter ants and termite populations. In some
cases I had even the impression that the abundant colonies of
different species of these animals were managing the forest for
their own benefit. They act therefore as an environmental factor
similar to man's impact upon vegetation in the temperate zone but
with an unusual distribution pattern. I wonder whether this fact
would influence the numerical analysis of vegetation data in
tropical rain forests.

H. VAN GROENEWOUD (Forest Research Lab, Fredericton, N.B.)

I would like to make a few remarks with regard to the papers presented by Messrs. Greig-Smith and Holgate.

    1. If the responses of some species to an ecological gradient
can be expressed by narrow bell shaped curves, the quantitative
relationships among these species and among these and others not
so curved, are curvilinear and these cannot be expressed effectively
by coefficients that are based on linear relationships. This
problem carries also over into similarity or distance measures among
vegetation sub-samples.

    These are some of the species that are creating some of the back-
ground noise Mr. Greig-Smith referred to. Those species may carry
ecological information but this information is lost in an analysis
based on a linear model.

2. Another difficulty in the ecological interpretation of vegetation data is caused by the absence of certain species. In the mathematical analysis of vegetation data, zero abundance or presence indicates only one point along the gradient. The absence of a species, however, can indicate any ecological condition outside the range of that species along the gradient.

3. If the response of a species to conditions along a gradient is bell shaped, a certain quantity of that species indicates either of two ecological conditions thus further confusing ecological interpretations.

P. GREIG-SMITH (University of North Wales, Bangor, U.K.)

I am most interested to hear of Knight's work in Panama, mentioned by Dr. Cooper. I agree with Dr. Cooper's emphasis on the need to use ecological intuition in the interpretation of analyses and with his remarks on the sensitivity of rain forest.

The difficulty of using the occurrence of rare species to remove some plots before analysis, suggested by Dr. Goodall, is that in sets of data like the Sabah one it would, I suspect, result in the elimination of most of the plots. Prior removal of plots would be more practical in more heterogeneous data.

I have not tried nearest neighbor methods to test for regularity because the data in question were on a plot basis. The practical difficulty of using such methods is that a map of species occurrence, as opposed to recording plot occurences, is very much more time consuming to obtain. Even if it is known in advance that particular species are of interest, it may take a prolonged search on the ground to identify the nearest neighbor of the same species of a particular individual.

I agree with Dr. Lieth that termites and leaf-cutting ants are important biotic factors in rain forest, though I do not agree with him in attributing overriding importance to them.

Dr. Van Groenewoud's point about the importance of species of narrow ecological amplitude in producing background noise is a valid one. In practice, however, it is inevitable that a set of data will extend beyond the range of such species. I have perhaps oversimplified the problem of background noise. Whether information is to be regarded as background noise or not depends in part on the context of the data.

MULTIPLE PATTERN ANAYLSIS, OR MULTISCALE ORDINATION: TOWARDS A VEGETATION HOLOGRAM?

IMANUEL NOY-MEIR
DEREK J. ANDERSON
Australian National University

SUMMARY

Multivariate methods for the analysis of vegetation (classification
and ordination) describe the patterns of covariation of species,
but only at a single predetermined scale. Block size variance
analysis (pattern analysis) describes variation of pattern over a
wide range of scales, but only for one species (or a pair of species)
at a time. A method is proposed which combines "information" from
all species at all scales to produce an integrated representation
of "total" pattern.

Data from all species are recorded from contiguous sites, and
grouped into successively larger blocks. Variance-covariance
matrices at each block size are calculated as for pattern analysis,
and added to form a combined covariance matrix. This is subjected
to principal component analysis to obtain a species ordination.
Subsequently each characteristic root (generalized variance) is
partitioned into contributions from the various block sizes, which
are plotted against block size as in pattern analysis.

The method has been applied to a model community and a limestone
grassland community, with encouraging results.

1. INTRODUCTION

Phytosociology--a Peter Pan among biological disciplines for so many
years--is undergoing a rapid transformation into a mature vegetation
science. This transformation is due in no small part to the injec-
tion of quantitative--and particularly multivariate--methodology
into the analysis of complex data, for which classical descriptive
methods alone become ineffective.

   Those multivariate methods which have been employed in the
analysis of vegetational relationships, whether labelled as classifications or ordinations, essentially seek a simplified description
of between-species relationships in a species (attributes) x sites
(individuals) matrix. While the species in this matrix are relatively well defined, the scale-factor involved in the definition of
sites, or "areal individuals," is often chosen arbitrarily. (The
converse situation applies in psychology, where the individuals are
well defined but the selection of attributes is subject to arbitrary
decision.) In most phytosociological work data are collected from
sites which are plots of a single, predetermined area. The inherent
difficulties (including potential loss of information) associated
with this arbitrary choice of sampling scale are not avoided by the
use of 'plotless' sampling schemes, since the size of the 'site'
then depends on the number of individual plants used to define it
(e.g., single or multiple nearest neighbors). Strictly, the results
of any multivariate analysis apple only to the scale (or 'block
size') at which the original data are enumerated; the same multivariate analysis carried out on data from a single area, but with
varying site-scales, may provide different solutions and therefore
interpretations(e.g., [8], [9]).

   In a sinse we are thus dealing not with a data matrix, but with
a "data cube" (or "data space") the dimensions of which are the
species (z), the sites (y) and the various site-scales (x) (Fig. 1).
A conventional multivariate analysis essentially entails a search-
procedure for "structure" on a single vertical species x sites plane
which is orthogonal to the scales axis and intersects it at the
chosen site-scale. Unless an investigator has an a priori reason
for being interested in variation at a single scale, (or at most a
narrow range of scales), the analysis is unduly restrictive; there
are doubtless many circumstances which warrant analysis of data
gathered from varying site-scales. One obvious way of doing this

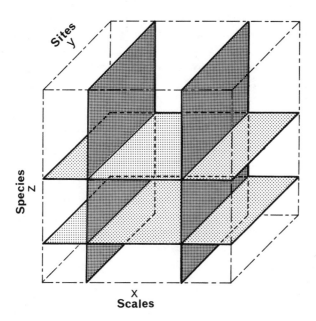

Figure 1. Schematical representation of species (z) x site (y) x
           scale (x) data cube

is to repeat the analysis independently at a number of different
scales; in other words describe the pattern on a number of parallel
vertical planes through the data space. The comparison of the solutions then provides some insight into the interaction of between-
species "pattern" with scale. The difficulty with this approach
arises from a lack of direct formal relationships between the
various solutions, which may well jeopardize a successful integrated
interpretation of the results.

By way of contrast, one quantitative approach which directly
examines scale variation is the pattern analysis technique developed
by Greig-Smith ([3], [4]). Pattern analysis (or block size variance
analysis) searches for the pattern of each sampled species at varying scales by relating the partitioned variance to block size. Thus
the method operates, in the context of the data space illustrated in
(Fig. 1), on a horizontal site x scale plane orthogonal to the
species axis and intersecting it at the position of the given

species (or if we consider only the species x scale plane, on a
horizontal line). The usual practice in dealing with multispecific
communities has been to carry out independent pattern analyses for
each species present in sufficient abundance and then to make empirical comparisons between the variance: block size relationships
so produced (e.g., [6]; [5]; [1]), thus examining pattern on a number of parallel horizontal planes through the data space. However,
this comparison can provide a meaningful interpretation only if the
correlations or covariances for each pair of species and their
dependence on block size are also examined ([7], [8], [4], [2]);
two species exhibiting pattern at the same block size may tend to
have a higher abundance in the same phase of the pattern (positive
covariance) or in alternate phases (negative covariance). The plots
of all variances and covariances against block size give a full
description of the pattern in the data cube. But comparison and
integration of the results into a coherent whole by inspection alone
becomes a cumbersome process if more than a few species are involved.

     It would appear then that a method which employs a search for,
and provides an integrated description of, the overall "pattern" in
a species x site x scale data cube will have considerable value in
extending the range and power of presently available quantitative
methods. The requirement is for a method which will combine information from all species in all sites and at all scales; such a
method requires compatible measures for "between-species" effects
on the one hand and scale "pattern" on the other. It may be regarded as either an extension of multivariate (= multispecific)
analysis to handle data taken from several scales or as a generalization of pattern analysis to a multispecific level. We shall in
what follows outline one such method and test it empirically by
application to two sets of data.

2. THE GENERAL METHOD

The method presented here utilizes the fact that the conventional
block size variance analysis employs an Euclidean variance-covariance model which is compatible with that used in principal component
analysis.* Thus, when data are given for a number of species over

---

* Similar compatibility could possibly be achieved by the use of a
  model based on information statistics.

a range of block sizes, the variance-covariance matrix at each block size may be simplified to a principal component solution. This solution will account for the variances and covariances of all species in terms of their contributions to a smaller number of generalized "composition variables" or components. However, the components from different block sizes will in general be neither identical nor simply related. The method proposed here consists essentially of adding, for each variance and covariance, the values over all block sizes into a combined variance (or covariance). The combined variance-covariance matrix is then subjected to a principal component analysis. The resulting characteristic vectors (eigen-vectors) can be plotted as a species ordination, which will show which species contribute (positively or negatively) to each of these main components of variation. The characteristic root (eigen-value) corresponding to each of these vectors expresses the magnitude of the variance accounted for by the vectors. Each root is subsequently partitioned into values contributed by the various block sizes and the usual variance: block size graph obtained. The interpretation of these graphs is similar in character to that of normal pattern analysis, peaks in the graph indicating contagion, and troughs indicating regularity. The only difference is that the variances (roots) are now generalized "multispecific" or "phytosociological" variances, the relationship of which to the species is defined by the species ordination.

Thus it will usually be sufficient to consider a small number of components which will express a large part of the heterogeneity present in a variance-covariance matrix. Additionally, the original species scores for each block at each size (as deviates from the total mean) can then be used to calculate component scores for these blocks. This calculation will result in a site ordination for each block size, the axes at the different scales coinciding both in direction and interpretation. The distribution of the sites at the various scales in this vegetation space may be examined for possible non-linear and/or clustered configurations, and may, hopefully, be related to environmental information.

A detailed mathematical exposition of the method is presented in the Appendix.

3. EMPIRICAL TESTING OF THE METHOD

The method as outlined has been tested using two sets of data from:

(i) a model community (sampled by a transect of 64 contiguous sites) composed of 13 "species" with different types of distribution, and correlations between species, at various scales;

(ii) a limestone grassland community of seven species from Monk's Dale, Derbyshire, England (sampled by one transect of 192 contiguous sites), for which conventional pattern analyses on cover data have already been reported [1].

3.1 Model community

The consciously inbuilt structure of the model is detailed in Table 1. The model data were constructed to exhibit random, contagious and regular distributions at different scales of pattern, differential levels of abundance between phases as well as simulated "competitive" relationships.

The results for the principal component analysis of the combined covariance matrix are presented in Table 2. The first four roots, which are readily interpretable, account for slightly more than 80% of the total variance present in the matrix.

The first component, with a strong peak at block size 16 (Fig. 2), reflects the alternating pattern of species A-B-E (high positive component loadings) versus C-D (high negative loadings) which was imposed at this scale. The third component very clearly represents the "competition effect" within each of the phases of this coarse pattern by separating A from B-E, and C from D, with most of the variance at block size 2.

The variance of the second component is fairly evenly distributed between block sizes, indicating a nearly random distribution. This component connects the basic "random" species F and species G-H which were derived from it by elimination (and which are positively correlated with it due to the common initial random set). Unexpectedly the "uniform" (I-J) and "regular" (K to M) species also showed similarity to F-G-H on this component; this indicated that a positive correlation between the random and the uniform/regular groups of species had inadvertently been built-in to the data (by using a similar set for elimination?). Similar "spurious" correlations must account for some cases in other components, where species behave in a way which cannot be fully explained by the consciously built-in features of the model.

Table 1. Structure of model community

Species	Mean score	Inbuilt pattern
A	2.31	contagious at block sizes 2 and 16
B	2.31	contagious at 2 and 16; correlation with A: – at 2, + at 16
C	1.41	contagious 2, 16; – correlation with A, B at 16
D	1.41	contagious 2, 16; as C at 16; – correlation with C at 2
E	1.56	as B; non-zero scores all reduced by 1
F	2.73	random (from random number tables)
G	2.00	scores as for F, with 30% random elimination
H	2.02	as F, with 30% regular elimination
I	4.47	± uniform distribution
J	2.97	as I, 30% random elimination as in G
K	2.77	regular at 8
L	1.80	as K, 30% elimination as in G, J
M	0.67	as L, with additional 30% elimination

Table 2. Principal component analysis for model community

| | | Components | | |
		1	2	3	4
Roots		74.56	63.38	31.45	27.05
% efficiency		30.9	26.3	13.0	11.2
Species	A	3.35	-1.58	-4.00	1.18
	B	5.01	-1.65	1.88	0.26
	C	-1.23	0.39	2.52	-0.57
	D	-2.26	0.22	-1.31	-0.16
	E	4.38	-1.36	1.51	0.22
	F	1.28	2.55	-0.64	-1.75
	G	1.93	3.93	0.00	-0.69
	H	1.04	2.46	-0.40	-1.98
	I	-0.06	0.62	0.21	0.22
	J	2.31	3.80	-0.31	0.58
	K	-0.77	1.11	0.67	3.17
	L	-0.16	3.08	0.53	2.66
	M	-0.18	1.61	-0.35	0.48

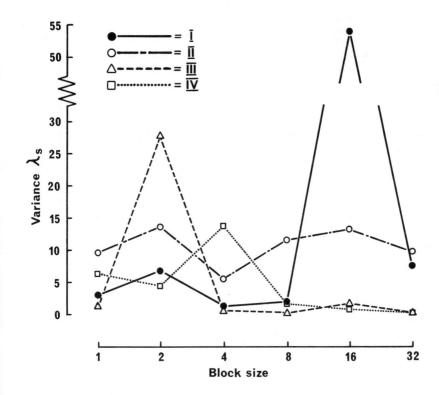

Figure 2.   Variance (λ_s) / block size graphs for first four components of the model community

The sharp fall in variance of the fourth component to a near-zero
level at block size 8 indicates regularity at that scale; it does
indeed separate species K-L-M, for which such a pattern had been
imposed, from the random species which lack it.

Species, the distributions of which were derived from that of
other, basic, species by systematic reduction or random elimination
of values, usually behave in a qualitatively similar way in the
ordination; the absolute values of loadings were either smaller or
larger, depending on whether the elimination decreased or increased
the variance.

On the whole, despite some blurring by "noise," the major features which were built into the model were recovered quite clearly
by the analysis.

3.2 Monk's Dale grassland

The results for the principal component analysis of the Monk's Dale data are presented in Figure 3. The first three roots extracted account for nearly 80% of the total variance present in the matrix. The first component, with Helianthemum and Festuca at opposite ends, has a marked variance peak at the largest block size, 64 (Fig. 4). The loadings of species on this component are closely and almost linearly related to their regression coefficients for cover on pH, calculated for this scale on a larger sample in the original study (Fig 5). This component is thus interpreted as the response to a pattern in the "pH-factor-complex" at block size 64; this is consistent with the original interpretation, which was arrived at by a combination of pattern analysis, between-species correlation and regression.

The second component shows a clear curvilinear relation to the first component (Fig. 3a). This suggests that the species of intermediate values for the first component are not just indifferent to pH but actually show a preference for medium pH values. The peak at block size 32 indicates that, not surprisingly, areas of soil with extreme levels of surface pH tend to be nearer to soils with intermediate pH status than to each other. This component has no counterpart in the original analysis, but perhaps (a posteriori) was reflected there in the lower pattern intensity for the intermediate species (particularly Hieracium and Helictotrichon) at block size 64.

The third component operates mainly within the "neutral-calcicolous" phase (Fig. 3b), separating Helianthemum from Briza at block size 8 (= 40 cm). Since no small-scale environmental variation was detected originally in the field, it is reasonable to interpret this effect as the result of competition between the two species. This effect was hinted at in the original analysis by the fact that both Helianthemum and Briza showed pattern at this same block size, and were negatively (but not significantly) correlated at this scale.

On the whole, the results of the integrated multiple pattern analysis are consistent with those of the earlier analysis, serving to reinforce the initial interpretation. But the new method presents more information in a simpler and clearer form, thereby allowing the detection of some effects which were overlooked originally.

It is hoped that the information from the ordination of sites at all scales (the program for which is in preparation) will contribute

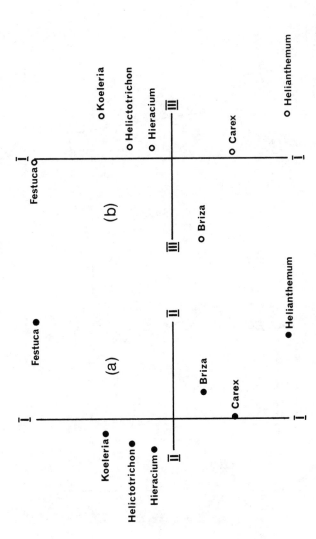

Figure 3. Principal component analysis for Monk's Dale grassland; (a) components I v. II, (b) components I v. III.

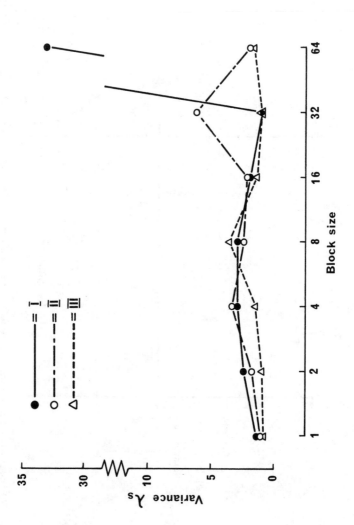

Figure 4. Variance (λ_s) / block size graphs for first three components of Monk's
Dale grassland community

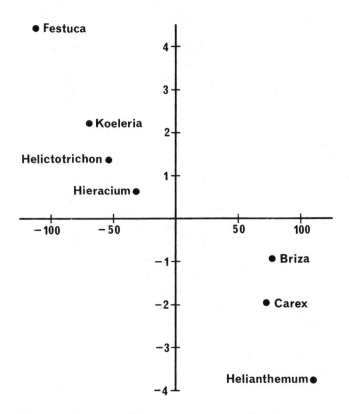

Figure 5.   Loadings on first component (ordinate) v. regression
            coefficients on pH (abscissa) for species in Monk's
            Dale community

further to the interpretation of the vegetational pattern in this
limestone grassland.

4.   DISCUSSION

The results so far obtained with the new method seem to justify the
hope that it would provide an interpretable description of 'total'
species-sites-scales pattern.   Further experience with different
types of data is necessary to increase confidence in the way the
results are being interpreted.

It may be expected that the edge of multiple pattern analysis over simpler techniques will be greatest in the study of complex communities, in which a large number of species contribute significantly to patterns of variation extending over a wide range of scales.

A number of problems still remain to be solved. The significance test for deviation from random distribution which can be used in univariate pattern analysis ([10]; [4]) is not applicable directly to the multivariate case, and a rigorous multivariate analogue may be hard to find. The choice between the summation of dispersion, covariance or correlation matrices (see Appendix) requires further consideration from both theoretical and ecological points of view. These two problems--of significance tests and standardization--are common to all multivariate methods.

To an extent we see a pertinent analogy to be drawn between multiple pattern anlaysis and the photographic technique known as holography. Although a hologram is simply a photographic record of a pattern of interfering wave fronts which show no direct resemblance to the recorded subject, a hologram nonetheless contains all the "information" about that subject. With a single hologram one can examine a subject from different points of view and even focus at different depths throughout the image without running the risk of distorting the original subject. In this context we see multiple pattern analysis as a tool which enables attention to be focused on to significant patterns which may be present on any plane (whether species or scale) of a data cube. It may be seen as a synthesis of the so far disparate approaches of pattern analysis and ordination, each of which may be misleading by restricting attention to a single plane. We hope that the product of this synthesis will prove a useful, generalized method for the analysis and interpretation of vegetational relationships, thereby contributing also to the matriculation of Peter Pan.

ACKNOWLEDGMENT

We are indebted to Mr. K. R. Slater for preparing the figures.

REFERENCES

[1] Anderson, D. J. 1965. Studies on structure in plant communities. I. An analysis of limestone grassland in Monk's Dale, Derbyshire. J. Ecol. 53:97-107.

[2]   Austin, M. P. 1968. Pattern in a Zerna erecta dominated
      community. J. Ecol. 56:197-218.

[3]   Greig-Smith, P. 1952. The use of random and contiguous
      quadrats in the study of the structure of plant communities.
      Ann. Bot. Lond. (N.S.) 16:293-316.

[4]   ——————. 1961. Data on pattern within plant communities.
      I. The analysis of pattern. J. Ecol. 49:695-702.

[5]   ——————. 1961. The use of pattern analysis in ecological
      investigations. Recent advances in botany. (Toronto) 2:
      1354-1358.

[6]   Kershaw, K. A. 1959. An investigation of the structure of a
      grassland community. II. The pattern of Dactylis glomerata,
      Lolium perenne and Trifolium repens. III. Discussion and
      conclusions. J. Ecol. 47:31-53.

[7]   ——————. 1960. The detection of pattern and association.
      J. Ecol. 48:233-242.

[8]   ——————. 1961. Association and co-variance analysis of
      plant communities. J. Ecol. 49:643-654.

[9]   Noy-Meir, I., Tadmor, N. H. and Orshan, G. 1970. Multivariate
      analysis of desert vegetation. I. Association analysis at
      various quadrat sizes. Israel J. Bot.: (In press).

[10] Thompson, H. R. 1958. The statistical study of plant distri-
      bution patterns using a grid of quadrats. Aust. J. Bot.
      6:322-342.

APPENDIX

The Partition of Characteristic Roots by Block Size

The score (density, cover, etc.) of the j-th (out of m) species in
the i-th (out of n) site of the smallest size ($b_1=1$), will be called
$x_{ij} = x_{ij1}$. These scores form the n x m matrix $X = X_1$. The "primary" sites are then successively grouped into blocks of size b_2,
$b_3, \ldots, b_s, \ldots, b_h, n$; there are h block sizes apart from the largest
one (n) which is the entire sample. The number of blocks of size
b_s is $n_s = n/b_s$. The scores, s_{ijs}, for each block size form the
n_s x m matrix X_s. For each block size b_s the "corrected scores"
Y_{ijs} are then calculated by subtracting from the score of each site
the mean for the block of next largest size in which it is located,
and dividing by b_s.

The corrected score matrix at each of the h sizes is then expanded to form an n x m matrix, Y_s; for each primary site the score in
each of these matrices will be the Y_{ijs} for the block of size b_s in

which it is located. The total deviation of the primary site score from the overall mean, $Y_{ij} = x_{ij} - \bar{x}_j$, is then equal to the sum of the Y_{ijs} over all block sizes; in matrix notation

$$Y = Y_1 + \ldots + Y_s + \ldots + Y_h = \sum_{s=1}^{h} Y_s \tag{1}$$

A between-species dispersion matrix G_s is now calculated for every block size

$$G_s = Y'_s Y_s \tag{2}$$

The elements of G_s (between species j, k) can be calculated by the short formula

$$g_{jks} = \frac{1}{b_s} \sum_{i}^{n_s} x_{ijs} x_{iks} - \frac{1}{b_{s+1}} \sum_{i}^{n_{s+1}} x_{ij(s+1)} x_{ik(s+1)} \tag{3}$$

The elements in the principal diagonal are the sums of squares of the species, corrected for block size b_s

$$g_{jjs} = \frac{1}{b_s} \sum_{i}^{n_s} x_{ijs}^2 - \frac{1}{b_{s+1}} \sum_{i}^{n_{s+1}} x_{ij(s+1)}^2 \tag{4}$$

The dispersion matrix is converted into a variance-covariance matrix C_s by dividing by the appropriate number of degrees of freedom ν_s

$$C_s = \frac{1}{\nu_s} G_s = \frac{1}{\nu_s} Y'_s Y_s = \left(\frac{Y_s}{\sqrt{\nu_s}} \right)' \left(\frac{Y_s}{\sqrt{\nu_s}} \right) \tag{5}$$

The value of ν_s can be calculated from the largest block size down, by the formulae

$$\nu_h = n_h - 1 \quad \text{and} \quad \nu_s = n_s - 1 - \sum_{t=s+1}^{h} \nu_t \tag{6}$$

The elements of the C_s matrices, particularly those on the principal diagonal (variances) are those used in the usual pattern analysis.

These elements are now added over all block sizes to give the combined variance-covariance matrix C

$$C = C_1 + \ldots + C_s + \ldots + C_h = \sum_{s=1}^{h} C_s \qquad (7)$$

The characteristic roots λ, and species vectors v, of C, are extracted by principal component analysis, and fulfil the equation

$$vC = \lambda v \qquad (8)$$

and are standardized so that

$$vv' = \lambda \qquad (9)$$

From (8) and (9) follows

$$\lambda^2 = vCv' \qquad (10)$$

and

$$\lambda = \frac{vCv'}{\lambda} \qquad (11)$$

By partitioning C back into the contributions from the various block sizes, a similar partition of λ is achieved

$$\lambda = \frac{v(\sum_s^h C_s)v'}{\lambda} = \sum_s^h \frac{vC_s v'}{\lambda} = \sum_s^h \lambda_s \qquad (12)$$

where

$$\lambda_s = \frac{vC_s v'}{\lambda} \qquad (13)$$

The values of λ_s thus calculated are then as usual plotted against log b_s for each of the principal components.

The vectors of component scores for the primary sites may be calculated, by analogy to ordinary principal component analysis, as:

$$u = \frac{1}{\sqrt{\lambda}} \, vY'$$

(14)

and partitioned into contributions from the various block sizes

$$u = \frac{1}{\sqrt{\lambda}} \, v \left(\sum_s^h Y_s \right)' = \sum_s^h \left(\frac{1}{\sqrt{\lambda}} \, vY_s' \right) = \sum_s^h u_s$$

(15)

where

$$u = \frac{1}{\sqrt{\lambda}} \, vY_s'$$

(16)

Each vector u_s provides the component scores for blocks of size b_s, corrected for larger block sizes. Total component scores for each size, u_s^*, are obtained by summing from the largest size down:

$$u_s^* = \sum_{t=h}^{s} u_t$$

(17)

or directly from equation (14) by using the total site scores (as deviations from the total mean) for each of the block sizes. A series of superimposable ordinations of "sites" at all block sizes is thus achieved.

Possible modifications of the method

The total dispersion matrix G is equal to the sum of the G_s matrices as defined in (2)

$$G = Y'Y = \left(\sum_s^h Y_s' \right) \left(\sum_s^h Y_s \right) = \sum_s^h Y_s'Y_s = \sum_s^h G_s$$

(18)

since the sums of cross products of y_s from different block sizes are all zero ($Y_s'Y_t = Z$ for $s \neq t$) due to the definition of y_s as a deviation from the mean for the larger size. However, the total variance-covariance matrix C* derived from the total dispersion matrix

$$C^* = \frac{1}{\nu}G = \frac{1}{n-1}G = \frac{1}{n-1}\sum_s^h G_s \tag{19}$$

is not identical to the combined variance-covariance matrix C as
defined in (7) because in the latter the contributions from the
various scales have been divided by different degrees of freedom.
It is possible to carry out principal component analysis, subsequent
partition of roots and derivation of site scores on C^* instead of C;
the division by ν_s can then be made at the final stage, on the λ_s.
In fact this solution is statistically more rigorous and mathemat-
ically neater, fulfilling as it does some additional simple
relationships which are lost through the "weighting" by $1/\nu_s$ prior
to principal component analysis. This weighting in combining elements of variation from various block sizes was adopted initially
because it was felt to be consistent with the way in which these
elements are normally compared in the variance block size graphs.

This was confirmed by a comparison of the results of the two
versions (C and C^*) when applied to the model community. The use
of the true total variance tended to emphasize pattern at the
smaller block sizes more than was considered useful. However this
point deserves further consideration and the decision may depend on
the ecological context of a particular study.

Another possible modification is standardization of scores by
species standard deviations, thus summing correlation matrices
instead of covariance or dispersion matrices. This would eliminate
the domination of the analysis by species with high variances
(usually abundant species), when this is considered a desirable
objective.

RECORD OF PREPLANNED AND SPONTANEOUS DISCUSSIONS

D. W. GOODALL (Utah State University)

This paper is concerned with the analysis of pattern displayed
jointly by a number of species within a relatively small area--such
as can be covered by a contiguous series of quadrats. The authors
refer to their approach in terms of a relationship to classification
or ordination, but this is somewhat misleading--these processes are
usually applied to sets of samples covering much larger areas--often
of the order of tens or hundreds of kilometers, where the samples

are perforce very far from contiguous. References to the individual quadrats of a transect as "sites" may also contribute to the potential for misunderstanding here.

Given, though, that the authors are concerned with variation within what would normally be regarded as a stand, rather than between stands, the approach proposed seems very promising. The joint variation over the area in quantity of the various species present is analyzed into principal components, and the variance attributable to successive axes is then partitioned by spatial scale. This enables evidence from different species on the existence of pattern at different scales to be mutually reinforcing, and also permits the interrelations among the species at each scale to be included in the same analysis.

This extension to the multivariate case of the study of pattern by analysis of variance (often called "grid analysis") will retain the drawbacks that have long been clear when the method is applied to each species separately--that the scales studied are constrained to follow an integral multiplicative series, usually powers of two; that there is a grossly unequal division of degrees of freedom, to the detriment of comparisons at larger scales; and that the variance estimates are independent only in the trivial case of random distribution, thus precluding effective tests of significance for null hypotheses of interest. Interpretation is therefore forced to rely in the last resort on subjective evaluation.

The authors maintain that the results of multivariate analysis apply only to the scale at which the data were enumerated, and that a change in scale may lead to different interpretations. This is true where the size of samples envisaged lies within the scale of patterned variation in the community. With small samples, for instance, classification may differentiate hummocks and hollows of bogs as primary classes, rather than different types of bog each with its own hummock and hollow characteristics. It should be remembered, however, that the majority of plant sociologists aim at using samples as homogeneous as they can find, at the scale with which they are concerned. If in fact the vegetation sampled is homogeneous at the scale of the sample size used, it implies that variation upwards from this sample size will have no effect on the classification or ordination results. It is well understood that distinct classification will result from different scales of sampling when overt mosaic patterns occur; this is in fact part of the reason for concern with minimal or representative area in sampling.

When the sample size chosen is large enough to cover the various elements of the mosaic, the results of classification may be relatively independent of the exact size of the samples. Pattern analysis--particularly in a form which takes simultaneous account of pattern in the various species, as in the present paper--may well have its greatest value in indicating how large samples need to be to meet this requirement.

The point to which the authors might well have given more attention is the potential value of a technique for pattern analysis such as they describe, not simply as a means of describing and quantifying a pattern in a particular stand, but as providing additional valuable information in comparisons of different stands and different vegetation types. Classification and ordination procedures normally use the presence of species in samples, or the quantities in which they occur, without reference to their spatial pattern (the Braun-Blanquet school admittedly includes assessments of "sociability" in its records, but they play no recognizable part in the classification process). Pattern analysis is in a position to provide extra, and largely independent, information to contribute to the classification process. It has been suggested, for instance, that different types of pattern may characterize different seral stages. Contrasts between the results obtained in different stands by the technique described here might well facilitate the recognition of such stages.

C. F. COOPER (University of Michigan)

As with most such classification and ordination techniques, the procedure outlined by Noy-Meir and Anderson is primarily valuable as an aid in establishing hypotheses about the structure of vegetation and about the nature of the processes leading to that structure. Ecologists have traditionally relied mostly on intuition in establishing such hypotheses; they have used statistical methods mostly to test the conformity of field data to a particular structural hypothesis. This application of statistics is conceptually identical to testing of effects in agricultural experimentation, and exact significance tests may be very important.

A basic problem in plant ecology, on the other hand, is to make some sense and order out of a mass of field measurements of various biotic and abiotic phenomena. If statistical or graphical methods are used to erect hypotheses, it is usually necessary to use all available data in the process. Additional data must then be

collected to test the hypotheses. The multiscale ordination
presented here seems to be of substantial value as an aid in organizing field data in an economical manner for hypothesis generation.

The authors state that a rigorous significance test may be hard
to find. If multiscale ordination is to be used primarily in
hypothesis generation, rigorous tests may not be required. Deriving
mathematical models of multivariate frequency distributions from
first principles may not be worth the effort if it is feasible at
all. As the authors point out, there may be so much uncertainty
about the underlying nature of the probability distributions that
a rigorous multivariate solution will be exceedingly difficult. As
an alternative, I suggest using Monte-Carlo methods for generating
tables of significance. With high speed computers, there is no
difficulty in running a hypothetical situation thousands of times,
using a random number generator to produce new values for each run.
The distribution of outcomes can then be used as a basis against
which to test observed data. For hypothesis generation, we are not
especially interested in cases of marginal significance where an
exact test may be important.

This approach is particularly appropriate in that a major aim of
ordination analysis should be to produce a biologically sound mathematical model of the actual processes that led to the observed
structural characteristics of the vegetation. Such a model should
be tested in simulation runs, using logical initial input values.
A simulation that consistently yields vegetation structures similar
to those observed in the field is likely to contain a high degree of
realism, and more important, to lead to new conclusions about ecological processes.

H. L. LUCAS, JR. (North Carolina State University)

Up to now, use of the analysis of variance and components of variance to study "aggregation" or "non-randomness" has not been noted
explicitly in this symposium. Conversations with participants
indicate, however, that it has been used and, in fact, was the
method of arriving at the results presented in a preceding paper
by D. J. Anderson. At the suggestion of D. W. Goodall, who has
employed the analysis of variance and has noted to me certain
associated problems, I shall make some comments. The ideas might
be useful to ecologists.

To be concrete, consider a situation like that discussed yesterday by N. G. Hairston. Suppose we have 2^4 contiguous plots. The
analysis of variance for the data might proceed as follows:

V.S.	D.F.	M.S.	E(M.S.)
Blks. of 8	1	s_8^2	$k_{81}\sigma_1^2 + k_{82}\sigma_2^2 + k_{84}\sigma_4^2 + k_{88}\sigma_8^2$
Blks. of 4 in Blks. of 8	2	s_4^2	$k_{41}\sigma_1^2 + k_{42}\sigma_2^2 + k_{44}\sigma_4^2$
Blks. of 2 in Blks. of 4	4	s_2^2	$k_{21}\sigma_1^2 + k_{22}\sigma_2^2$
Units in Blks. of 2	8	s_1^2	$k_{11}\sigma_1^2$
Mean		$\bar{X}$	$\bar{\sigma}^2$

The k's depend on the number of units per block and, if applicable, finite population corrections; $\bar{X}$ with expectation $\bar{\sigma}^2$ under the Poisson assumption is included to make certain tests. The components of variance (σ^2's) are estimated by equating the mean squares to their expectations and solving.

The expectations given are those for the usual analysis of variance model. A model reflecting the general nature of field patterns can, however, be used and the variance components and the expectations can be written in those terms. Such is the approach of statistical geneticists.

One of the problems apparently has been that the mean squares, although they may have been computed to be orthogonal are, like the estimated σ^2's not in general uncorrelated. This poses problems in testing. Estimates of the variances and covariances of the mean squares and of the estimated variance components can, however, be worked out and approximate tests and confidence limits for various items of interest can be constructed.

Another problem is that variances of the mean squares and the estimated variance components for the larger block sizes are relatively poorly estimated. This can be overcome by using unbalanced sampling designs constructed in such a way as to distribute the information available more evenly over the several variance components. Optimal designs are in part dependent on the magnitudes of the components of variance.

The field design and the analysis of variance do not have to be of the completely "nested" type as shown. Two-or-more-way classification schemes can be invoked to distinguish variance patterns in

different directions (e.g., down slope vs. across slope). In the
excellent paper we have just heard, a multivariate counterpart of
the analysis of variance was employed. It yielded quantities
analogous to mean squares. Perhaps an extension along the lines
indicated here for the univariate case would be useful.

D. J. ANDERSON (Australian National University, Canberra, Australia)

I agree with Dr. Cooper that the question of a rigorous significance
test for the multivariate case may be of academic interest only; as
I emphasized in an earlier discussion it is the magnitude and direction of the effects which need to be judged for their ecological
relevance. If I may be heretical in so distinguished a company, I
would emphasize that ultimately, the results of any statistical
analysis can only be interpreted in terms of their ecological
utility.

I cannot agree with Professor Goodall that the use of "sites"
rather than "quadrats" is misleading in our description of multiple
pattern analysis. In fact Noy-Meir has analyzed data collected by
Dr. Costin from solifluxion terraces in the Kosciusko region of New
South Wales, in which the sites or quadrats are contiguous (10mx|m)
sections of a belt transect extending over several hundred meters.
The analysis provided a successful basis for interpreting variation
which has an essentially between-, rather than within-stand basis.

I very much hope that this method will help those workers who are
interested in defining minimal or representative area, although I
personally remain sceptical of its utility in this regard. If we
examine the minimal area (as defined by species/area curves) for
the Australian shrublands I discussed in an earlier paper, the
minimal area so defined will vary between about $2m^2$ and $200m^2$,
depending only on whether the summer or winter facies of the same
community is examined.

In these circumstances, and knowing too the general influence of
spatial pattern on estimates of "minimal area," I am doubtful if
multiple pattern analysis will provide the clear answer that Professor Goodall is seeking.

I. NOY-MEIR (Australian National University)

The range of application of the method as presented here, as of
univariate pattern analysis, is indeed limited by the length of
contiguous transects it is technically possible to record. However,
the recording of simple quantity estimates in transects hundreds to

thousands of meters long is quite feasible, and is often done in surveys. Most ecologists would consider variation along these transects as including "between-community" variation. The range could probably be extended by aerial survey methods. In any case, one of the main lessons that seems to emerge from the application of simple and multiple pattern analysis to vegetation is that it is impossible to distinguish clearly "within-community" ("mosaic") pattern from "between-communities" variation. This distinction is purely a matter of convention and it may be more useful to talk of variation, patterns and communities at various scales (e.g., macro-, meso-, and micro-). When the variance/scale plot in multiple pattern analysis has a single well-defined peak-and-trough it can be used to determine the size of the "representative area" for sampling and describing the vegetation, as suggested by Professor Goodall. However the results so far show that this size is usually different for different components (i.e., different aspects of phytosocio-logical variation) and sometimes there is more than one such scale even for a single component. Thus while a "representative area" can be defined in a limited context (e.g., for the first component and within a range of scales from 10 to 100 meters), there is usually no single overall "representative" (or "minimal") size. Moreover in many cases the peaks and troughs are broad and diffuse--a situation which may arise when patches vary strongly in size. Then there are no well-defined optimal quadrat-sizes.

With some limitations, multiple pattern analysis may also be used when sampling is not contiguous, for instance multi-stage nested sampling, as can be used for large areas. Variances and covariances for the various stages can be calculated in a similar way, with sub-sequent PCA of a combined matrix and repartition of the roots. The interpretation of the change in variance with scale will however be complicated by the effect of the distance between centers of quad-rats for each sampling stage, which now (in contrast to contiguous sampling) is not equal to quadrat length.

The use of "pattern" as an attribute in classification, ordina-tion or model validation, the application of Monte-Carlo methods in judging significance and the extensions suggested by Dr. Lucas, are certainly interesting possibilities.

ON THE STATION-SPECIES CURVE

V. V. KRYLOV
All-Union Research Institute of Marine Fisheries
and Oceanography
Moscow, U.S.S.R.

SUMMARY

Approximation for the exact mean length of the list of species (N_m) for m station.

1. INTRODUCTION

Exact mean length of the list of species (N_m) for m stations
$(m = 1,2,3,...,r;$ r is the total number of stations in a given
survey) can be determined from the formulae:

$$N_m = \frac{\sum\limits_{i=1}^{m} (-1)^{i+1} C_{r-i}^{r-m} \sum\limits_{i=m}^{r} x_i C_i^m}{C_r^m} \text{, or}$$

$$N_m = \sum\limits_{i=1}^{r} x_i - \sum\limits_{i=1}^{r-m} x_i \frac{C_{r-m}^i}{C_r^i} \text{,}$$

where x_i is the number of species, each of them found at i stations
$(i = 1,2,3,...,r)$. $x_0=0$.
   N_m can be approximated by the function:

$$N_m' = N_{max} \exp[-\alpha/m^\beta] \text{ .}$$

Coefficient N_{max} is interpreted as the maximum number of species
which can be obtained in the studied region if $m \to \infty$ and the same
methods of collecting and, particularly, of examination of the
material are used. N_{max} and, possibly, β do not depend on the
catching capacity of the sampling gear.
   N_{max} must be used instead of n if the function of hypergeometric
distribution is used as an association criterion for pairs of stations, i.e., if two stations are considered to be associated when

$$\sum\limits_{j}^{b} \frac{C_a^j C_{n-a}^{b-j}}{C_n^b} \leq \alpha'$$

here α' is the significance level, a and b are the numbers of species found at stations "A" and "B" $(a \geq b)$, j is the number of species,
each of them obtained at both stations, n is the total number of
species identified in the material.

2. REPORT

Values of N_m have been computed for plankton collected in the East
China Sea during two research cruises at 23 stations (February-March
1962) and at 42 stations (April-May 1963) distributed approximately
evenly over the area of the sea.   The samples were taken in vertical
hauls at standard layers between 200 m or the bottom and the surface
with the Juday net (diameter 37 cm, gauze 38 meshes per linear cm).
It was found that

$$N'_m = 272.5 \, \exp\left(-\frac{1.7946}{m^{0.4820}}\right) \qquad \text{(February-March 1962)},$$

$$N'_m = 315.0 \, \exp\left(-\frac{2.0806}{m^{0.4000}}\right) \qquad \text{(April-May 1963)}.$$

Correlation coefficients between N_m and N'_m are more than 0.9999 for
both surveys.   The application of the hypergeometric distribution
to the material collected in the East China Sea made it possible to
reveal the groups of stations located within the main planktonic
biocoenoses.

A TEST OF THE DIFFERENCE
BETWEEN CLUSTERS

M. D. MOUNTFORD
The Nature Conservancy
19, Belgrave Square
London, England

SUMMARY

A probability model describing the joint distribution of indices of
similarity is proposed. An exact test of significance is devised
applicable to clusters defined by an external criterion; a conservative test of the significance of clusters defined by internal
criteria is obtained.

1. INTRODUCTION

Indices of similarity are used by ecologists and systematists as a means of comparing and classifying biological communities and organisms. The indices have the property of measuring, in some sense, the degree of resemblance of two biological units. Various indices of similarity have been proposed by many writers: lists of the more widely used are given by Sokal and Sneath [9]. More recently, Goodall [1] has proposed a similarity index which has the advantage of a probabilistic interpretation. The indices are constructed from the information on a number of characteristics or variates each observed on the two units. For example, the similarity between two plant communities in which k species are represented with relative frequencies $P_1,\ldots,P_k$ and $\pi_1,\ldots,\pi_k$ may be measured by the index,

$$J = \frac{\sum P_i \pi_i}{\sqrt{[(\sum P_i^2)(\sum \pi_i^2)]}} \tag{1}$$

J is the direction cosine of the angle between the normalized vectors representing the two communities in k dimensional space; J varies from 0, in the case in which the two sites have no species in common, to 1, the value obtained when the two sites have the same species composition and the same set of relative abundances.

In comparative and classificatory studies the biological units are combined by various methods to form groups of units. One point of interest in the formation of the groups is the relative magnitude of the indices of similarity of pairs of units within the same group and the indices of similarity of units in different groups. In comparative studies the units are allotted to groups usually by some external characteristic; e.g., a set of plant communities are classified according to their geological character. Alternatively in cluster analysis, the grouping procedure is based on the observed values of the indices of similarity. The method of grouping is such that the indices of similarity of pairs of units in the same group are on the average greater than the indices of pairs of units in different groups. Following a procedure commonly adopted, the groups already formed are divided into sub-groups, which are in turn sub-divided--or, the alternative procedure of agglomerating groups into larger groups is followed--to form a hierarchical tree or dendrogram.

A problem posed by these procedures, both comparative and classificatory, is that of testing the reality of the grouping. Is the
similarity of units within the same group significantly greater than
that of units in different groups; or are the units effectively
members of the same single aggregate, the observed differences
between the indices of similarity being the result of sampling
variation? To answer this question it is, of course, necessary to
formulate a probability model against which the significance of the
observed differences can be gauged. The model proposed here, though
an extremely simple representation, may be appropriate to some
practical situations.

2. PROBABILITY MODEL

Suppose there are n units to be classified according to the information on p variates observed on each unit. It is assumed that the
n units are random variables in p-dimensional space such that either
x_{ij}, the index of similarity of the ith and jth unit, or a suitable
transform of x_{ij}, is normally distributed with mean μ_{ij} and with
variance σ^2 independent of i and j. If the distribution of x_{ij} is
non-normal, an appropriate normalizing transform may be indicated
by the form of the index.

The indices x_{ij} and $x_{k\ell}$ may be assumed to be distributed independently if $i,j \neq k,\ell$. However the indices x_{ij} and x_{ik}, having the unit
i in common, will be correlated. It will be assumed that this
correlation is described by the covariance function,

$$\text{cov}[x_{ij}, x_{k\ell}] = \sigma^2; \ i=k \text{ or } \ell, \ j=\ell \text{ or } k$$

$$= \rho\sigma^2; \ i=k \text{ or } \ell, \ j \neq k \text{ or } \ell$$
$$i \neq k \text{ or } \ell, \ j=k \text{ or } \ell$$

$$= 0 \qquad i,j \neq k,\ell.$$

Let g_i, (i=1,2,...,n) and e_{ij}, ($1 \leq i < j \leq n$) be a set of independent
normal variates with zero means and variances σ_g^2 and σ_e^2. Consider
the derived set of variables

$$y_{ij} = g_i + g_j + e_{ij}$$

These are nultinormally distributed with zero means, variance $2\sigma_g^2 + \sigma_e^2$ and covariances

$$\text{cov}[y_{ij}, y_{ik}] = \sigma_g^2$$

$$\text{cov}[y_{ij}, y_{k\ell}] = 0 \qquad i,j \neq k,\ell. \tag{2}$$

Thus the model of (2) can be recast in the form

$$x_{ij} = \mu_{ij} + g_i + g_j + e_{ij}$$

where

$$\sigma_g^2 = \rho\sigma^2$$

and

$$\sigma_e^2 = (1-2\rho)\sigma^2 . \tag{3}$$

This second model is that used to describe the simple dialled cross. By analogy, the index of similarity x_{ij} may be considered as the sum of the unit effects g_i and g_j added to the specific combining effect e_{ij} of the units i and j.

If a number of assumptions are made, the work of Gower [2] can be shown to be relevant to the model defined in (3). He showed that many indices of similarity x_{ij} are such that point representations P_i and P_j of the units i and j can be found whose distance d_{ij} apart is $\sqrt{[2(1-x_{ij})]}$; i.e., x_{ij} varies in the same manner as

$$d_{ij}^2 = r_i^2 + r_j^2 - 2r_i r_j \cos\theta_{ij}$$

where the r_i, r_j are the radius-vectors, and θ_{ij} is the angle between the two points. Let us suppose that the units i and j have evolved by a process of random diffusion from an initial state of complete resemblance. Taking the origin at the point representation of the initial state of identity, then

$$E[r_i] = 0; \quad E[2r_i r_j \cos \theta_{ij}] = 0$$

and

$$E[r_i^2] = \mu/2 \quad (\text{say}).$$

Writing

$$g_i = r_i^2 - \mu/2$$

and

$$e_{ij} = -2r_i r_j \cos \theta_{ij}$$

Then

$$E[g_i] = E[g_i g_j] = E[g_i e_{ij}] = 0$$

and x_{ij} varies as

$$d_{ij}^2 = \mu + g_i + g_j + e_{ij} .$$

If the g_i and e_{ij} are normally distributed, then the model defined by (3) is obtained.

3. TEST OF SEPARATION OF TWO CLUSTERS

Consider a separation of a set of n nunts into two groups. Group 1 and Group 2, with n_1 and $n-n_1=n_2$ units respectively. The two groups will be said to form two distinct clusters if the indices of similarity of pairs of units within the same group are significantly greater on the average than the indices of pairs of units belonging to different groups.

Let x_{ij}; $1 \le i < j \le n_1$ be the $n_1(n_1-1)/2=N_1$ indices of similarity formed from the pairs of units belonging to Group 1; $x_{n_1+k,n_1+\ell}$; $1 \le k < \ell \le n_2$ the $n_2(n_2-1)/2=N_2$ indices belonging to Group 2; and x_{i,n_1+k}; $1 \le i \le n_1$, $1 \le k \le n_2$ the $n_1 n_2$ indices formed by pairing a unit in Group 1 with one in Group 2. First, for simplicity, it is assumed that the indices of similarity within each group are distributed about a common mean; $E[x_{ij}]=\mu_{11}$; $E[X_{n_1+k, n_1+\ell}]=\mu_{22}$ say. Furthermore it is assumed that the indices of similarity x_{i,n_1+k} are distributed about a common mean μ_{12}. The variance-covariance function is taken to be that defined by (2) or (3).

The presence or absence of clustering is tested against the null hypothesis

$$\mu_{11} = \mu_{22} = \mu_{12} = \mu$$

It has been shown by a number of authors, among them Yates [10] and McGilchrist [3], that the sum of squares

$$S_n^2 = \sum_{1 \le i < j \le n} (x_{ij} - \bar{x} - \hat{g}_i - \hat{g}_j)^2 \tag{4}$$

where

$$\hat{g}_i = \frac{1}{n-2} \left\{ \sum_{j \ne i} x_{ij} - (n-1)\bar{x} \right\}$$

is distributed as $\sigma_e^2 \chi_{N-n}^2$, where $N=n(n-1)/2$. For the alternative hypothesis

$$\mu_{11} \ne \mu_{22} \ne \mu_{12}$$

it follows (see [7], p. 147) that S_n^2 is distributed as a non-central χ^2 with $N-n$ degrees of freedom, and with the non-centrality parameter λ equal to the value of S_n^2 when expected values are substituted for the variables.

When i and j both belong to Group 1,

$$E[x_{ij}-\bar{x}-\hat{g}_i-\hat{g}_j] = \mu_{11} - \mu - \frac{1}{n-2} \left[2(n_1-1)\mu_{11}+2n_2\mu_{12}-2(n-1)\mu \right]$$

where

$$\mu = [N_1\mu_{11}+N_2\mu_{22}+n_1n_2\mu_{12}] / N$$

that is,

$$E[x_{ij}-\bar{x}-\hat{g}_i-\hat{g}_j] = \frac{n_2(n_2-1)}{(n-1)(n-2)} [\mu_{11}+\mu_{22}-2\mu_{12}]$$

Similarly when i and j belong to different groups

$$E[x_{ij}-\bar{x}-\hat{g}_i-\hat{g}_j] = - \frac{(n_1-1)(n_2-1)}{(n-1)(n-2)} [\mu_{11}+\mu_{22}-2\mu_{12}]$$

Hence the non-centrality parameter is

$$\lambda = \frac{n_1(n_1-1)n_2(n_2-1)}{2(n-1)(n-2)} [\mu_{11}+\mu_{22}-2\mu_{12}]^2 \tag{5}$$

The quantity

$$Q = \bar{x}_{11} + \bar{x}_{22} - 2\bar{x}_{12} \tag{6}$$

is distributed normally with mean $\mu_{11}+\mu_{22}-2\mu_{12}$ and with variance, in terms of model (3), equal to

$$(\frac{1}{N_1} + \frac{1}{N_2} + \frac{4}{n_1 n_2})\sigma_e^2$$

$$= \frac{2(n-1)(n-2)}{n_1(n_1-1)n_2(n_2-1)} \sigma_e^2 = v^2\sigma_e^2 \quad \text{(say)}. \tag{7}$$

Hence

$$s_n^2 = Q^2/v^2 + \sigma_e^2\chi_{N-n-1}^2 ,$$

where χ_{N-n-1}^2 is distributed as a central χ^2 with N-n-1 degrees of freedom. Furthermore Q^2/v^2 is distributed independently of $s_n^2-(Q^2/v^2)$ and the ratio

$$t = \left[\frac{(N-n-1)Q^2/v^2}{s_n^2-(Q^2/v^2)} \right]^{\frac{1}{2}} \tag{8}$$

considered as a t-variate with N-n-1 degrees of freedom provides a test of the deviation from zero of $(\mu_{11}+\mu_{22}-2\mu_{12})$.

The quantity $(\mu_{11}+\mu_{22}-2\mu_{12})$ is a reasonable measure of the degree of clustering of the two groups; it is the difference between the indices of similarity of pairs within the same group and those of pairs of units belonging to different groups. If the grouping is determined by an external criterion or prior consideration, the ratio(8), used as a one-sided test, measures the significance of the grouping.

Now consider the general model in which the individual $\mu_{ij}=E[x_{ij}]$ are allowed to be different.

Let $\bar{\mu}_{11}$ be the arithmetic mean of the pairs of units occurring in Group 1; and let $\mu_{ij}=\bar{\mu}_{11}+\delta_{ij}$ for $1\leq i<j\leq n_1$. $\bar{\mu}_{22}$ and $\bar{\mu}_{12}$ and the resulting deviations δ_{ij} are similarly defined.

When the expected values μ_{ij} are substituted for x_{ij} in (4) it is easily seen that the non-centrality parameter equals

$$\frac{n_1(n_1-1)n_2(n_2-1)}{2(n-1)(n-2)} \; [\bar{\mu}_{11}+\bar{\mu}_{22}-2\bar{\mu}_{12}]^2 \; + \; \left[s_n^2\right]_{x_{ij}=\delta_{ij}}$$

Furthermore the quantity

$$Q = \bar{x}_{11} + \bar{x}_{22} - 2\bar{x}_{12}$$

is distributed normally with mean $\bar{\mu}_{11}+\bar{\mu}_{22}-2\bar{\mu}_{12}$. Hence for the general alternative hypothesis in which the μ_{ij} are allowed to be different, the test ratio of (8) undervalues the significance of the deviation from the null hypothesis. It gives a lower bound to the significance point of most types of clustering.

4.   GROUPING DETERMINED BY AN INTERNAL CRITERION

4.1 Test Statistic

If, as is usual in classificatory studies, the grouping procedure is based on the internal evidence provided by the observed indices of similarity, the value of the test-statistic t of the previous section, will be conditioned by the procedure actually used. In so far as most grouping procedures strive to maximize the within-group similarity and minimize the similarity of units in different groups, it is clear that the test-statistic will be greater than that expected in a grouping determined by an external and independent criterion. A valid test of a grouping obtained from the observed values of the indices of similarity would take into account the various steps taken, and criteria, used in obtaining the final grouping. As there are 2^{n-1} different ways of forming two groups from n units, and as each grouping can be arrived at by different routes, it is hardly practicable to devise a separate test for each grouping produced by the classificatory procedure. However there is an over-all test which can be applied to all possible splits of the

n units into two groups, with at least two units in each group. The
basic idea is the same as that of Scheffé's [8] method of multiple
comparison. The value of a test-statistic for the division into two
groups produced by the classificatory procedure is tested against
the probability distribution of the maximum of the $2^{n-1}-n-1$ values
obtained from every possible distinct decomposition of the n units
into two groups, with at least two units in each group. The test
is conservative in the sense that whenever the observed grouping is
not the configuration that produces the maximum value of the test-
statistic, the significance of the deviation from the null hypothesis
is under-valued.

4.2 Distribution Function of Maximum Statistic

Instead of using t, defined by (8), as the test statistic it is more
convenient to work with the derived quantity

$$b = \frac{Q}{vS_n} \tag{9}$$

where S_n^2 is defined by (4). The method used to obtain the distribution of the maximum value of b is similar to that of Quesenberry
and David [6]. Let the units be numbered 1 to n. For each split
of the n units into two groups let $b_{i,\ldots,\xi}$ be the value of b when
the smaller of the two groups consists of the units numbered $i,\ldots,\xi$.
If the two groups both consist of n/2 units it is immaterial which
of the two receives the denotation $i,\ldots,\xi$. There are $2^{n-1}-n-1=T$,
say, distinct sets $i,\ldots,\xi$ corresponding to the T distinct ways of
splitting the n units into two groups, with at least two units in
each group. Let β be the maximum of the T distinct $b_{i,\ldots,\xi}$. Then

$$\Pr(\beta>h) = \sum \Pr(b_{i,\ldots,\xi}>h) - \sum \Pr(b_{i,\ldots,\xi}>h; \ b_{j,\ldots,\eta}>h)$$
$$+ \sum \Pr(b_{i,\ldots,\xi}>h; \ b_{j,\ldots,\eta}>h; \ b_{k,\ldots,\zeta}>h) \text{ etc.} \tag{10}$$

where the summations are over the T distinct splits. Thus the
second summation, of b's taken two at a time, is made up of $T(T-1)/2$
terms.

This method of obtaining the distribution of the maximum b works
satisfactorily if the third and higher summations are of negligible
magnitude. This was the case for values of n up to 10. The first
summation on the right-hand side of (10) is soon computed; it equals

$$T \ Pr\left\{ t_{N-n-1} > \left[\frac{h^2 (N-n-1)}{1-h^2} \right]^{\frac{1}{2}} \right\}$$

where t_{N-n-1} denotes the t-variate with N-n-1 degrees of freedom.

The second summation of (10), $\sum Pr(b_{i,\ldots,\xi} > h; \ b_{j,\ldots,n} > h)$, is derived by decomposing the sum of squares S_n^2 defined in (4), into three independent components.

Let $Q_{i,\ldots,\xi}$ be the value of $\bar{x}_{11} + \bar{x}_{22} - 2\bar{x}_{12}$ corresponding to the set $1,\ldots,\xi$, and let $\sigma_e^2 v_{i,\ldots,\xi}^2$ be its variance.

Let

$$R_{i,\ldots,\xi} = \frac{Q_{i,\ldots,\xi}}{v_{i,\ldots,\xi}}$$

Then

$$S_n^2 = R_{i,\ldots,\xi}^2 + \frac{(R_{j,\ldots,n} - \lambda R_{i,\ldots,\xi})^2}{1-\lambda^2} + \chi_{N-n-2}^2 \sigma_e^2$$

where λ is the correlation between $Q_{i,\ldots,\xi}$ and $Q_{j,\ldots,n}$. It follows, by Theorem 1 of Mosiman [4], that

$$b_{i,\ldots,\xi} = R_{i,\ldots,\xi} \ / \ S_n$$

and

$$b'_{j,\ldots,n} = \frac{R_{j,\ldots,n} - \lambda R_{i,\ldots,\xi}}{[\sqrt{(1-\lambda^2)}] \ S_n}$$

have the joint density function

$$f(b_{i,\ldots,\xi}; \ b'_{j,\ldots,n}) = \frac{N-n-2}{2\pi} \left[1 - b_{i,\ldots,\xi}^2 - b'^2_{j,\ldots,n} \right]^{\frac{N-n-4}{2}}$$

Therefore $b_{i,\ldots,\xi}$ and $b_{j,\ldots,n}$ have the joint density function,

$$f(b_{i,\ldots,\xi}; \ b_{j,\ldots,n}) = \frac{N-n-2}{2\pi[\sqrt{(1-\lambda^2)}]} \left[1 - \frac{1}{1-\lambda^2} (b_{i,\ldots,\xi}^2 \right.$$
$$\left. - 2\lambda b_{i,\ldots,\xi} b_{j,\ldots,n} + b_{j,\ldots,n}^2) \right]^{\frac{N-n-4}{2}} \quad (11)$$

As $f(b_{i,\ldots,\xi};\ b_{j,\ldots,\eta})$ is symmetric about $b_{i,\ldots,\xi}=b_{j,\ldots,\eta}$ then

$$I_2 = Pr\{b_{i,\ldots,\xi}>h;\ b_{j,\ldots,\eta}>h\} = 2Pr\{b_{i,\ldots,\xi}>h;\ b_{j,\ldots,\eta}>b_{i,\ldots,\xi}\}.$$

Write

$$u_1 = \frac{b_{i,\ldots,\xi}}{\sqrt{[2(1+\lambda)]}} + \frac{b_{j,\ldots,\eta}}{\sqrt{[2(1+\lambda)]}}$$

$$u_2 = \frac{b_{i,\ldots,\xi}}{\sqrt{[2(1-\lambda)]}} - \frac{b_{j,\ldots,\eta}}{\sqrt{[2(1-\lambda)]}}$$

Then

$$I_2 = \int\int \frac{N-n-2}{\pi}\left(1-u_1^2-u_2^2\right)^{\frac{N-n-4}{2}} du_1 du_2$$

the integral being taken over the region

$$\left(\frac{1+\lambda}{2}\right)^{\frac{1}{2}} u_1 + \left(\frac{1-\lambda}{2}\right)^{\frac{1}{2}} u_2 > h \quad \text{and} \quad u_2 < 0$$

If $u_1 = r\cos\theta$, $u_2 = r\sin\theta$, then

$$I_2 = \int_{h\sqrt{\left(\frac{2}{1+\lambda}\right)}}^{1} \frac{N-n-2}{\pi}\left(1-r^2\right)^{\frac{N-n-4}{2}} r\left(\cos^{-1}\frac{h}{r} - \cos^{-1}\left(\frac{1+\lambda}{2}\right)^{\frac{1}{2}}\right) dr$$

Integrating by parts

$$I_2 = \frac{h}{\pi} \int_{h\sqrt{\left(\frac{2}{1+\lambda}\right)}}^{1} \frac{\left(1-r^2\right)^{\frac{N-n-2}{2}}}{r\sqrt{(r^2-h^2)}} dr \tag{12}$$

The integral was evaluated numerically for different values of n, h and λ.

4.3 Computation of Correlation Coefficients

The values of the correlation coefficient λ between $Q_{i,\ldots,\xi}$ and $Q_{j,\ldots,\eta}$ were obtained as follows. For the group consisting of only the two units i and j

$$Q_{ij} = x_{ij} + \frac{2}{(n-2)(n-3)} \sum_{k,\ell \neq i,j} x_{k\ell} - \frac{1}{(n-2)} \sum_{k \neq i,j} (x_{ik}+x_{jk}) \quad (13)$$

i.e.,

$$\sum_{j \neq i}^{n} Q_{ij} = 0 \qquad (14)$$

Multiplying (14) by Q_{ik} and taking expectations,

$$(n-2)\rho' v^2 + v^2 = 0 \qquad \text{i.e., } \rho' = - \frac{1}{(n-2)} \qquad (15)$$

where

$$v^2 = \text{var}[Q_{ij}]; \quad \rho' v^2 = \text{cov}[Q_{ij}, Q_{ik}] .$$

Multiplying (14) by $Q_{k\ell}$, $(k,\ell \neq i,j)$, and taking expectations,

$$(n-3)\rho'' v^2 + 2\rho' v^2 = 0$$

where

$$\rho'' v^2 = \text{cov}[Q_{ij}, Q_{k\ell}] .$$

Hence

$$\rho'' = \frac{2}{(n-2)(n-3)} \qquad (16)$$

It is easily verified that the statistic $Q_{i,\ldots,\xi}$ may be expressed as

$$Q_{i,\ldots,\xi} = \frac{2(n-2)(n-3)}{k(k-1)(n-k)(n-k-1)} \sum_{i,\ldots,\xi} Q_{ij}$$

where the summation is taken over all distinct pairs from the set of k (say) units $i,\ldots,\xi$. If the second set $j,\ldots,\eta$ consists of ℓ units, and if h units are common to both sets, then the covariance between $Q_{i,\ldots,\xi}$ and $Q_{j,\ldots,\eta}$ is equal to

$$\frac{[2(n-2)(n-3)]^2}{k^{(2)}(n-k)^{(2)}\ell^{(2)}(n-\ell)^{(2)}}\ C$$

where

$$C = E\left[\left(\sum_{i,\dots,\xi} Q_{ij}\right)\left(\sum_{j,\dots,\eta} Q_{ij}\right)\right]$$

$$= v_{ij}^2\left\{\frac{h^{(2)}}{2} + \left[h(k-1)(\ell-1)-h^{(2)}\right]\rho' + \left[\frac{h^{(2)}}{2}-h(k-1)(\ell-1)+\frac{k^{(2)}\ell^{(2)}}{4}\right]\rho''\right\}$$

$$= \frac{v_{ij}^2}{2(n-2)(n-3)}\left\{(n-1)(n-2)h^{(2)}-2(n-1)(k-1)(\ell-1)h+k^{(2)}\ell^{(2)}\right\}$$

From (7),

$$\frac{\mathrm{var}[Q_{ij}]}{\mathrm{var}[Q_{i,\dots,\xi}]} = \frac{k^{(2)}(n-k)^{(2)}}{2(n-2)(n-3)}$$

Hence the correlation between $Q_{i,\dots,\xi}$ and $Q_{j,\dots,\eta}$ is

$$\lambda(h,k,\ell) = \frac{(n-1)^{(2)}h^{(2)}-2(n-1)(k-1)(\ell-1)h + k^{(2)}\ell^{(2)}}{\sqrt{[k^{(2)}(n-k)^{(2)}\ell^{(2)}(n-\ell)^{(2)}]}} \tag{17}$$

4.4 Computational Procedure for n Up to 10

The procedure used for computing the 5% and 1% significance points
of β was much the same as that of Quesenberry and David [6]. A
first approximation and an upper bound D_1 to the significance level,
α, was obtained by setting the first summation of (10) equal to α
i.e., D_1 is such that

$$\mathrm{Pr}\left\{t_{N-n-1}>\left[\frac{-D_1^2}{1-D_1^2}(N-n-1)\right]^{\frac{1}{2}}\right\} = \frac{\alpha}{2^{n-1}-n-1}\ .$$

The next approximation and lower D_{20} is given by

$$(2^{n-1}-n-1)\mathrm{Pr}\,(\beta>D_{20}) = \alpha + \sum\mathrm{Pr}\left\{b_{i,\dots,\xi}>D_1;\ b_{j,\dots,\eta}>D_1\right\}. \tag{18}$$

On replacing D_1 in (18) by D_{20} a second approximation D_{21} is
obtained and the process can be continued until sufficient accuracy

is obtained. These bounds, for values of n up to 10, are given in
Table 1. When only one value is quoted the bounds agree to the
third decimal place.

Table 1. Bounds for upper 5% and 1% points of max. b

n	5%	1%
4	.998	.99993
5	.917	.963
6	.816	.879
7	.735	.798
8	.668; .772	.730
9	.619; .622	.674; .675
10	.570; .582	.626; .629

4.5 Approximate Upper Significance Points

The relatively wide bounds, .570 to .582, of the 5% point for n=10
indicate the substantial effect of the second, pairwise, summation
on the right-hand side of equation (10). It is therefore to be
expected that the third and higher summations will have a measurable
effect for values of n greater than 10.

To avoid the labor of attempting to compute the higher summations,
approximate solutions to the upper significance levels were sought.

As n increases, the joint density function (11) tends to the
bivariate normal

$$f(b_{i,\ldots,\xi}; b_{j,\ldots,\eta}) = \frac{N-n-3}{2\pi\sqrt{(1-\lambda^2)}} \exp\left\{-\frac{(N-n-3)}{2(1-\lambda^2)}\left(b_{i,\ldots,\xi}^2\right.\right.$$

$$\left.\left. -2\lambda b_{i,\ldots,\xi}b_{j,\ldots,\eta} +b_{j,\ldots,\eta}^2\right)\right\}$$

As the number of units n increases it is to be expected that, on
average, the correlations between the Q's will lessen. Histograms of
values of λ for n=10 and n=20 are shown in Figure 1. For n=10, 52%
of the values of λ lie between ±.1 and 88% between ±.2; for n=20,
89% lie between ±.1 and 97% between ±.2. As n increases the joint

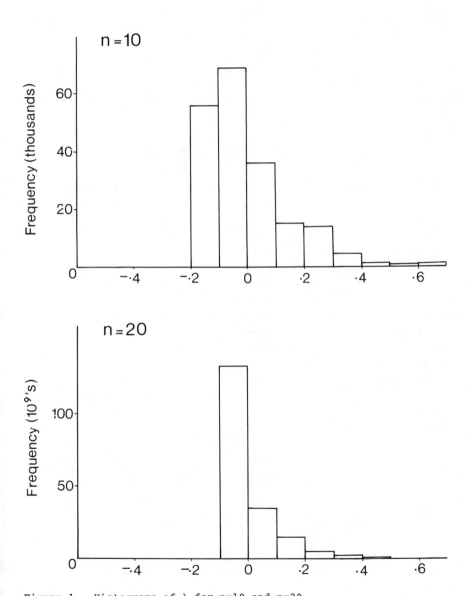

Figure 1.   Histograms of λ for n=10 and n=20

distribution of the b's tends to the independent normal; the tendency is enchanced by the mutually cancelling effects of the negative and positive correlations. Hence for large values of n the distribution function of the maximum b is

$$Pr(\beta > h) = 1 - [1-Pr(b>h)]^T$$

where b is normally distributed with zero mean and variance 1/N-n-3. The upper significance levels given by this approximation are shown in Table 2.

Table 2. Approximate upper 5% and 1% points of max. b

n	5%	1%
10	.581	.629
11	.547	.590
12	.518	.557
13	.494	.529
14	.472	.505
15	.454	.483
16	.437	.464
17	.422	.447
18	.409	.432
19	.397	.418
20	.385	.405
25	.342	.356
30	.309	.321
35	.285	.294
40	.266	.273
45	.249	.256
50	.237	.242
60	.216	.220
70	.199	.203
80	.187	.190
90	.176	.178
100	.166	.169

5. NUMERICAL EXAMPLE

Table 3 is reproduced from Mountford [5]. It gives the similarity values of the soil fauna found in 10 Lake-District Woodland sites.

Table 3. Values of $I \times 10^4$ for data of soil animals in Lake-district woodlands

Site	Yew			Oak				Larch		
	Limestone			Slate				Lime-stone	Slate	Lime-stone
	1	2	3	4	5	6	7	8	9	10
1		180	103	36	44	42	41	90	56	108
2			141	59	63	80	71	108	83	128
3				42	53	50	60	100	57	92
4					133	106	166	77	139	86
5						133	185	97	116	81
6							211	82	109	60
7								108	116	76
8									83	106
9										86

The index of similarity I is defined as the solution of

$$e^{aI} + e^{bI} = 1 + e^{(a+b-j)I}$$

where a and b are the numbers of species found in samples drawn from two sites and j is the number of species found in both samples. I increases with j; it also has the merit of being independent of sample size for samples drawn from a population with a logarithmic distribution of species numbers.

Unfortunately I have been unable to derive the probability distribution of I; for the purpose of this numerical demonstration, I is assumed to be normally distributed and the probability model described in the previous sections is taken to be a realistic description.

Following the procedure described in Mountford [5] the dendrogram shown in Figure 2 was produced: the numbers at the branch-points are the values of $\bar{x}_{12}$.

The overall nature of the test justifies its application to any bifurcation of the dendrogram. Also if a group is made up of distinct sub-groups the test will be made only yet more conservative.

The ten sites fall into two main groups; one group composed of the five limestone sites and the second group made up of the five slate sites.

Designating the five limestone sites as Group 1,

$$\bar{x}_{11} = 115.6; \qquad \bar{x}_{22} = 141.5; \qquad \bar{x}_{12} = 66.92$$

$$s^2_{10} = 61253.9$$

Hence

$$Q = \bar{x}_{11} + \bar{x}_{22} - 2\bar{x}_{12} = 123.26$$

and the variance of Q,

$$\sigma^2 v^2 = \sigma^2 \left(\frac{1}{10} + \frac{1}{10} + \frac{4}{25} \right) = 0.36\sigma^2$$

$$b^2 = \frac{Q^2}{v^2 s^2_{10}} = 0.689$$

$$b = 0.830$$

For n=10, this value is greater than the 1% significance level of β. Hence the two groups form significantly distinct clusters.

The dendrogram in Figure 2 indicates the possibility of further subdivision of Group 1 into the two clusters (5, 6, 7) and (4, 9). For this division of the five sites,

$$\bar{x}_{11} = 139.0; \qquad \bar{x}_{22} = 176.\dot{3}; \qquad \bar{x}_{12} = 124.\dot{3}$$

$$s^2_5 = 4266; \qquad Q = 66.\dot{6}; \qquad v^2 = 2$$

Hence

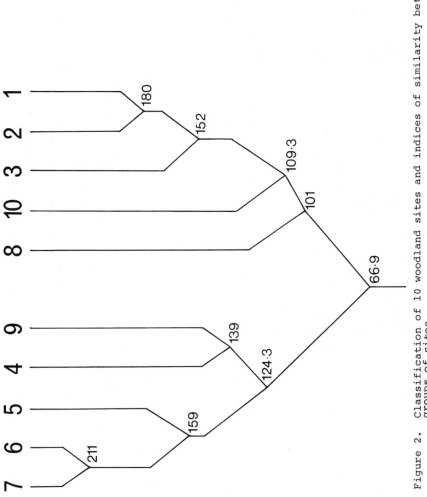

Figure 2.   Classification of 10 woodland sites and indices of similarity between
groups of sites

$$b^2 = 0.5209$$

$$b = 0.72 \quad .$$

The 5% significance level of β for $n=5$ is 0.917:   and hence the two sub-clusters are not significantly distinct.

6. DISCUSSION

The usefulness of the method hinges on the descriptive accuracy of the probability model. In particular the model predicates that the index of similarity, or a transform of the index, is normally distributed. The sampling distributions of most of the various indices are so algebraically complex that they as yet remain underived. Of course the problem of their derivation can be surmounted--perhaps by Monte-Carlo studies--and a transform of the index, meeting the requirements of the probability model, may be available. The method is more readily applicable to indices of similarity constructed according to probability considerations. The index proposed by Goodall [1] has, under certain conditions, a rectangular sampling distribution: appropriately transformed it is immediately available for cluster analysis.

Because of the very large number of ways of dividing a set of units into two groups, the overall test is likely to be extremely conservative. It would be made less so if it were tailored to the method of obtaining the grouping. Because of the complex of conditional probabilities involved, this would entail heavy algebraic manipulation, the results of which would be applicable only to that particular configuration. On the other hand the overall test can be used on all types of classification formed by repeated splitting of groups into two. It is especially suited to the analysis of a dendrogram, to which it can be applied at all levels.

ACKNOWLEDGMENTS

I am grateful to Miss G. McCall for her valuable assistance and to Mr. P. Holgate and Mr. J. G. Skellam for their helpful comments.

REFERENCES

[1]   Goodall, D. W. 1966.   A new similarity index based on
      probability. Biometrics. 22:882-907.

[2]   Gower, J. C. 1966. Some distance properties of latent roots
      and vector methods used in multivariate analysis.
      Biometrika. 53:325-338.

[3]   McGilchrist, C. A. 1965. Analysis of competition experiments.
      Biometrics. 21:975-985.

[4]   Mosimann, J. E. 1962. On the compound multinomial distribution,
      the multivariate B-distribution, and correlations among
      proportions. Biometrika. 49:65-82.

[5]   Mountford, M. D. 1962. An index of similarity and its appli-
      cation to classificatory problems.   In Murphy, P. W. (ed),
      Progress in Soil Zoology: 43-50.

[6]   Quesenberry, C. P. and David, H. A. 1961. Some tests for
      outliers. Biometrika. 48:379-390.

[7]   Rao, C. R. 1965. Linear Statistical Inference and its Appli-
      cations. John Wiley and Sons, Inc., New York.

[8]   Scheffé, H. 1953. A method for judging all contrasts in the
      analysis of variance. Biometrika. 40:381-400.

[9]   Sokal, R. R. and Sneath, P. H. 1963. Principles of Numerical
      Taxonomy. W. H. Freeman, San Francisco and London.

[10] Yates, F. 1947. Analysis of data from all possible reciprocal
      classes between a set of parental lines. Heredity. I:
      287-301.

RECORD OF PREPLANNED AND SPONTANEOUS DISCUSSIONS

R. R. SOKAL (State University of New York, StonyBrook)

I would like to compliment Dr. Mountford on this pioneering development of significance tests for clusters. Until I have studied the
mathematical part of his paper in detail I am not prepared to
comment on it, but I would like to question some of the general
philosophy underlying it.

I wonder what taxonomic or ecological meaning can be ascribed to
differences in mean similarity values between clusters. On the
whole it would seem to me that an overall, external criterion of
goodness of a classification is preferable to significance tests of
subsets. Also when tests of differences between clusters are made,
those should involve not only mean similarities but also measures
of density, shape and possibly other characteristics.

INFORMATION THEORY TECHNIQUES FOR CLASSIFYING PLANT COMMUNITIES

LASZLO ORLOCI
Department of Botany
University of Western Ontario
London, Canada

SUMMARY

This paper is concerned with the classification of stands of
vegetation based on the use of the notions of information theory.
Relating to different definitions of information, two distinct
families of classification techniques are considered. The first
includes a sorting of stands, according to their best fit, between
classes of specified parameters. Fitness in a class is measured
in terms of the minimum discrimination information. The second
family embraces cluster seeking in general in new collections of
stands, with no class parameters specified a priori. The clustering
techniques, that are described in this paper, are agglomerative and
hierarchic, utilizing the information, generated by paired comparisons, as the decision parameter.

1.   INTRODUCTION

The term information is used in this paper in a strictly technical
sense. It is conceived as a physical, quantitatively measurable,
property of events related to probability in accordance with Shannon's [13] definition. It is a direct consequence of this definition
that events of low probability are regarded as those of high information content. Information in this context thus relates more
closely to surprisal than to either knowledge or meaningfulness of
the ordinary speech.

Information techniques are noted for their great flixibility in
being easily applied to ecological problems in the most diverse
situations. This flexibility resides in the fact that the information techniques are not necessarily burdened by restrictive
assumptions about the data, the underlying probability distribution,
or the properties of the sample space, which characterize the
conventional multivariate techniques.

2.   DEFINITIONS

2.1   Collection

The term collection is used in a general sense that may equate to
sample or population depending on the nature of the problem at hand.
The collection is presented in the form of an r x c matrix of values,
designated by the symbol X, whose elements x_{hj}, define the value of
the different species in the individual stands. The r rows of X,
referenced by the first subscript of x_{hj}, represent species
(characters, attributes) and the c columns correspond to stands
(individuals). The term stand implies a plant community within
given boundaries.

2.2   Total Information

Let F_h represent an s_h-valued distribution of frequencies, corresponding to the hth species (row) vector X_h, and let F_j represent
an s_j-valued distribution of frequencies relating to the jth stand
(column) vector X_j of X. The elements in F_h and F_j are such that:

$$\sum_{j=1}^{s_h} f_{hj} = c, \ p_{hj} = f_{hj}/c \text{ and } \sum_{j=1}^{s_h} p_{hj} = 1 ,$$

and

$$\sum_{i=1}^{s_j} f_{ij} = r, \ p_{ij} = f_{ij}/r \ \text{and} \ \sum_{i=1}^{s_j} p_{ij} = 1 \ .$$

The total information in F_h can be conveniently measured by the c multiple of Shannon's [13] entropy function. In this function information is related to probability in the following manner:

$$I(F_h) = -\sum_{j=1}^{c} \ln p(x_{hj}) = -\sum_{j=1}^{s_h} f_{hj} \ln p_{hj}$$

$$= c \ln c - \sum_{j=1}^{s_h} f_{hj} \ln f_{hj} \ . \tag{1}$$

To define the total information in stand F_j, i is substituted for j and r for c in the summation in equation (1). After such substitutions we have:

$$I(F_j) = r \ln r - \sum_{i=1}^{s_j} f_{ij} \ln f_{ij} \ . \tag{2}$$

When the elements of X represent frequencies, or counts within sampling units, the total information in stand X_j can be defined directly in terms of the x_{hj} such that

$$I(X_j) = N_j \ln N_j - \sum_{h=1}^{r} x_{hj} \ln x_{hj} \ . \tag{3}$$

In this expression N_j represents the jth column total. The definition of total information in equation (3) is used by Pielou [11] for the measurement of species diversity in plant communities. Note that equations (2) and (3) are not equivalent. According to Basharin [2] the sampling distribution of Shannon's information is approximately normal. The goodness of the approximation, however, depends on sample size. Pielou [11] has used the approximate normality of the distribution to test specific hypotheses about species diversity.

2.3 Mutual and Discrimination Information

Any n species (or any n distributions in general) and the relation-

ship between them can be represented in an n-dimensional frequency
table.   In such a frequency table the n individual frequency distributions $(F_h; h=1,2,\ldots,n)$ constitute the principal marginal distributions and the body of the table contains the frequencies of the
joint observations between the individual distributions taken 2,3,
$\ldots,n-1$, or n at a time.   The information relating the different frequency distributions is the information which they commonly possess.
In the simplest case of paired comparisons the common or mutual
information between the hth and the ith species is defined by:

$$I(F_h; F_i) = I(F_h) + I(F_i) - I(F_h, F_i) . \tag{4}$$

The individual terms in this expression represent the total information in the two species individually, defined according to equation
(1), and the joint information given by:

$$I(F_h, F_i) = - \sum_{\ell=1}^{s_h} \sum_{m=1}^{s_i} f_{h\ell im} \ln p_{h\ell im} . \tag{5}$$

Similar expressions can be found for any n way comparison.   The
symbol $f_{h\ell im}$ and $p_{h\ell im}$ represent the frequency and relative proportion of the pair in which one member $x_{h\ell}$ is the class value or class
symbol in the ℓth class of F_h and the other member x_{im} is the class
value or class symbol in the mth class of F_i.   The number of
realized combinations is c.   The unrealized combinations, $s_h s_i - c$ in
total, are represented by zeros in the frequency table.

Equations (4) and (5), after altering their subscripts, are
directly applicable to stands.   The appropriate expressions relating
the jth and kth stand are given as follows:

$$I(F_j; F_k) = I(F_j) + I(F_k) - I(F_j, F_k) \tag{6}$$

and

$$I(F_j, F_k) = - \sum_{\ell=1}^{s_j} \sum_{m=1}^{s_k} f_{\ell jmk} \ln p_{\ell jmk} . \tag{7}$$

In equation (7) the symbols $f_{\ell jmk}$ and $p_{\ell jmk}$ represent the frequency
and relative proportion of the pair in which one member $x_{\ell j}$ is the
class value or class symbol in the ℓth class of F_j and the other

member x_{mk} is the class value or class symbol in the mth class
of F_k.

When species are evaluated in terms of counts of individuals
per sampling unit, or as frequencies, the data matrix X is itself
a contingency table. In such a table the observations are sorted
according to two criteria of classification, namely species
affiliation and stand of occurrence. The information posessed in
common by the two criteria can be defined as the amount of information discriminating between the observed values of x_{hj} and the
random expectations $E[x_{hj}]$. The mutual information of the two
criteria then is equivalent to the discrimination information
defined by:

$$I(\text{species; stands}) = I[X; E(X)] = \sum_{h=1}^{r} \sum_{j=1}^{c} x_{hj} \ln[x_{hj}/E(x_{hj})]. \quad (8)$$

The expected values are defined according to $E(x_{hj}) = N_h N_j/N$, where
N_h, N_j and N represent the hth row and jth column total, and the
grand total of X respectively. In terms of the total and joint
information equation (8) can also be written in a computationally
simple form:

$$I(\text{species; stands}) = \sum_{h=1}^{r} \sum_{j=1}^{c} x_{hj} \ln x_{hj} + N \ln N$$

$$- \sum_{h=1}^{r} N_h \ln N_h - \sum_{j=1}^{c} N_j \ln N_j. \quad (9)$$

With a signum transformation of the x_{hj}, equation (8) or (9) is
equivalent to the function given by Macnaughton-Smith [7] in connection with the evaluation of 2 x 2 contingency-tables. The
sampling distribution of the discrimination information in general,
is considered by Kullback [5] and Kullback etc. [6]. These authors
show that twice the discrimination infromation is approximately
distributed as chi-square under the null hypothesis of no departure
from expectation, implying that the observations are proportional
to the marginal totals in the contingency-table. The reference chi-
square is given at the (number of rows -1)(number of columns -1)
degrees of freedom.

Phytosociological problems frequently require comparison between
an observed frequency distribution (F) and a reference distribution

of a given type $E(F)$. In the general case, the information discriminating between the two distributions, each with s elements, is defined by:

$$I[F; E(F)] = \sum_{e=1}^{s} f_e \ln[f_e/E(f_e)] , \qquad (10)$$

such that $\sum f_e = \sum E(f_e)$. Twice the discrimination information of equation (10) is approximately distributed as chi-square at s-1 degrees of freedom under the null hypothesis of no departure of the observations from the specified values.

2.4 Equivocation Information

Abramson [1] gives a diagram relating the total information and joint information to mutual information. In terms of his diagram the equivocation information is defined by the difference:

$$d(h; i) = I(F_h, F_i) - I(F_h; F_i) \qquad (11)$$

for the hth and ith species, or

$$d(j; k) = I(F_j, F_k) - I(F_j; F_k) \qquad (12)$$

for the jth and kth stand.

2.5 Relative Measures of Information

In the expressions so far presented information is weighted by frequencies. This weighting introduces a size effect which may be undesired. There are several alternative measures, free from the effect of size, that may interest phytosociologists:

   (i) Average information. Information when expressed as an average value per observation is called entropy. It is equal to the total information divided by the total number of observations. Since it is not dependent on absolute frequencies it is directly comparably between different collections. An alternative possibility is to attach a probability to information. The probability can then be used as a standard measure free from the effect of size.

   (ii) Rajski's metric and relative coherence. A relative measure may also take the form of a metric or relative coherence [12]. The metric is defined by:

$$d(h;i) = 1 - I(F_h; F_i) / I(F_h, F_i) , \tag{13}$$

which can also be represented by $[I(F_h, F_i) - I(F_h; F_i)] / I(F_h, F_i)$. The value of $d(h;i)$ varies between zero and unity indicating a perfect correlation, or alternatively, the total lack of relatedness between F_h and F_i. The coherence coefficient and Rajski's metric are related in the following manner:

$$R(h;i) = \sqrt{[1 - d^2(h;i)]} . \tag{14}$$

The value of the coherence coefficient also varies from zero to unity. The limits indicate total difference or perfect correlation respectively. When h is replaced by j and i is replaced by k, equations (13) and (14) represent the distance or the relative coherence between the jth and kth stands:

$$d(j;k) = 1 - I(F_j; F_k) / I(F_j, F_k) \tag{15}$$

and

$$R(j;k) = \sqrt{[1 - d^2(j;k)]} . \tag{16}$$

3. SORTING BETWEEN CLASSES OF SPECIFIED PARAMETERS

3.1 'Centroid' Sorting

The problem in 'centroid' sorting amounts to finding the class among a number of classes of given mean distributions to which the stand whose affiliation is in question is most closely related. If a class mean vector is specified by the frequency distribution $E(F)$, and F represents the stand to be assigned, a suitable measure of relatedness is given by the information which discriminates between F and $E(F)$ as defined in equation (10). The stand in question then is assigned to the class from the mean distribution of which it is separated by the least amount of discrimination information.

3.2 Group Sorting

In specific circumstances all members of a given class may be specified by their frequency distributions. When such is the case

it may be desired to base the sorting on the pooled information
collectively discriminating between the members of a given class
E(F) and the individual stand F whose affiliation is in question.
This technique of sorting can also be broadened to include situations where the entity F is itself a class of several stands.

4. CLUSTER SEEKING IN NEW COLLECTIONS

4.1 Clustering by Equivocation Information

The techniques described relate to centroid sorting (Ducker etc. [3])
and average linkage clustering (Sokal and Michener [14]), but in the
present case the average value of the equivocation information, or
alternatively Rajski's metric, is used as the decision parameter.

   At the first step in the analysis a dissimilarity matrix D is
generated with a characteristic element d(j;k) relating the jth and
kth stands in terms of equation (12) or (15). In the first clustering pass through D all local minima are located and the corresponding
pairs of stands are united. An element of D represents a local
minimum if it is the minimum value in both its row and its column.
The continuation of the analysis may take one of two alternative
routes:

   (i) Centroid clustering. A residual D is generated from the
original observations by comparing the mean distributions of the
subsets, some of which may contain only a single stand. The
residual D then is scanned for local minima and the subsets are
revised. The labels and essential parameters of the revised subsets
are stored in a register. The process is repeated until the order
of D is reduced to one. A dendrogram is then plotted from the labels
and parameters stored in the register.

   (ii) Average linkage clustering. The order of D is reduced by
simple or weighted averaging of the corresponding elements between
pairs of columns and the equivalent pairs of rows whose union is
indicated in the previous clustering pass. For example, if the
relation d(B;C) represents a local minimum then the element in the
intersection of the Ath row and BCth column in the reduced matrix
is defined by

$$d(A;BC) = [d(A;B) + d(A;C)] / 2 \quad .$$

The ds may also be defined as weighted averages (e.g., [4]) in which

the size of the subsets is taken into consideration. The residual
D then is scanned and the subsets are revised. The labels and
essential parameters of the revised subsets are stored in a register.
The process is terminated when the order of D is reduced to one.
A dendrogram is then plotted.

4.2 Clustering by Total and Discrimination Information

Let X_A and X_B represent two subsets of stands of X. The heterogeneity information that separates X_A from X_B may in this instance
be defined as the information increase in the system resulting from
the union X_{A+B} of the two subsets. The specific definitions relate
to total information and discrimination information:

(i) Clustering by heterogeneity information relating to total
information. The increase of total information with regard to a
single species is

$$\Delta I(F_h)_{AB} = I(F_h)_{A+B} - I(F_h)_A - I(F_h)_B .$$ (17)

The different values in equation (17) are defined by equation (1).
The increase is zero when the two subsets possess identical values
of x_{hj} in identical proportion. The values of the heterogeneity
information relating to the individual species are pooled to obtain
an expression of total heterogeneity,

$$d(A;B) = \sum_{h=1}^{r} \Delta I(F_h)_{AB} .$$ (18)

Such a total, however, only qualifies as a conservative approximation to the actual value of joint information, unless it is shown
that the individual species do not possess mutual information. In
the clustering process two subsets are united if they are mutually
nearest in terms of the heterogeneity information, implying that
the corresponding d(A;B) represents a local minimum.

A similar definition of heterogeneity relating to total information has been used by Macnaughton-Smith [7] in subdivisive clustering
of a collection of individuals on the basis of presence/absence data.
The phytosociological applications in this connection relate to
agglomerative clustering of stands utilizing species presence
(Williams etc. [15]) and quantitative data (Orloci [9], [10]).

(ii) Clustering by heterogeneity information relating to discrimination information. An alternative definition of heterogeneity

(Orloci [10]) relates to information as defined in equation (8) in connection with contingency tables. The appropriate expression of heterogeneity in this case is given by

$$d(A,B) = I(\text{species; stands})_{A+B} - I(\text{species; stands})_A$$
$$- I(\text{species; stands})_B . \tag{19}$$

Equation (19) measures the amount of increase of the relatedness of the two criteria of sorting of the observations in the contingency-table X_{A+B}, arising from the union of the two subtables X_A and X_B in accordance with equation (8). The two subtables are united if the corresponding $d(A;B)$ is a local minimum. The properties of the related sampling distribution are considered by Kullback [5] and Kullback etc. [6]. According to these authors twice the hetero-geneity information is approximately distributed as chi-square under the null hypothesis that the two subtables are from the same popula-tion. The reference chi-square distribution is given at the number of species - 1 degrees of freedom. Only species occurring in the two subtables are considered.

Regarding the technique of agglomeration, the two alternatives given in section 4.1 can also be used, after minor alterations, with the definitions of d given in section 4.2.

5. DISCUSSION

No attempt is made to present in this paper an enumeration of all the information based classification techniques that may conceivably interest phytosociologists. The phytosociological applications of information analysis are still in an initial state. In the initial applications, naturally, the basic problem is concerned with the choice of technique. Regarding priorities, cluster analysis usually takes precedent. But once the new collection is clustered into a set of disjoint classes the problem of assigning external individuals to the established classes often arises.

Phytosociological collections usually are unique in as much as they have not previously been encountered. The initial problem, therefore, frequently amounts to cluster analysis. The clustering models described in this paper are based on the use of different measures of heterogeneity. Of these, equivocation information or

Rajski's metric may be compared to absolute distance or relative distance. Relative distance, by removing all proportional differences between stands from the analysis (see [8]), seems to amplify the appearance of a clustered pattern of the collection. This may also apply to Rajski's metric.

The heterogeneity functions may greatly differ in their effect on the appearance of dissimilarity. The equivocation information, and also Rajski's metric, has a zero value if two stands have identical information content, and the distribution of their joint frequencies follows the same pattern as the individual frequency distributions. A zero value of heterogeneity in terms of the total information, on the other hand, indicates that the two subsets within species have identical values occurring with identical relative frequencies. In the most extreme case, such as a subset consisting of two stands, the heterogeneity information is zero if in the two the corresponding values are identical. And in the case of the discrimination information, a zero heterogeneity indicates that in the two subtables of stands the relationship between the observed values and their random expectations is proportional.

A distinction is made between two cases of assignments. In the first case only the class mean distributions are specified. The decision parameter then is the information which discriminates between the frequency distribution, whose assignment is considered, and the mean distributions of the reference classes. Often, however, it is desired to base the assignments not on the mean distributions but on the total data available for the definition of the properties of the reference classes.

Finally, it may be noted that evaluation of the different clustering techniques must take into account that each may relate to a different information function, and thus, to a different aspect of the structure in the collection at hand. Each technique, therefore, potentially leads to different final results.

ACKNOWLEDGMENT

The studies leading to the present paper were supported by a National Research Council of Canada grant.

REFERENCES

[1]   Abramson, N. 1963. Information theory and coding. New York,
      London, McGraw-Hill.

[2]   Basharin, G. P. 1959. On a statistical estimate for the
      entropy of a sequence of independent random variables.
      Theory Probab. Applic. 4:303-336.

[3]   Ducker, S. C., Williams, W. T. and Lance, G. N. 1965.
      Numerical classification of the pacific forms of Chlorodes-
      mis (Chlorophyta). Aust. J. Bot. 13:489-499.

[4]   Gower, J. C. 1967. A comparison of some methods of cluster
      analysis. Biometrics. 23:623-637.

[5]   Kullback, S. 1959. Information theory and statistics.
      New York, Wiley; London, Chapman and Hall.

[6]   ——————, Kupperman, M. and Ku, H. H. 1962. Test for
      contingency-tables and Markov chains. Technometrics.
      4:573-608.

[7]   Macnaughton-Smith, P. 1965. Some statistical and other
      numerical techniques for classifying individuals. London,
      H.M.S.O.

[8]   Orloci, L. 1967. An agglomerative method for classification
      of plant communities. J. Ecol. 55:193-205.

[9]   ——————. 1968. Definitions of structure in multivariate
      phytosociological samples. Vegetatio. 15:281-291.

[10] ——————. 1968. Information analysis in phytosociology:
      partition, classification and prediction. J. Theoret.
      Biol. 20:271-284.

[11] Pielou, E. C. 1966. The measurement of diversity in different
      biological collections. J. Theoret. Biol. 13:131-144.

[12] Rajski, C. 1961. Entropy and metric spaces. pp. 41-45 in
      "Information theory" edited by C. Cherry.

[13] Shannon, C. E. 1948. A mathematical theory of communication.
      Bell System Tech. J. 27:379-423; 623-656.

[]4] Sokal, R. R. and Michener, C. D. 1958. A statistical method
      for evaluating systematic relationships. University of
      Kansas Sci. Bull. 38:1409-1438.

[15] Williams, W. T., Lambert, J. M. and Lance, G. N. 1966.
      Multivariate methods in plant ecology. V. Similarity
      analysis and information analysis. J. Ecology. 54:427-445.

CONTINGENCY-TABLE ANALYSIS OF RAIN FOREST VEGETATION

W. H. HATHEWAY*
Department of Botany
North Carolina State University
Raleigh, North Carolina

Smithsonian Institution
Washington, D. C.

SUMMARY

Williams [11] gave a method for the use of scores in the analysis of association in contingency-tables. Inasmuch as these methods constitute a theoretical basis for and an extension of techniques used by Whittaker [10] in the analysis of vegetational gradients, they deserve wider attention from plant ecologists than they have received. Williams' method, applied to complex rain forest data, detects variations in the vegetation that apparently correspond to a moisture or drainage gradient. Scores along this and other gradients are provided for species and plots which can be utilized in the preparation of vegetation maps.

*Present address: Center for Quantitative Science
and
College of Forest Resources
University of Washington
Seattle, Washington

1.   INTRODUCTION AND PROCEDURES

In the early stages of analysis of complex ecosystems a working classification of the vegetation is useful. Its separation into conceptually distinct entities allows the work to be subdivided logically and prevents throwing together into the same study unit entities which are qualitatively different, such as bogs and upland vegetation. Once such a broad classification has been made, it is often of interest to determine whether the vegetational types thus recognized are homogeneous or whether regular patterns of variation exist within them. A common pattern within a relatively homogeneous vegetational type such as a slope forest is a regular but perhaps scarcely perceptible change in species composition along an environmental gradient within the type, such as a drainage catena. In the present paper we shall be concerned with the application of two methods for detecting continuous variation in a complex equatorial rain forest.

From the point of view of the forest botanist ordered tabulations of stands and species along a gradient, such as the ones suggested by Curtis and McIntosh [2] or Whittaker [10] are informative because they place close to one another those species which grow together as well as the stands in which they occur. If the number of species or the number of stands is large, construction of an ordered species-stand table is time consuming, and errors are hard to avoid. For these reasons it is important to search for consistent rules for ordering species and stands that are readily translated into mathematics and hence into computer routines. The computer then produces a repeatable, objective ordered tabulation of species and stands.

Accordingly, we assume that it is possible to arrange stands in correct order along an environmental gradient and look for methods for ordering the species along the same gradient. If the species and stands are ordered correctly, then the resulting tabulation will exhibit correlation between the species and the stands. For example, Curtis and McIntosh found continuous variation in the vegetation of southern Wisconsin from forests dominated by Black Oak (Quercus velutina) to others dominated by Sugar Maple (Acer saccharum). Black Oak was uncommon in the Sugar Maple forest type and Sugar Maple rare in the Black Oak type. These relations are exhibited in Table 1. Of course, ecologists usually deal with much larger tables that contain quantitative observations on several species in many stands. In fact, in Wisconsin Curtis and McIntosh found that White Oak

(Q. alba) and Red Oak (Q. rubra) occupied positions intermediate
between those of Black Oak and Sugar Maple and that the entire
series represented essentially continuous variation along a moisture gradient.

Table 1. Commonness and rarity of two species in two vegetational
types

Species	Stands	
	Black Oak Type	Sugar Maple Type
Black Oak	Common	Rare
Sugar Maple	Rare	Common

A slightly different approach was reported by Whittaker [10],
termed by him a composite weighted average technique. He assigned
scores to species according to the location of their maximum population levels along a moisture gradient--e.g., mesic: 1; submesic: 2;
subxeric: 3; xeric: 4. To obtain corresponding scores for the
stands, the frequency counts of the species present in any stand
were multiplied by their moisture gradient scores and averaged.
This method, discovered independently by Whittaker and other authors
listed by him, had earlier been described more fully by Fisher [3].

Fisher showed further that given the set of scores calculated
for the stands it is possible to obtain new scores for the species
in exactly the same way. After a few such iterations the relative
scores for both species and stands become stable if correlations
are strong between species and stands along the gradient.

The method was generalized by Yates [12] and Williams [11], in
papers discussing the use of scores in the analysis of association
in contingency tables. Williams showed how optimum scores for the
species and stands may be calculated directly from a frequency table
of stands and species. He assumes that the best arrangement of
species and stands is the one which exhibits the strongest correlation between scores for species and stands.

Because Williams' paper is not familiar to many ecologists, his
main arguments are repeated here. We suppose that we are given
frequency counts of tree stems in plots. If

n_{ij} is the number of trees of species i in plot j, $(i=1,2,\ldots,p)$, $(j=1,2,\ldots,q)$,

$n_{i.}$ is the total frequency in the study area of species i,

$n_{.j}$ is the total frequency in plot j,

$n_{..}$ is the total frequency in the table,

x_i is the score of the ith species, and

y_j is the score of the jth plot,

then the following restrictions are placed on the scales used for the scores:

$$\sum_i n_{i.}x_i = 0 \qquad\qquad \sum_j n_{.j}y_j = 0 \qquad\qquad (1)$$

$$\sum_i n_{i.}x_i^2 = n_{..} \qquad\qquad \sum_j n_{.j}y_j^2 = n_{..} \quad . \qquad\qquad (2)$$

The correlation coefficient between the x_i's and the y_j's is then

$$R = \sum_i \sum_j n_{ij}x_iy_j \ / \ n_{..} \quad .$$

In Whittaker's case scores for the species are given along a moisture gradient and we are required to obtain corresponding scores for the plots. Williams maximizes $n_{..}^2 R^2$ for variations in the y_j subject to the restrictions (1) and (2), obtains

$$y_j = \sum_i n_{ij}x_i \ / \ Rn_{.j}$$

and shows that the constant factor R may be obtained from the relation

$$n_{..}R^2 = \sum_j (\sum_i n_{ij}x_i)^2 \ / \ n_{.j} \quad .$$

He suggests several alternative tests of significance for R^2. It will be observed that apart from the constant factor R the plot scores y_j are calculated essentially as suggested by Whittaker.

In case scores for the species are not available, then scores for both species and plots may be estimated from the data of the species-plot frequency table. Williams again maximizes

$$n_{..}^2 R^2 = (\sum_i \sum_j n_{ij} x_i y_j)^2$$

for variations in both the x_i and y_j, subject to the scaling restrictions (1) and (2) and obtains equations

$$\sum_k t_{kj} (m_k y_k) - R^2 (m_j y_j) = 0 \qquad\qquad (3)$$

where

$$m_j = \sqrt{(n_{.j}/n_{..})}$$

and

$$t_{kj} = \sum_i \frac{n_{ik} n_{ij}}{n_{i.} \sqrt{(n_{.k} n_{.j})}} .$$

Therefore R^2 is a latent root of the matrix $T = (t_{kj})$. The corresponding latent vector is $(m_j y_j)$.

Williams observes that $T = GG'$, where G is a matrix the elements of which are

$$g_{ij} = n_{ij} / \sqrt{(n_{i.} n_{.j})} .$$

It can be shown that the latent vectors $(m_j y_j)$ of T are uncorrelated and of unit length. If $y_1 = (1,1,\ldots,1)$ is substituted in equation (3), it will be found that m_1 is a latent vector corresponding to the latent root $R^2 = 1.00$. This root and its vector correspond to the expected values of the frequencies in the original contingency table and could have been eliminated from the analysis by starting with the matrix

$$G_d = G - \bar{G}, \text{ where the elements of } \bar{G} \text{ are}$$

$$\bar{g}_{ij} = \sqrt{(n_{i.} n_{.j})} / n_{..} .$$

The matrix T is an association matrix in the sense of Gower [4], because the element t_{kj} is a coefficient of association between plots k and j. Moreover, this coefficient is obtained by the so-called Q-technique, discussed by Gower and by Orloci ([6], [7]).

Data centering in the restricted sense of Orloci is not appropriate
in analysis of contingency-tables. Orloci and Gower both discuss
the duality of Q and R methods in principal components analysis.
Gower arranges the latent roots and vectors of a Q matrix in a table
(Table 2) where the latent roots λ_j are arranged in order of de-
scending size from left to right. In the table $(c_{1j}, c_{2j}, c_{3j}, \ldots, c_{nj})$
is the latent vector corresponding to the root λ_j. If the latent
vectors are normalized, so that the sums of squares of their elements are equal to the corresponding latent roots, then the points
$Q_i = (c_{i1}, c_{i2}, c_{i3}, \ldots, c_{in})$ are representations of the original
individuals in the new coordinate system. In fact, when the first
two vectors are used as coordinates for a two-dimensional scatter
diagram, Q_i is assigned the coordinates (c_{i1}, c_{i2}), and the number
of dimensions is somewhat arbitrarily reduced. In common practice
the coordinates of the vectors corresponding to latent roots the
values of which are nearly zero are ignored in the analysis and in
fact are unstable from one sample to another.

Table 2. Latent roots and vectors of an association matrix

	λ_1	λ_2	λ_3		λ_n
Q_1	c_{11}	c_{12}	c_{13}	$\cdots\cdots\cdots$	c_{1n}
Q_2	c_{21}	c_{22}	c_{23}	$\cdots\cdots\cdots$	c_{2n}
$\cdot$	$\cdot$	$\cdot$	$\cdots\cdots\cdots$		$\cdot$
Q_n	c_{n1}	c_{n2}	c_{n3}	$\cdots\cdots\cdots$	c_{nn}
$\bar{Q}$	$\bar{c}_{\cdot 1}$	$\bar{c}_{\cdot 2}$	$\bar{c}_{\cdot 3}$	$\cdots\cdots\cdots$	$\bar{c}_{\cdot n}$

Gower observed that in Q analyses of association matrices the
vector corresponding to the largest latent root often has more or
less constant elements and can be regarded as corresponding to a
mean value of the elements of the association matrix. Since this
vector is essentially irrelevant to the interpretation of the results
of a principal components analysis, he suggests that it be eliminated by means of a change of origin and possibly an orthogonal
transformation. The results of this transformation lead to an

unexpected correspondence between the two-stage principal components analysis proposed by Gower and the method of canonical variates for the analysis of contigency-table data suggested by Williams.

A transformation of an association matrix A with elements a_{ij} which preserves distances between the original observed points has elements

$$\alpha_{ij} = a_{ij} - \bar{a}_{i.} - \bar{a}_{.j} + \bar{a}_{..} \quad .$$

In the Williams contingency-table analysis this transformation has its counterpart in the matrix

$$T_d = G_d G_d'$$

$$= (G-\bar{G})(G-\bar{G})'$$

$$= T - \bar{G}\bar{G}'$$

with elements

$$t_{kj} - \frac{\sqrt{(n_{.k} n_{.j})}}{n_{..}} \quad .$$

Since the rank of the matrix α is one less than that of the association matrix A, it has a zero root. The corresponding vector has components $(1,1,\ldots,1)$. Gower notes that the latent vectors of α are orthogonal. If $\underline{v}$ is any latent vector of α, then its inner product with the vector $(1,1,\ldots,1)$ is $\underline{1}'\underline{v}=0$, so that the elements of $\underline{v}$ sum to zero. This result has a counterpart in the contingency-table analysis, for the latent vectors of the matrix T are $(m_j y_j)$. On division by m_j these give the required plot scores y_j. But as has already been noted, one of the latent vectors of the matrix T has elements (m_j), which on division by m_j yields the vector $(1,1,\ldots,1)$. The associated latent root unity, which corresponds to the expectations (mean values) of the matrix G, is eliminated by beginning the analysis with the matrix $G_d=G-\bar{G}$.

This procedure also leads directly to the scores for the species and plots, since the elements of the latent vectors of T_d are (y_j), not $(m_j y_j)$. Because of the restrictions placed on the scales of the scores

$$\sum_j n_j y_j = 0 \ ,$$

the weighted sum of the scores is zero.

Consequently contingency-table analysis leads to a matrix T (or T_d) which is an analogue for enumeration data of Gower's matrix α, and can therefore be regarded formally as an optimum principal components representation of the data as well as one which, because it maximizes the correlations between species and plots, leads to scores which represent the relationship between the latter in its simplest form. Because enumeration data are assumed, the Williams technique is especially appropriate for vegetational frequency counts.

Williams developed a test of statistical significance for only the largest of the correlations between sets of variates and suggests that to account for the association in a contingency-table in terms of more than one pair of variates is a somewhat artificial concept. But vegetation, the product of the reaction of a flora over long periods of time to a complex external environment, ordinarily reflects simultaneously the influence of several environmental and historical factors. Thus a contingency-table of species and plots may be expected to show meaningful association between several sets of scores.

The degree of this association is measured in the contingency-table analysis by the correlation coefficient between the two sets of scores. The total association in the table, on the other hand, is measured by the usual chi-square statistic. Williams shows that the total chi-square in the table may be partitioned into components that correspond to the squares of the several correlation coefficients. Thus for practical purposes it is possible to estimate the proportion of the total association in the table that may be attributed to each of the successive pairs of calculated canonical variates. In addition, there exist methods in the multivariate analysis of normally distributed variables, discussed by Bartlett [1], Kendall [5], and Seal [9] for determining how many of the roots can be assigned zero values and accordingly ignored. For the purposes of this paper, however, we prefer to regard canonical analysis as simply a mathematical technique for summarizing data describing the distribution of trees in one forest. Nevertheless, it is important to be able to make some judgment of the usefulness of these results.

In practice two situations commonly occur. In the first, only one of the calculated latent roots is large relative to the remaining ones. If a test of significance is required, the method suggested by Williams may be applied. If the results are not significant, the degree of association in the table may have been small or perhaps not readily represented on a linear scale.

If often happens in applications in vegetational work, however, that several roots are large, so that none of them accounts for a considerable proportion of the total association in the table. In such cases it is useful to study separately the scores corresponding to each root. It is possible, for example, to reconstruct the species-plot contingency-table in the order indicated by the scores. All species and plots with large positive scores are thrown together at one corner of the table, those with large negative scores at the other; species and plots with intermediate scores are represented in the cells of the table along the main diagonal connecting opposite corners. Comparison of plots at the ends of this scale or some knowledge of the requirements of the species may suggest an environmental gradient which it represents. If environmental data are available, correlations of these with plot scores are often suggestive. For comparisons in the field, maps of the scores such as Figure 2 are valuable.

If unrelated types of vegetation are included in the contingency-table, so that there is little overlap in species composition from one plot to another, much of the measured association in the table may reflect such contrasts and appear in the analysis as a large correlation coefficient. These puzzling situations are often clarified by scatter diagrams of the scores of stands corresponding to two canonical variates. Stands (or groups of stands) which fall far from the main directions of variation obviously contribute little to an understanding of the environmental gradeints affecting most of the vegetation under study and might well be eliminated from the analysis.

Because the theory of significance tests of latent roots and vectors calculated from observational data is difficult, empirical observations on the stability of these results are useful. In these studies the original data from which the scores were calculated are modified, for example by the addition or deletion of stands. Latent roots are relatively little affected by small modifications in the input data. The components of latent vectors--and thus in the present case the scores of the plots and species--are relatively

stable if their latent roots are large but very unstable if the
roots are small. Accordingly, no reliance should be placed on the
scores corresponding to roots whose values are zero or nearly so.

In summary, statistical significance is expected if considerable
association exists in the contingency-table and if it can be accounted for satisfactorily in terms of variation along a single
environmental gradient. In many practical cases, several environmental gradients and their interactions may be involved, or the
stands may be heterogenous. Then several correlation coefficients
are large, and statistical tests are less straightforward. Graphical analysis, mapping, and rearrangement of species and plots in
the contingency-table in the order of the scores are suggested.

In the case described below, significance tests are not meaningful, because the area studied is not regarded as a sample drawn from
a larger universe. Since the scores merely describe the existing
situation in one forest, their greatest value is in suggesting
relationships which may deserve further careful study. It will be
shown that soil moisture and drainage affect the distributions of
tree species on the study area importantly. Average positions of
all tree species are calculated along this catena. These scores
can be applied in a provisional way outside the study area because
what has been learned about the behavior of these trees in a single
location is a contribution to knowledge of their autecology over a
broader range.

2. THE STUDY AREA

The data for the present study are derived from an inventory of a
5.5 hectare tract of upland Amazonian rain forest, called the
Mocambo Reserve, near Belem, Brazil. This site has been under continuous study by Dr. J. M. Pires and his colleagues for more than 15
years. In 1956 Pires and Nilo T. da Silva collaborated with S. A.
Cain and G. M. de Oliveira Castro in a phytosociological study of
the Mocambo forest. A few years earlier Th. Dobzhansky, in papers
written with Pires and other Brazilian botanists, had discussed the
great diversity of tree species in this reserve. The Mocambo forest
is presently an important part of the Guama Ecological Research Area
(APEG), at which studies sponsored jointly by the Brazilian Government, the Smithsonian Institution, and the Rockefeller Foundation
are carried out. The Mocambo Reserve is situated about two

kilometers north of the Guama River, on land belonging to the
Instituto de Pesquisas e Investigacao Agropecuarias do Norte (IPEAN),
an agricultural experiment station. Mean annual temperature at
station headquarters is 26°C, and mean annual precipitation is about
2500 mm, rather evenly distributed throughout the year. In terms of
the Holdridge life zone system, the Mocambo Reserve is classified
as Tropical Moist Forest (climatic association); according to
Professor Paul W. Richards, who visited the study area in October
and November 1967, Tropical Rain Forest. The forest is of the tall,
multistratal evergreen type. Mean canopy height is about 35 meters,
and the upper canopy is about 80 percent closed, except in patches
where blowdown has occurred. The canopy trees--species of Goupia,
Vochysia, Eschweilera, Qualea, Parkia, Piptadenia, and Anacardium
are among the most important--have trunks 50 to 120 cm in diameter
and about 20 meters long to the first branch. The leafy parts of
the crowns are 10 to 15 meters above the points at which the main
limbs branch off the trunks. These crowns are mostly 15 to 20 m.
in diameter. Degree of buttressing is variable. The large Qualeas,
Vochysias, Anacardiums, and Hymenaeas are essentially unbuttressed,
but high buttresses do occur in Eschweilera odora, the most common
tree in the forest, and in species of Piptadenia and Sloanea. On
the other hand, Eschweilera corrugata, almost as common as E. odora
and about as large, lacks buttresses. Bark characters are correspondingly variable. Qualea and Hymenaea have smooth, thin bark,
but that of Vochysia flakes off in papery scales. There is
considerable variation in leaf types and sizes. Parkia, Enterolo-
bium, Piptadenia, and Pentaclethra are bipinnate legumes. The
leaves of Vochysia, Qualea, and most other canopy species are simple
and elliptic, with entire margins, somewhat coriaceous, about 5 cm
long and 2 cm wide.

The shrub layer in the forest is well marked and consists chiefly
of miniature single-stemmed dicot trees or young trees of canopy
species 1 to 1.5 m. tall. Palms of the genus Geonoma are common but
seldom dominant, as they usually are in wetter lowland forests in
the American tropics. Broadleafed monocot herbs, such as species of
Calathea and Ischnosiphon, are on the whole more common in the shrub
layer than dwarf palms. This shrub layer is dense. Kneeling, one
can see the trunks of canopy trees no further than 20 m. away. At
eye level, visibility increases to about 50 m.

Between the shrub stratum and the forest canopy 30 meters higher
is a layer of scattered smaller trees 10 to 20 m. tall, with slender

trunks and often somewhat oblong crowns. Among the most common
trees of the forest are several species of Protium, on slightly
buttressed stilt roots, Tetragastris pilosa, and Theobroma subin-
canum, all with slender, horizontal branches and drooping,
elliptical, entire-margined leaves, some up to 25 cm long, with
acuminate tips.

The ground layer is relatively open, as in most tropical moist
to wet forests, with the usual scattered groups of tree seedlings
and beds of Selaginella, the latter especially along trails. Semi-
epiphytic herbaceous vines (chiefly Araceae and Cyclanthaceae) are
common but not very abundant, mostly climbing the smaller trees to
heights of about 10 m. The number of these vines in the crowns of
the trees as well as bromeliads, orchids, and other similar compact
epiphytes is modest. The forest has a conspecuous complement of
giant bush ropes, some as thick as a man's thigh, with parts lying
in loose coils on the ground, the rest ascending to the roof of the
forest. A thin layer of mosses, leafy liverworts, and algae covers
much of the lower parts of the trunks of the trees, partially ob-
scuring the bark, but becoming only a millimeter or two thick.
Higher than 10 m. the moss layer is spotty. Epiphylls cover the
older leaves of some Calatheas and palms in the shrub layer.

Beside each tree 10 cm or over in diameter on the Mocambo Reserve,
Dr. Pires has placed a numbered stake. From this reference number
we can look up its geographical coordinate, its scientific name, its
stem diameter, total height to the first branch, crown diameter,
buttress height and width, and number of buttresses. Dr. Pires
arranged his work in 10 x 10 m. plots. Since the number of woody
species with which we are dealing is very large and since the
starting point for all our calculations is a two-way tabular array
of species and plots, I prepared a computer program which finds
among Dr. Pires' 555, 10 x 10 plots the maximum possible number of
30 x 30 m. plots and identifies their coordinates. I wish to
acknowledge the advice of Dr. John Graham in developing this key
program.

The first map (Fig. 1) shows the spatial arrangement of Dr. Pires'
original 10 x 10 m. plots, each represented by a dot, and the 51,
30 x 30 m. plots constructed within this arrangement by the computer
program. In my large plot No. 1, the 9 original 10 x 10 m. subplots
are illustrated. The unusual shape of the study area is related to
its topography and drainage. The region outside the plot is swampy.

Much of it is, in fact "Igapo": black-water swamp forest with
standing water up to at least 20 cm deep in some places. The
Mocambo Reserve is thus a low island in the middle of this swamp,
and it is largely because of its former inaccessibility that it is
still preserved today.

3. RESULTS

I carried out the analysis of Dr. Pires' data on an IBM 360-75
computer, located at the Research Triangle Institute, North Carolina,
and connected to the North Carolina State University Computing
Center by teleprocessing devices. The computer program forms a
two-way tabulation of frequency counts of 295 woody species in the
51 plots, a total of 2644 trees. It should be noted that some of
these trees were blown down in windstorms some time after Dr. Pires
had completed his inventory. Two association matrices are formed
from the species-plot frequency table by Q techniques: a standard
principal components matrix (Orloci, [7]) and the Williams T
matrix. Latent roots and vectors of both 51 x 51 matrices are
extracted by the QR method. Scores y_j for the plots in the contingency-table analysis are obtained by dividing the components of the
vectors by $(m_j) = \sqrt{(n_{.j}/n_{..})}$. Scores for the species are obtained
from the formula

$$x_i = \sum_j n_{ij} y_j / n_i . R \quad .$$

Scores for the plots are shown on the base map of the Mocambo
Reserve in Figures 2 and 3. It is perhaps not surprising that the
first canonical variate in both contingency-table and principal
components analyses should present similar patterns on the map.
Plot scores decrease from a maximum value in the center of the maps
to lower values on the periphery, somewhat more smoothly in the case
of the contingency-table technique. There is, of course, no assurance that these scores correspond directly to observable gradients
in the environment, since scores for both species and plots were
calculated from the frequency table data. It was possible to
interpret some of these sets of scores in a practical way, however,
because it was possible to compare the map produced by the Williams
technique with the vegetation on the ground in November 1968, and
certain correlations were immediately obvious.

The center of plot 11 is roughly the highest point in the forest,
about 5 meters above the small stream shown in Figure 4. The land
slopes off gradually in all directions from this point, and the
areas in the upper panhandle and right sides of the map--that is,
the eastern part of the reserve--are relatively low and flat. The
crosses on the map show the locations of eight soil pits, in four
of which there was standing water at the time of my visit. The
numbers indicate the depth of this standing water, measured from
ground level, on November 14. There was no water in the pits
labeled with the symbol for infinity. Depths to apparent ground
water levels of 25, 50, and 100 cm have been connected by dashed
lines. This chart is therefore a crude first approximation to a
contour map of the Mocambo forest. Obviously a more accurate map
is needed, and the necessary survey is now under way. I would
like to emphasize that before the computer studies were carried
out and the soils pits dug, it was not thought that the small
variation in topography warranted the preparation of a very accurate
topographic map.

Independently of the vegetational studies, Ing. Italo Falezi, of
the IPEAN station, initiated a detailed soil survey of the APEG
reserve. A few weeks before our visit he had completed the Mocambo
map, which he very kindly permitted me to copy, and which is reproduced in Figure 5. Ing. Falezi informed me that soil Type 3, a
yellow latosol, and Type 4 were the best drained soils in the
Mocambo Reserve, followed in order by Types 1, 2, 5, and 6. Types
2 and 5 are ground-water laterities. The heavy clays of these soils
are strongly mottled in the region of the fluctuating ground-water
table, and I observed that much of the spoil in the soil pits,
having been dried out by exposure to the air, had become hard and
even brick-like. Ing. Falezi informed me that Type 5, in the region
of the upper panhandle, is complex. Apparently there has been an
accumulation of silts on the surface in some places, so that here
this soil has some of the characteristics of alluvial floodplain
("varzea") types.

Table 3 shows the index scores on this first scale for species
with total densities of 10 or more occurrences on the Mocambo tract.
According to this table, Pentaclethra macroloba, Licania spp.,
Carapa guianensis, Sagotia racemosa, and so forth are species of
relatively low, wet situations; Mouriri sagotiana, Tetragastris
pilosa, and Tovomita stigmatosa and others are trees of high, well

Table 3.   Scores of species with densities of 10 or more.   First
canonical variate, E. J. Williams procedure

Density	Score	Species
12	16.18	Mouriri sagotiana
12	13.02	Tetragastris pilosa
42	12.14	Tovomita stigmatosa
14	11.69	Osteophloeum platyspermum
12	11.47	Ptychopetalum olacoides
53	11.32	Protium paraense
22	10.85	Pouteria laurifolia
13	10.40	Thyrsodium paraensis
49	9.96	Protium guayacanum
21	9.79	P. aracouchini
10	8.95	Piptadenia suaveolens
27	7.93	Protium polybotrium
11	7.85	Couepia leptostachya
11	7.18	Pourouma melinonii
22	6.89	Protium puncticulatum
45	6.74	Iryanthera paraensis
32	6.69	Vantanea parviflora
42	6.62	Micropholis acutangula
16	6.41	Pouteria virescens
25	6.03	Dendrobangia boliviana
88	5.92	Vochysia guianensis
179	5.92	Tetragastris trifoliolata
17	5.84	Qualea albiflora
19	5.52	Eschweilera sp.
41	4.86	Theobroma subincanum
12	4.20	Helicostylis pedunculata
19	3.77	Piptadenia psilostachya
28	3.50	Symphonia globulifera
21	3.44	Poraqueiba guianensis

Table 3 (continued).

13	3.22	Caryocar glabrum
29	2.91	Tapura singularis
11	2.83	Parahancornia amara
19	2.54	Sterculia pruriens
11	1.23	Inga myriantha
261	0.01	Eschweilera odora
18	-0.16	Pithecellobium jupunba
12	-0.57	Protium pernervatum
34	-0.69	Iryanthera paraensis
22	-1.38	Chrysophyllum prieuri
20	-1.49	Anacardium giganteum
50	-1.60	Protium cuneatum
20	-2.23	Tovomita schomburgkii
12	-3.77	T. choisyana
56	-5.41	Protium decandrum
59	-5.62	Vouacapoua americana
70	-6.17	Goupia glabra
16	-6.66	Licania heteromorpha
15	-7.35	Swartzia racemosa
14	-7.37	Licania micrantha
10	-7.42	Eugenia patrisii
11	-7.44	Tachigalia myrmecophila
243	-7.86	Eschweilera corrugata
23	-8.32	Rinorea flavescens
13	-10.92	Elvasia elvasioides
10	-15.99	Crudia oblonga
21	-18.01	Licania macrophylla
14	-18.20	Sagotia racemosa
10	-19.19	Carapa guianensis
13	-26.34	Licania longistyla
20	-32.51	Pentaclethra macroloba

drained ground. An important advantage of the Williams technique
is that numerical socres are assigned to the species along each
gradient, so that the results are more informative than a simple
ranking.

Figures 6, 7, 8, 9, and 10 illustrate the distributions of some
of these species on the Mocambo Reserve. The correspondence between
the distribution of Licania longistyla and that of Falezi's soil
Type 6 (a somewhat humid gley) is especially striking. Sagotia
racemosa, which is abundant on alluvial floodplains in the APEG
reserve, is restricted to the upper panhandle, in the complex
portion of Falezi's soil Type 5. This small tree may be an important indicator of relatively rich and potentially arable land. The
very important large tree Goupia glabra has a ring-like distribution
which seems clearly related to some aspect of soil moisture or
drainage, and the understory trees Tovomita stigmatosa and Mouriri
sagotiana are certainly species of upland well drained situations,
as suggested by their scores. These partial results do not imply
that all species on the Mocambo are highly specialized with respect
to drainage requirements. The very abundant Eschweilera odora is
almost universally and evenly distributed over the entire Mocambo
Reserve, and many other species exhibit what appear to be essentially random distributions.

On occasions a strong contrast is exhibited by the scores in two
or three plots, which may be very large, and those that remain, with
values clustered near zero. In contingency-table analysis this
pattern of scores is usually associated with discontinuous variation
in some aspect of the vegetation. Figure 11 provides a convenient
example for discussion. Plots 51, 41, and 15, with spotty distributions on the low-lying periphery of the study area, had scores
which contrasted strongly with the nearly neutral values of the
other plots. What appears to be the correct interpretation of this
situation was provided by Dr. Philip S. Humphrey of the Smithsonian
Institution, who pointed out to me that plot 51, with a highly
aberrant score, was evidently recovering from severe wind damage
that had occurred several years earlier. Pentaclethra macroloba,
a species of floodplains and other wet areas, was especially abundant in this plot. Since Pentaclethra reproduces vegetatively from
fallen logs, often striking root in places with ample moisture, it
is able rapidly to increase its numbers after destructive storms.
There was similar evidence of local blowdown near plots 15 and 41.

It is my impression that the third set of scores obtained by the contingency-table technique, which also exhibits striking variations in the low-lying areas of the forest (Fig. 12), is related to soil fertility. The contrast between the good growth of trees in the region of plots 40 and 41 on the one hand and the depauperate aspect of plot 15 was especially striking. Another useful contrast occurred in the best drained upland area. Plots with large negative scores on this index had relatively many dwarf palms and large monocot herbs (Calathea, Ischnosiphon) in the shrub layer.

Analysis by principal components led to similar maps. The set of scores corresponding to the third largest latent root, shown in Figure 13, is similar to the set corresponding to the second root in the contingency-table analysis. Plots 51 and 41 receive large positive scores, and plot 15 is of the same sign. Plots 11 and 12 also receive large positive scores by the principal components technique and somewhat smaller positive scores in the contingency-table analysis.

The scores corresponding to the second largest root in the principal components analysis exhibit a strong contrast between plots 11 and 12 (Fig. 14). This contrast also shows up in the third contingency-table analysis. In both maps plot 41 has a large score in the same direction as plot 12.

4. DISCUSSION

Although both techniques evidently identify the same environmental gradients and produce similar vegetation maps, certain differences may be noted. Where variation seems essentially continuous, as along a moisture gradient, the contingency-table maps are somewhat smoother than their principal components counterparts. Unfortunately, we lack environmental data which would tell us which of the two methods leads us closer to the true state of affairs. On walking through the forest, however, the impression is generally one of smooth transition, not abrupt change.

On the other hand, in cases of discontinuous variation, such as windthrow, the contingency-table analysis tends to exaggerate differences, most of the variation in a variate being concentrated in the contrast between two or three similar, aberrant plots and the remaining ones. Similar results have occurred in the analysis of certain other data which exhibit continuous variation along a gradient but in which the floristic composition of two or three plots happens to be very similar and differs from the rest in the presence in both of a few uncommon species. A large proportion of the total chi-square in the contingency-table is "accounted for" by the contrast between the floristically similar plots and the others.

In practical applications an understanding of the ecological situation on the ground is desirable. If the cause of the aberrant behavior of a few plots can be identified, the analysis can be repeated, eliminating the deviant plots. If this procedure is followed, however, it must be clearly understood that what is required of the analysis is an insight into the causes of spatial variation in the vegetation, rather than an objective simplification of a complex situation, such as a vegetation map. As a general rule, it is not desirable to throw a heterogeneous assemblage of data into the analysis, and of course this is the risk one runs if he tries to produce a vegetation map.

If the goal of a completely objective classification of the vegetation is abandoned, it may be asked what other justification can be made for this kind of analysis. Among the most valuable results of these calculations are the sets of scores of species along gradients, especially if it has been possible to measure the latter independently. If deviant samples are not eliminated in gradient studies, the scores are blurred to the extent that the scales along which they are measured are affected by the deviant factor. Expressed in somewhat different terms, elimination of deviant samples results in an increasing idealization of the ecological situation, and the practical goal of description of the vegetation as it exists is replaced by an attempt to identify and measure the principal environmental factors affecting its variation.

It is the common experience of botanists trained in temperate regions that tropical forests are overwhelming in their species diversity. Even if one is able, with the help of a guide, to name the trees, he is still faced with the problem of very great diversity if he wishes to understand the ecology of these species--their life histories, their physiological tolerances, and their interactions with one another and with other organisms that belong to this community. A gradient analysis is useful if it enables the botanist to identify the main patterns of variation in the vegetation and to locate the average position of each species along important environmental gradients. Formerly this kind of orientation was obtained only after years of experience in the tropical rain forest. Understanding of the spatial variation of vegetation, of course, is essential in planning studies of productivity, mineral cycling, and other processes of the ecosystem.

Difficulties in mapping tropical vegetation are also very largely associated with the great species diversity of tropical ecosystems. Since tropical forests are rarely dominated by one or even a few species, systems of classification based on an ability to recognize a few important trees are not practicable. Another alternative, a system based on the physiognomy of the vegetation, usually does not distinguish among types occurring contiguously on different types of soil.

It is obvious to woodsmen and experienced naturalists that the distributions of many tropical trees are largely restricted to certain physiographic situations. These observations are confirmed by gradient analyses. As Richards [8] pointed out when commenting on Ashton's [13] gradient studies in the complex Dipterocarp forests of Borneo, availability of water is the "master factor" in the development of tropical vegetation. In Borneo as in Brazil differences in moisture and drainage are associated with striking and consistent changes in floristic composition.

These observations suggest that if reliable scores can be obtained for a number of readily identified species along moisture gradients, vegetational maps similar to the one prepared for the Mocambo Reserve can be constructed over much larger areas. Small modifications of the computational procedure, such as weighting the contribution of each species according to its ecological amplitude (variability) would be desirable. Characterization of complex vegetation by some form of weighted average of the scores of conspecuous species produces objective maps that are predictors of certain features of the physical environment such as moisture or drainage. Accordingly, their accuracy and utility can be rigorously evaluated.

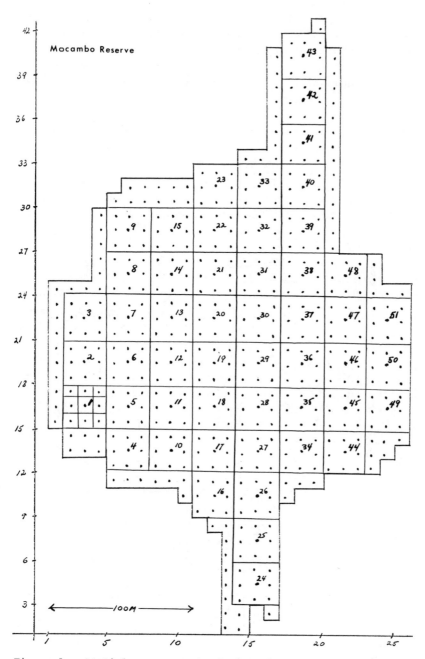

Figure 1.   Spatial arrangement of plots in the Mocambo Reserve

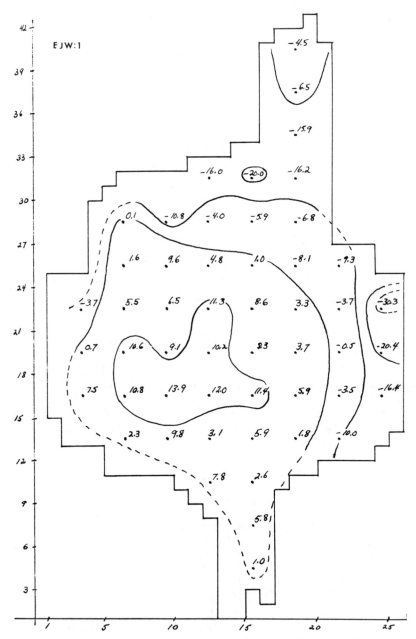

Figure 2.   Plot scores corresponding to the first latent root:
E. J. Williams procedure

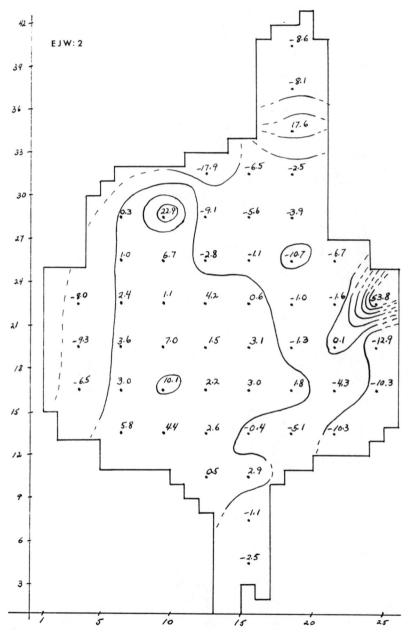

Figure 3.   Plot scores corresponding to the second latent root:
E. J. Williams procedure

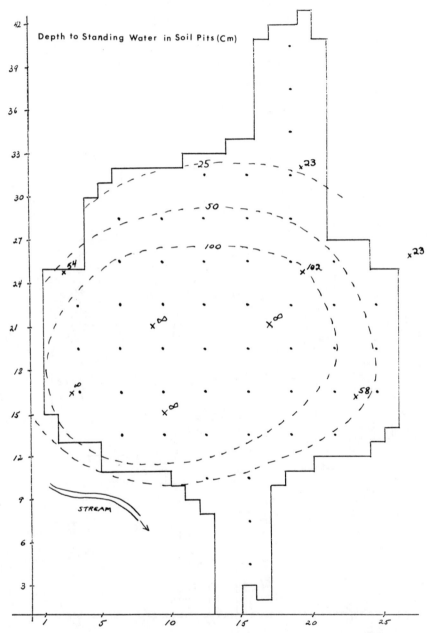

Figure 4.   Depth to standing water in soils pits on November 14,
            1968.   Pits labeled with symbol for infinity were dry.

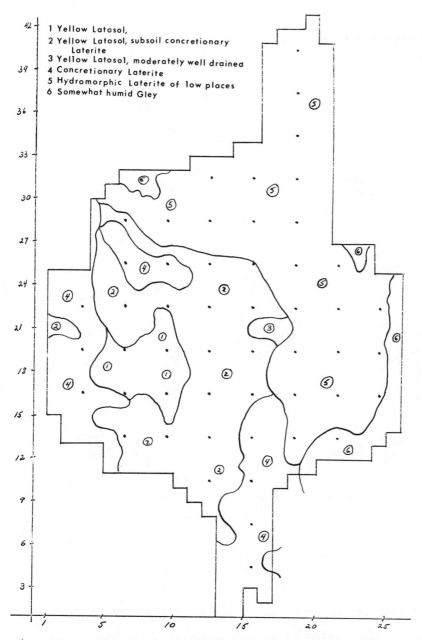

Figure 5.   Soils map of the Mocambo Reserve (Courtesy I. Falezi)

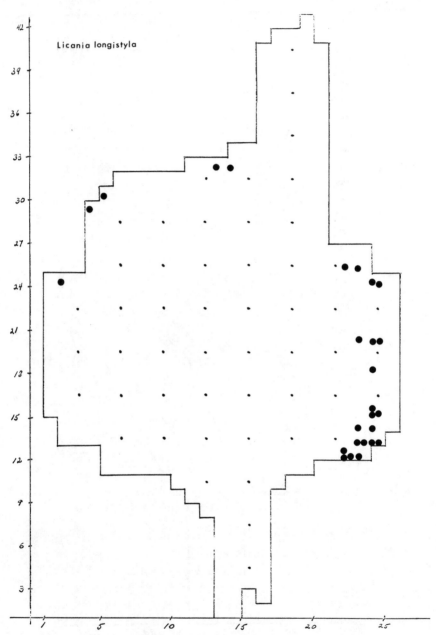

Figure 6.   Distribution of <u>Licania</u> <u>longistyla</u> (Rosaceae), a canopy
species

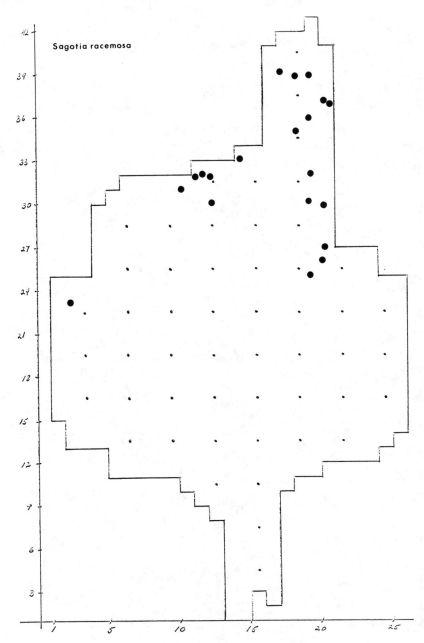

Figure 7.   Distribution of <u>Sagotia</u> <u>racemosa</u> (Euphorbiaceae), an
understory tree

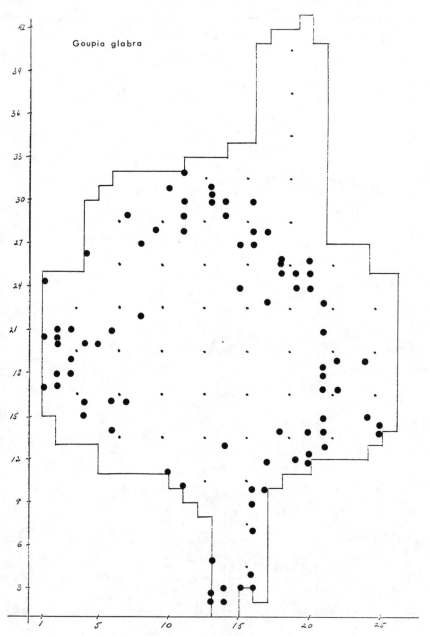

Figure 8.   Distribution of <u>Goupia glabra</u> (Celastraceae), an upper
canopy series

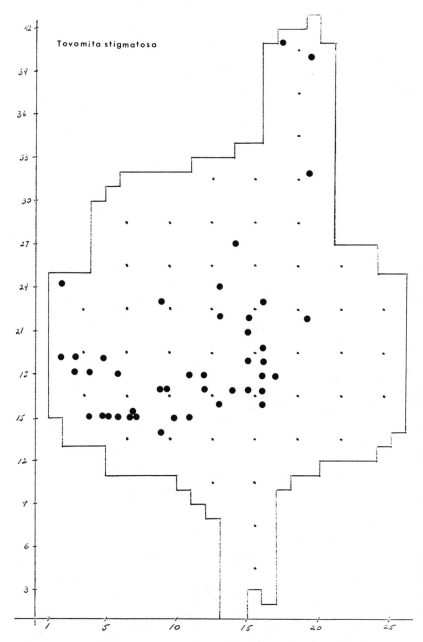

Figure 9.   Distribution of <u>Tovomita</u> <u>stigmatosa</u> (Guttiferae), an
understory species

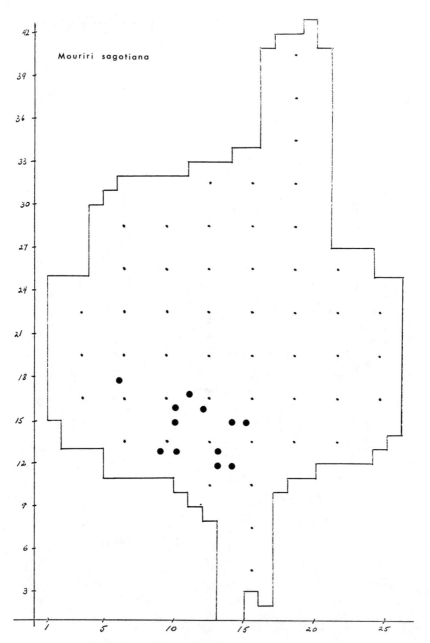

Figure 10.   Distribution of <u>Mouriri</u> <u>sagotiana</u> (Melastomataceae),
             an understory species

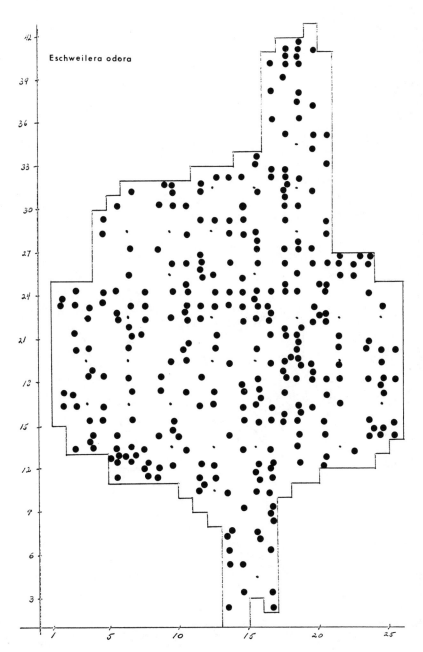

Figure 11.   Distribution of <u>Eschweilera</u> <u>odora</u> (Lecythidaceae),
a canopy series

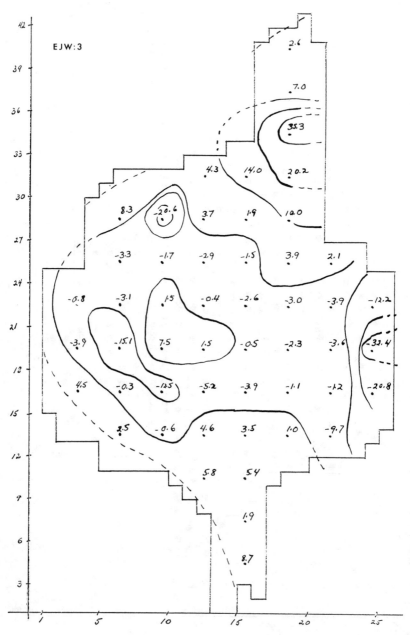

Figure 12.   Plot scores corresponding to the third latent root,
E. J. Williams procedure

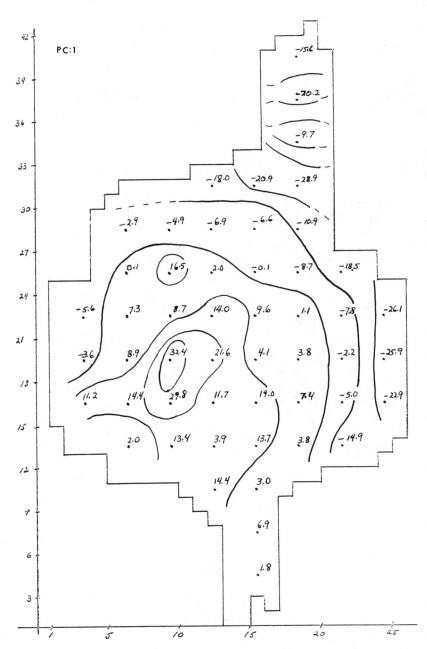

Figure 13.   Plot scores corresponding to the first latent root,
Principal Components procedure

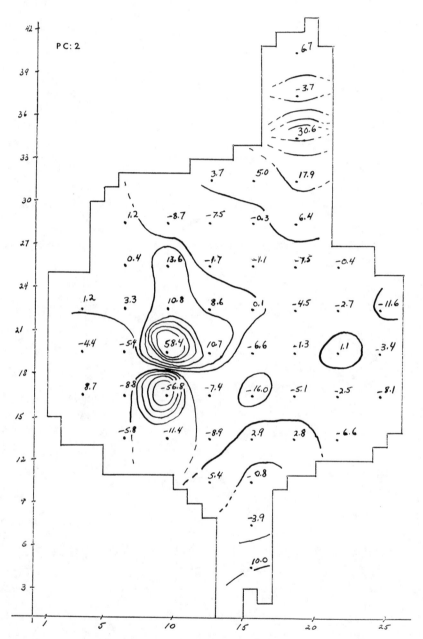

Figure 14.   Plot scores corresponding to the second latent root,
Principal Components procedure

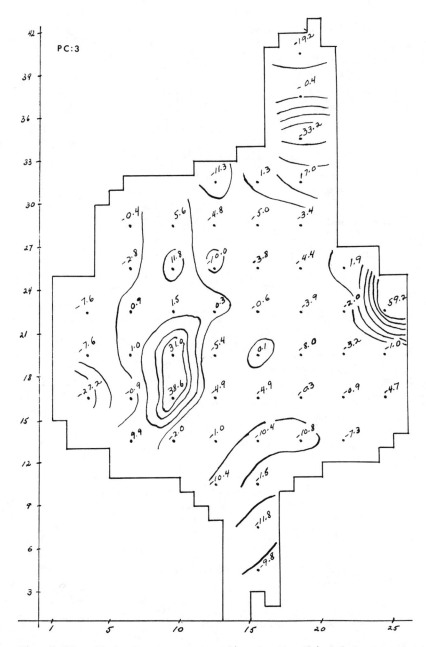

Figure 15. Plot scores corresponding to the third latent root,
Principal Components procedure

5.   ACKNOWLEDGMENT

I am very greatly indebted to J. M. Pires for permission to use his Mocambo tree data. Dr. Pires has devoted most of his career to the study of the Mocambo forest and by patient attention to detail has accumulated a large body of information which he has generously made available to scientists with whom he has collaborated. Although I was fortunate to come to the Mocambo forest after several months of intensive study in similar forests in Costa Rica, I would have been bewildered by its complexity without the help of Dr. Pires and his staff. Despite the fact that many generalizations in tropical forest ecology have been made from successful attacks on isolated problems, an appreciation of the forest as a functioning system requires a detailed knowledge of its major components. I reached many of my conclusions about the development of the Mocambo forest during informal discussions with Dr. Pires, and I had hoped that he would regard these ideas as an additional contribution to this paper. I regret that Dr. Pires has asked not to be named a joint author because he does not feel sufficiently at home in statistical ecology. In any case, it should be clear that my interpretation of the results of the contingency table analysis has depended on Dr. Pires's detailed knowledge of the Mocambo forest and of the distributions and life histories of the tree species in it, and it is obvious that in the absence of such field knowledge the interpretation of indices produced in a multivariate analysis would have been a sterile and nearly meaningless exercise.

REFERENCES

[1]   Bartlett, M. S. 1951. The goodness of fit of a single hypothetical discriminant function in the case of several groups. Ann. Eugen. Lond. 16:199-214.

[2]   Curtis, J. T. and McIntosh, R. P. 1951. An upland forest continuum in the prairie-forest border region of Wisconsin. Ecology. 32:476-496.

[3]   Fisher, R. A. 1940. The precision of discriminant functions. Ann. Eugen. Lond. 10:422-429.

[4]   Gower, J. C. 1966. Some distance properties of latent root and vector methods used in multivariate analysis. Biometrika. 53:325-338.

[5]   Kendall, M. G. 1957. A Course in Multivariate Analysis. Hafner, New York.

[6]   Orloci, L. 1966. Geometric models in ecology. I. The theory
      and application of some ordination methods. J. Ecol. 54:
      193-215.

[7]   —————————. 1967. Data centering: a review and evaluation
      with reference to component analysis. Syst. Zool. 16:
      208-212.

[8]   Richards, P. A. 1963. What the tropics can contribute to
      ecology. J. Ecol. 51:231-241.

[9]   Seal, H. 1964. Multivariate Statistical Analysis for
      Biologists. Wiley, New York.

[10] Whittaker, R. H. 1960. Vegetation of the Siskiyou Mountains,
      Oregon and California. Ecol. Monogr. 30:279-338.

[11] Williams, E. J. 1952. Use of scores for the analysis of
      association in contingency-tables. Biometrika. 39:274-289.

[12] Yates, F. A. 1948. The analysis of contingency-tables with
      groupings based on quantitative characters. Biometrika.
      35:176-181.

Added in the proof

[13] Ashton, P. S. 1964. Ecological studies in the mixed Diptero-
      carp forests of Brunei State. Oxford Forestry Memoirs
      No. 25.

RECORD OF PREPLANNED AND SPONTANEOUS DISCUSSIONS

L. P. LEFKOVITCH (Statistical Research Service, Ottowa, Canada)

The pointing out by Hatheway and Pires in this paper of the resemblance between the Williams' analysis of contingency-tables and
Gower's principal coordinate analysis is very illuminating. There
is one point that I would like to suggest which may be of some
interest in analyzing data of this kind and which has been used in
taxonomic studies rather than in ecology.

A key concept in the Williams, Gower and Orloci techniques is the
elimination of the expectation of the variables. In this way, the
pattern of the points in n dimensional space, the latter two authors
claim, is more clearly revealed. With this I agree, but this, however, is not sufficient. Let us suppose, for example, that we have
a set of values for an external variable, say, moisture content
appropriate to each plot, we may find it of interest to examine the
differences between the plots in a direction orthogonal to the
moisture contents, possibly so that a pattern which may relate to
some other external factor may be revealed. Such a technique was

used by Delaney and Healy [1] who considered the differences between
populations of mice after allowance was made for the age of the
individuals. For their measure of age, they took the size of a
certain tooth, performed a multiple regression of that size on the
variables of primary interest and used the computed linear combination of those variables as a direction to indicate age. In other
words, they found canonical axes, which in addition to satisfying
the usual conditions, were obliged to be orthogonal to the tooth
size direction. Gower [2] has reviewed this and a number of other
techniques in this context and has showed that the linear combinations arising from canonical correlation are to be preferred when
there is more than one concomitant variable.

Let us consider the external variables on which the authors have
remarked. Among them we have height, the depth of standing water,
soil type, soil fertility, windthrow, etc. An investigation of,
say, the influence of soil type could best be performed with the
common within-population covariance matrix adjusted so that it is
orthogonal to the remaining external variables. It is dangerous to
assume, that a certain principal component of the covariance matrix
represents, say, a gradient of moisture content and nothing else.
Such an inference need not be true and need not be considered.

Suppose that by some method or other a number of diredtions have
been found which can be attributed to a set of external variables.
Let us further suppose that all reasonable external variables have
been considered; what would the principal components of the adjusted
covariance matrix represent, assuming that some are distinct? Since
we have excluded external variables, it seems to me that we could be
getting at the interrelationships between the internal variables
themselves; in the example here, we could possibly infer that plots
may be grouped because a particular species happened to have been in
one place at a certain time and ha spread into adjacent regions by
the normal processes of reproduction or, alternatively, that certain
plots may be linked because of the biological dependence of one
species on another independently of any external factor.

In conclusion, I have found this paper to be of especial interest
since it suggests further lines of thought.

References

[1] Delaney, M. J. and Healy, M. J. R. 1964. Variation in the
     long-tailed field mouse (Apodemus sylvaticus (L.)) in
     northwest Scotland II. Proc. Roy. Soc. B. 161:200-207.

[2] Gower, J. C. (unpublished)   Growth free canonical variates
     and generalized inverses.

P. GREIG-SMITH (University College of North Wales, Bangor, U.K.)

The authors have presented an interesting procedure of ordination,
on which I shall not comment in detail--there are others present
far more competent to do so than I am. From the user viewpoint it
is evidently a satisfactory one though, judged on the evidence we
have heard, no more satisfactory than some others currently in use.
I wish to make two points.

   (i) It is clear that the authors were dealing with an unusually
simple topographical situation, in which environmental gradients are
likely to be more readily detected.

   (ii) How far is it useful to map extracted entities? We are
dealing essentially with a set of relationships between vegetational
composition and position on the ground and another set of relationships between environmental factors and position on the ground. Of
the three variates position on the ground is irrelevant to the main
probelms, and, moreover, its relationship to the other two is liable
to be very complex. Is it not more useful to compare directly
vegetation composition and environmental factors? Admittedly, in
the relatively simple situation with which the authors were dealing,
it was possible to draw useful conclusions from maps, but this
approach is likely to fail in the commoner situation of more complex
topography.

D. W. GOODALL (Utah State University)

It is commonly an embarrassment to statistical ecologists that their
statistical and ecological confreres use the work "frequency" in
quite different senses. So it should be pointed out first of all
that "frequency" in the present paper refers to data consisting of
counts of individuals, not the proportion of samples in which a
species occurs.

   The procedure derived in this paper from Williams' treatment of
contingency-tables is an interesting addition to the armoury of the
statistical ecologists, but is relevant only to count data. It
amounts to deriving canonical sets of variates relating simultaneously to the stands or samples of vegetation, and to the species
occurring in them. It has close analogies with principal components
analysis, as is illustrated by the similarities between Figures 2
and 3, Figures 11 and 14, and Figures 12 and 13. It is a pity that

the authors did not also derive the species scores corresponding
with the first principal component, for comparison with the species
scores from Williams' procedure in Table 3, which would have emphasized further the analogy between the two approaches.

Both these procedures work best where there is some overriding
factor or complex of factors causing differences in the composition
of the vegetation, and where the quantities of the various species
are more or less linearly related to this factor or complex. In
such cases, by far the greatest part of the variance will be accounted for by the first component or set of variates, and interpretation is straightforward. This corresponds with the situations with
which both Whittaker and Curtis were concerned when they used the
rather intuitive and informal approaches mentioned. Where relations
are more complex, several components or variate sets may need to be
taken into account, and interpretation becomes much more difficult.
This is particularly the case if relations between species and the
underlying environmental variables are non-linear where, for instance, some of the species have optima in intermediate parts of the
range but non-linear models might be conceptually more satisfying
and facilitate interpretation.

The author suggests that "Stands...which fall far from the main
directions of variation obviously contribute little to an understanding of the environmental gradients affecting most of the vegetation
under study, and might well be eliminated from the analysis." This
seems very sound, provided one is willing for the time being to
restrict attention to the "gradients affecting most of the vegetation;" this was in fact the rationale underlying the procedure for
successive rejection of outliers which I published recently in
Vegetatio. Very often, though, these outlying stands provide valuable ecological insights not available in the main corpus of data,
and consequently their rejection should only be temporary, while one
is proceeding to simplify the rest of the data. They should after-
wards be re-examined to see how their deviant position arose, and
what their relations with the main body of observation may be. A
multi-stage procedure like this may provide the means of approaching
the goal of an objective classification, which the authors felt it
necessary to abandon because of the difficulties of a single-stage
analysis of a heterogeneous assemblage.

J. GURLAND (Mathematics Research Center, University Of Wisconsin)
There is another technique for analyzing contingency-tables, which

involves the logit transformation. A presentation of the method, by Grizzle, based on maximum likelihood, appeared in Biometrics [2], and a presentation by Berkson, based on minimum chi-square, appeared in Biometrics [1]. According to this method a test of fit of the underlying model can be carried out, and the effects of rows, columns, and interactions can be tested as in a two-way layout in analysis of variance. The maximum likelihood technique required iteration but the minimum chi-square procedure leads directly to weighted least squares. The method based on logits has a further advantage in that it is general and can also be applied to contingency-tables of higher order than two. The authors of this paper might wish to try this method on their data in addition to the ones already employed.

References

[1]   Berkson, J. 1968. Application of minimum logit χ^2-estimate to a problem of Grizzle with a notation on the problem of no interaction. Biometrics. 24:75-95.

[2]   Grizzle, J. E. 1961. A new method of testing hypotheses and estimating parameters for the logistic model. Biometrics. 17:372-385.

C. I. BLISS (Connecticut Agricultural Expt. Station, New Haven, U.S.A.)

Analysis in terms of the inverse sine would simplify the weights, in contrast with the logit which requires in addition an estimate of the expected logit when computing the weights.

J. GURLAND (Mathematics Research Center, University of Wisconsin)

In reply to Dr. Bliss' questions I offer the following comments. Other possibilities besides the logit transformation are the probit and the inverse sine transformations. These are all used in biological assay and for a large class of situations lead to results which agree closely. In some situations some of these transformations may work better than others depending upon the context, but it is not convenient to go into this in detail here.

Although, as Dr. Bliss states, the weights for the inverse sine transformation depend only upon sample size. I believe there must be a good reason why this transformation has not been considered for the purpose of analyzing contingency-tables. I suggest the following one: In bioassay, for example, we would require

$$\sin^{-1} \sqrt{p} = \alpha + \beta x$$

where p is the observed proportion, x is the dosage, and α, β are
parameters to be estimated. For some combinations of parameter
values and x we could have $\sin(\alpha+\beta x) < 0$ which contradicts $p > 0$.
Similarly, in the case of a contingency-table, the estimated parameters might lead to negative probabilities. A similar difficulty
in a different context arises in analysis of variance when estimates
of components of variance turn out negative.

J. T. SCOTT (State University of New York, Albany)

I agree with Dr. Greig-Smith that we should examine the vegetation-
environment relation directly. But I was tremendously impressed by
the plots of vegetation in space given in this paper. The first
component of the ordination showed a very regular pattern along the
ground for an elevation change of only a few meters. This means
that the vegetation changes very uniformly along whatever factor or
factor complex is changing. It also seems that the region must have
been without disturbance for a long time. When we tried to map
ordination number for our case in the Adirondacks, New York we got
a mess.

I was even more impressed with pattern of the Goupia species which
produced no seedlings. The species formed a ring between the wet
region and the top of this 5 meter high knoll. How did this pattern
get that way? It shows a very narrow range of environment and
therefore the vegetation composition in tropical regions may be very
sensitive "environmental instruments," much more sensitive than in
mid-latitudes. If the species does not seed except after disturbance
then why was it successful only in such a narrow range of site
(environment)? The questions regarding this species seem to me to
be tremendously important.

C. I. BLISS (Connecticut Agricultural Expt. Station, New Haven)

Professor Gurland's concern over negative angles in the inverse sine
transformation overlooks one important application. In terms of p,
the biological interpretation of a negative inverse sine, or of one
larger than 90°, would be indirect at best. Values outside this
observable range, however, can arise in terms of working angles by

maximum likelihood, as described by Cochran (1940, _Ann_. _Math_. _Stat_.)
or by Fisher and Yates in "Statistical Tables (6th Ed., 1963).
They might occur when the bulk of the observed values fall within
a range of approximately 7 to 93%, where conclusions from percentages transformed to angles would differ only negligibly from the
more meaningful but more complex solutions in probits or logits.
Unbiased means of maximum likelihood angles may then include oc-
casional negative values. They have occurred in contingency-tables
that I will describe in volume 3 of my book.

HATHEWAY

Dr. Scott's question about the ring-like distribution of _Goupia_
glabra can be answered only with some knowledge of the ecology of
that species. It is one of the largest of the Mocambo trees, but
it seems not to reproduce under the canopy of that forest. Of the
70 _Goupia_ trees over 10 cm in diameter present in the study area,
only three were not dominant--that is, their crowns were below the
level of the forest canopy. I spent a morning searching the forest
floor for _Goupia_ seedlings, with negative results. Since the fruits
are small-seeded orange berries less than a centimeter in diameter,
I suspect that the species is bird distributed. I have been told by
Dr. Pires and others that _Goupia_ is a colonizer of relatively open
ground.

Near the Guamá River small patches of forest are disturbed from
time to time by local windstorms. If the birds which eat _Goupia_
berries visit such tangled patches--toucans, tityras, nunbirds,
boat-billed flycatchers, and various tanagers are obvious possibili-
ties--then it is reasonable to suppose that _Goupia_ seedlings may
become established occasionally in temporary openings over much of
the forest. If in addition _Goupia_ seedlings grow faster than those
of other species over only a rather narrow range of soil moisture
conditions and if the tree is long-lived, as it seems to be, then it
would be expected to exhibit a ring-like distribution on the Mocambo
forest.

The width of the band in which _Goupia_ outgrows and outpersists
its competitors depends, of course, on how similar are the environmental preferences of the latter. If it is true that the greater
the diversity of the woody flora the more likely that many species
will be closely competitive, then sharp zonation along moisture
gradients should be especially pronounced in floristically rich
forests like the Mocambo.

COMMENTS ON THE DISTRIBUTION OF INDICES OF DIVERSITY

K. O. BOWMAN
Oak Ridge National Laboratory

K. HUTCHESON
E. P. ODUM
L. R. SHENTON
University of Georgia

SUMMARY

Several measures of diversity in populations (and samples) have been used by ecologists and information theoreticians. For example, ecologists have used Shannon's information statistic

$$h_s = - \textstyle\sum p_i \ln p_i \quad ,$$

Simpson's distance measure

$$h_{si} = \textstyle\sum p_i^2 \quad \text{(or its root)},$$

and Brillouin's combinatorial type measure

$$h_b = n^{-1} \ln n! \, / \, n_1! \, \ldots \, n_k! \quad .$$

Distributional properties of these are studied from three points of view: (a) the exact distribution is found by sample configuration enumeration, (b) approximate distributional properties are found through Monte-Carlo simulations, (c) asymptotic moments are developed and their uses in assessing departures from normality considered. It is brought out that no one method is adequate for the parameter space in general, and that asymptotic moment developments are generally markedly unstable. However, the distribution of h_s is remarkably close to normal when the category probabilities follow MacArthur's model.

The study illustrates method of gaining insight into some of the complex distributional problems associated with multinomial structures, many of which defy neat mathematical formulation.

Diversity in Sodom

'There is diversity in Sodom, simple-hearted queens,
sour cats, even one who would rather not think that
there is anything queer about him, and there is a
possibly normal guest who has blundered into the
party on a wrong night.' Philip Hope-Wallace,
Manchester Guardian Weekly, 2/20/69.

1.   INTRODUCTION

Several indices of dispersion have been suggested by ecologists, the
most commonly used being the statistic corresponding to the entropy
concept defined by

$$H = \sum_{i=1}^{k} p_i \ln(1/p_i) \qquad (1)$$

where $p_i \geq 0$, $\sum p_i = 1$. Thus the statistic is

$$h = -\sum_{i=1}^{k} (n_i/n) \ln(n_i/n) \qquad (2)$$

and this can be regarded as an estimate of H being given a random
sample of n observations from data falling into the k categories
c_i (i=1,2,...,k) with probabilities p_i (i=1,2,...,k). It is well
known that h and H lie between zero and ln k, the latter being
achieved when the observations (or category probabilities) are
uniformly distributed (since x ln x→0 as x→0 it is customary to
take n_i ln n_i as zero when n_i=0).

Another index [5] is based on setting up the distribution of
species by ranks, the ranks being associated with the logarithmic
series

$$n_1, \ n_1\theta/2, \ n_1\theta^2/3,..., \qquad (0<\theta<1)$$

where the number with one representative is n_1, with two representa-
tives $n_1\theta/2$, and so on. The number of species is $k=(n_1/\theta)\ln[1/(1-\theta)]$,
and the total number of observations is $n_1+2(n_1\theta/2)+ \ldots = n/(1-\theta)$.
The index of diversity is now defined as n_1/θ.

Simpson [18], using the fact that the probability of drawing
successive members belonging to the same class (or category) in

random sampling is p_i^2 suggested the statistic

$$i_k = \sum_{i=1}^{k} (n_i/n)^2 \, , \tag{3}$$

or its equivalent $1-i_k$.

A variant of this is a diversity index suggested by McIntosh [10], namely $i_m = \sqrt{i_k}$, or one due to Wigert* taking the form $i_w = 1 - \sqrt{i_k}$.

Brillouin [3] has used a rather more efficient index of diveristy, namely

$$i_B = \frac{1}{n} \ln \frac{n!}{n_1! \, n_2! \, \cdots \, n_k!} \, , \tag{4}$$

and if all n_i are large this is approximately equal to h [13].

Good [6] has suggested the following generalized measures of heterogeneity, namely

$$c_{m,n} = \sum_{i=1}^{k} p_i^m (- \ln p_i)^n \qquad (m,n=0,1,2,\ldots) \tag{5}$$

with an obvious statistical version. This reduces to H when $m=n=1$, and to Simpson's index when $m=2$, $n=0$.

Pielou [15] has suggested the use of "evenness" (the ratio of the entropy measure h to its maximum value with respect to the cell frequencies n_i/n $(i=1,2,\ldots,k)$) or $h/\ln k$ and this clearly lies between zero and unity. Similarly, Lloyd and Ghelardi [8] have introduced an equitability measure $e_q = k^1/k$, where k is the number of species and k^1 is the number of species which associated with MacArthur's [9] model produce the observed entropy measure h.

A comparison of some of these indices for several sample configurations is given in Table 7. We use the "normed" versions:

Simpson $\hat{h}_s = [1 - \sum_{i=1}^{k} (n_i/n)^2] / (1 - 1/k)$

McIntosh $\hat{h}_m = [1 - \sqrt{\sum (n_i/n)^2}] / (1 - 1/\sqrt{k})$

---

* Oral communication.

Shannon $\hat{h} = - \sum (n_i/n) \ln (n_i/n) / \ln k$

Brillouin $\hat{h}_b = h_b / (\max h_b)$

A clear and concise account of the uses and properties of diversity measures is to be found in Pielou ([13], [14]). From the ecological point of view, much useful information is to be found in Odum [12].

Our main concern here is to investigate some of the distributional properties of $\hat{h}_s$, $\hat{h}_m$, $\hat{h}$ and $\hat{h}_b$. We give most attention to $\hat{h}$, partly because of its uses in information theory and also because of its closeness to $\hat{h}_b$; the latter seems especially difficult to handle analytically. Since for some purposes it is sufficient to know the mean and variance of a statistic, and its approximate distributional form for certain regions of the parameter space, we pay particular attention to the associated moment aspects.

Again to define the distribution of these indices of diversity we have to make some assumption with respect to the underlying population probabilities p_i, and the number of categories k. We assume that k is known prior to sampling, and that the category probabilities are either equiprobable or follow MacArthur's [9] model. In any case, our asymptotic expansions for the moments are valid in general, and those for the 2nd, 3rd and 4th moments probably new. Lastly we have done a small scale investigation of a method of determining the diversity h by cumulative sampling (and this method is very similar to that suggested by Pielou [15].

2.   GENERAL REMARKS ON THE DISTRIBUTIONAL PROBLEM

2.1   There are many problems which arise from a consideration of the exact distribution of these measures of diversity. In passing we merely mention (i) the difficulty surrounding the elusiveness of the number of species, and (ii) the conceptual problems associated with the existence (or otherwise) of a probability population structure. As for (i) it is possible that the introduction of the concept of random number of species might be acceptable to biologists, yet this would create serious problems for the statistical analyst. Comments on (ii) have appeared in the literature (see e.g., [13]

[14]) and those pointing towards the existence of a theoretical underlying population are of more interest to us than those which regard the diversity index as an unrepeatable measurement.

2.2 Very little is known about the exact distribution of any of the measures of diversity in common usage. There are of course asymptotic results, which for example, show that the entropy statistic h is approximately normally distributed for large samples n and fixed probabilities $(p_1, p_2, \ldots, p_k)$, unless the probabilities are uniform in which case the distribution becomes that of the so-called "goodness of fit" statistic chi-squared. These results become intuitive from elementary statistical theory when we write,

$$h = \sum_i p_i \ln p_i - \sum_i \varepsilon_i \ln p_i - \frac{1}{2} \sum \frac{\varepsilon_i^2}{p_i} + \ldots \tag{6a}$$

where

$$\varepsilon_i = \frac{n_i}{n} - p_i \quad , \quad \text{and} \quad \sum \varepsilon_i = 0.$$

For convenience, we write

$$h = \phi_0 + \phi_1 + \phi_2 + \ldots$$

where

$$\phi_0 = \sum_i p_i \ln(1/p_i) \quad ,$$

$$\phi_1 = - \sum_i \varepsilon_i \ln p_i \quad , \tag{6b}$$

$$\phi_r = \frac{(-1)^{r-1}}{r(r-1)} \sum_i \varepsilon_i^r / p_i^{r-1} \quad (r=2,3,\ldots).$$

In the equiprobable case, we lose the normality producing linear term in the random variables (ε_i) and

$$h^* = \ln k - \frac{k}{2} \sum \varepsilon_i^2 + \ldots \quad , \tag{7}$$

so that the chi-squared result is not surprising. Indeed the first four moments of a χ^2-variate with ν degrees of freedom (see e.g., [7]) are

$$\mu_1'(\chi^2) = \nu ,$$

$$\mu_2(\chi^2) = 2\nu ,$$

$$\mu_3(\chi^2) = 8\nu , \tag{8}$$

$$\mu_4(\chi^2) = 48\nu + 12\nu^2$$

and using (7), we have approximately

$$h^* = \ln k - \frac{\chi^2_{k-1}}{2n} \tag{9}$$

where the χ^2-variate has k-1 degrees of freedom. Thus for large n, $2n(\ln k-h^*)$ is approximately χ^2. It may be verified directly, using (6) and its extension (see [2]) that the dominant terms in the moments of h reduce in the equiprobable case to

$$\mu_1'(h^*) = \ln k - \frac{(k-1)}{2n} + \ldots$$

$$\mu_2(h^*) = \frac{(k-1)}{2n^2} + \frac{(k^2-1)}{6n^3} + \ldots$$

$$\mu_3(h^*) = - \frac{(k-1)}{n^3} + \ldots \tag{10}$$

$$\mu_4(h^*) = \frac{3(k-1)(k+3)}{4n^4} + \ldots$$

which agree with those derived from (8) and (9).

In the non-equiprobable case, we have for large samples [2]

$$\text{Mean: } \mu_1'(h) = \sum p_i \ln(1/p_i) - \frac{(k-1)}{2n} + \frac{1-\sum p_i^{-1}}{12n^2} + \frac{\sum(p_i^{-1}-p_i^{-2})}{12n^3} + \ldots \tag{11a}$$

Variance:
$$\mu_2(h) = \frac{\sum p_i \ln^2 p_i - (\sum p_i \ln p_i)^2}{n} + \frac{(k-1)}{2n^2}$$

$$+ \frac{(-1 + \sum p_i^{-1} - \sum p_i^{-1} \ln p_i + \sum p_i^{-1} \sum p_i \ln p_i)}{6n^3} + \ldots \tag{11b}$$

Third Moment:
$$\mu_3(h) = \frac{-[\sum p_i (H + \ln p_i)^3 + 3 \sum p_i (H + \ln p_i)^2]}{n^2} + \ldots \tag{11c}$$

Fourth Moment:
$$\mu_4(h) = 3 \frac{[\sum p_i (H + \ln p_i)^2]^2}{n^2} + \ldots \tag{11d}$$

$$(H \equiv - \sum p_i \ln p_i) \ .$$

These can be used to define the asymptotic normal distribution for fixed k and large n.

The equiprobable case needs separate treatment since some of the low order terms in the moments are zero. To set up the moments of h in this case we evaluate first of all the crude moments

$$\nu_r = E[h - \phi_0]^r$$

and then apply the usual correction formulae to derive the central moments. For example,

$$\mu_2 = \nu_2 - \nu_1^2 \tag{12a}$$

$$\mu_3 = \nu_3 - 3\nu_2\nu_1 + 2\nu_1^3 \tag{12b}$$

$$\mu_4 = \nu_4 - 4\nu_3\nu_1 + 6\nu_2\nu_1^2 - 3\nu_1^4 \ . \tag{12c}$$

Expectations of ϕ-products are found following the methods given in [17] and [2]. The products which must be considered to establish moments as far as a specified power of n^{-1} (n=sample size) being:

Moment	Power of n^{-1}				
	n^{-1}	n^{-2}	n^{-3}	n^{-4}	n^{-5}
ν_1	ϕ_2	ϕ_3, ϕ_4	ϕ_4, ϕ_5, ϕ_6	$\phi_5, \phi_6, \phi_7, \phi_8$	$\phi_6, \phi_7, \phi_8, \phi_9, \phi_{10}$
ν_2		ϕ_2^2	$\phi_2^2, \phi_2\phi_3,$ $\phi_2\phi_4, \phi_3^2$	$\phi_2\phi_3, \phi_2\phi_4, \phi_3^2$ $\phi_2\phi_5, \phi_2\phi_6, \phi_3\phi_4,$ $\phi_3\phi_5, \phi_4^2$	$\phi_2\phi_4, \phi_3^2, \phi_2\phi_5,$ $\phi_2\phi_6, \phi_3\phi_4, \phi_3\phi_5,$ $\phi_4^2, \phi_2\phi_7, \phi_2\phi_8,$ $\phi_3\phi_6, \phi_3\phi_7, \phi_4\phi_5$ $\phi_4\phi_6, \phi_5 \ .$
ν_3			ϕ_2^3	$\phi_2^3, \phi_2^2\phi_3,$ $\phi_2^2\phi_4, \phi_2\phi_3^2 \ .$	$\phi_2^3, \phi_2^2\phi_3, \phi_2^2\phi_4, \phi_2\phi_3\phi_4,$ $\phi_2\phi_3, \phi_2^2\phi_5, \phi_2^2\phi_6, \phi_2\phi_3\phi_5,$ $\phi_2\phi_4^2, \phi_3^3, \phi_3^2\phi_4$
ν_4				ϕ_2^4	$\phi_2^4, \phi_2^3\phi_3, \phi_2^3\phi_4, \phi_2^2\phi_3^2$

As examples, we have, using $E_a g$ to denote the coefficient of n^{-a} in the expectation of g,

$$E_1\phi_2 = -\tfrac{1}{2}(k-1), \tag{13a}$$

$$E_2\phi_3 = (k-1)(k-2)/6, \tag{13b}$$

$$E_2\phi_4 - -(k-1)^2/4 \tag{13c}$$

$$E_3\phi_4 = -(k-1)(k^2-6k+6)/12 \tag{13d}$$

$$E_3\phi_5 = (k-1)^2(k-2)/2 \tag{13e}$$

$$E_4\phi_5 = (k-1)(k-2)(k^2-12k+12)/20 \tag{13f}$$

$$E_3\phi_6 = -(k-1)^3/2 \tag{13g}$$

$$E_4\phi_6 = -(k-1)^2(5k^2-26k+26)/6 \tag{13h}$$

$$E_5\phi_6 = -(k-1)(k^4-30k^3+150k^2-240k+120)/30 \tag{13j}$$

$$E_4\phi_7 = 5(k-1)^3(k-2)/2 \tag{13k}$$

$$E_5\phi_7 = (k-1)^2(k-2)(4k^2-33k+33)/3 \tag{13L}$$

$$E_6\phi_7 = (k-1)(k-2)(k^4-60k^3+420k^2-720k+360)/42 \tag{13m}$$

$$E_2\phi_2^2 = (k^2-1)/4 \tag{13.1a}$$

$$E_3\phi_2^2 = -(k-1)/2 \tag{13.1b}$$

$$E_3\phi_2\phi_3 = -(k-1)(k-2)(k+5)/12 \tag{13.1c}$$

$$\dot{E}_4\phi_2\phi_3 = (k-1)(k-2)/2 \tag{13.1d}$$

$$E_3\phi_2\phi_4 = (k-1)^2(k+3)/8 \tag{13.1e}$$

$$E_4\phi_2\phi_4 = (k-1)(k^3+7k^2-72k+78)/24 \tag{13.1f}$$

$$E_5\phi_2\phi_4 = -(k-1)(7k^2-36k+36)/12 \tag{13.1g}$$

$$E_3\phi_3^2 = (k-1)(k-2)/6 \tag{13.1h}$$

$$E_4\phi_3^2 = (k-1)(k-2)(k^2+15k-52)/36 \tag{13.1i}$$

$$E_5\phi_3^2 = -(k-1)(k-2)(3k-8)/6 \tag{13.1j}$$

$$E_4\phi_2\phi_5 = -(k-1)^2(k-2)(k+7)/4 \tag{13.1k}$$

$$E_4\phi_2\phi_6 = (k-1)^3(k+5)/4 \tag{13.1L}$$

$$E_4 \phi_3 \phi_4 = -(k-1)(k-2)(k^2+22k-35)/24 \tag{13.1m}$$

$$E_4 \phi_3 \phi_5 = (k-1)^2(k-2)/2 \tag{13.1n}$$

$$E_4 \phi_4^2 = (k-1)(3k^3+23k^2-63k+45)/48 \tag{13.1o}$$

$$E_3 \phi_2^3 = -(k-1)(k+1)(k+3)/8 \tag{13.2a}$$

$$E_4 \phi_2^3 = (k-1)(k+13)/4 \tag{13.2b}$$

$$E_4 \phi_2^2 \phi_3 = (k-1)(k-2)(k+5)(k+7)/24 \tag{13.2c}$$

$$E_4 \phi_2^2 \phi_3^2 = -(k-1)(k-2)(k+5)/12 \tag{13.2d}$$

$$E_5 \phi_2^2 \phi_3 = -(k-1)(k-2)(k+77)/12 \tag{13.2e}$$

$$E_6 \phi_2^2 \phi_3 = -(k-1)(k-2)(k-10)/2 \tag{13.2f}$$

$$E_4 \phi_2^2 \phi_4 = -(k-1)^2(k+3)(k+5)/16 \tag{13.2g}$$

$$E_5 \phi_2^2 \phi_4 = (k-1)(k^4+22k^3+27k^2-870k+1020)/48 \tag{13.2h}$$

$$E_4 \phi_2^4 = (k-1)(k+1)(k+3)(k+5)/16 \tag{13.3a}$$

$$E_5 \phi_2^3 \phi_3 = -(k-1)(k-2)(k+5)(k+7)(k+9)/48 \tag{13.3b}$$

$$E_5 \phi_2^4 = (k-1)(k+5)(k-17)/4 \tag{13.3c}$$

$$E_6 \phi_2^4 = (k-1)(2k^2+53k-261)/4 \tag{13.3d}$$

$$E_5 \phi_2^3 \phi_4 = (k-1)^2(k+3)(k+5)(k+7)/32 \tag{13.3e}$$

$$E_5 \phi_2^2 \phi_3^2 = (k-1)(k-2)(k+5)(k+7)/24 \tag{13.3f}$$

From the like of which we derive, after simplification,

$$\nu_1 = -\frac{(k-1)}{2n} - \frac{(k^2-1)}{12n^2} - \frac{k^2(k-1)}{12n^3} + \cdots$$

$$\nu_2 = \frac{(k^2-1)}{4n^2} + \frac{(k-1)(k+1)^2}{12n^3} + \frac{(k-1)(13k^3+37k^2-k-1)}{144n^4} + \cdots$$

$$\nu_3 = - \frac{(k^2-1)(k+3)}{8n^3} - \frac{(k-1)(k+1)^2(k+3)}{16n^4} + \ldots$$

$$\nu_4 = \frac{(k-1)(k+1)(k+3)(k+5)}{16n^4} + \frac{(k-1)(k+1)^2(k+3)(k+5)}{24n^5} + \ldots ,$$

and from these, using (12), we have

$$\mu_1'(h) = \ln k - \frac{(k-1)}{2n} - \frac{(k^2-1)}{12n^2} - \frac{k^2(k-1)}{12n^3} - \frac{(k-1)(19k^3-11k^2-k-1)}{120n^4}$$

$$- \frac{(k-1)k^2(27k^2-30k+5)}{60n^5} + \ldots , \tag{14a}$$

$$\mu_2(h) = \frac{(k-1)}{2n^2} + \frac{(k^2-1)}{6n^3} + \frac{k^2(k-1)}{3n^4} + \ldots , \tag{14b}$$

$$\mu_3(h) = - \frac{(k-1)}{n^3} - \frac{(k^2-1)}{2n^4} - \frac{3k^2(k-1)}{2n^5} + \ldots , \tag{14c}$$

$$\mu_4(h) = \frac{3(k-1)(k+3)}{4n^4} + \frac{(k^2-1)(k+3)}{2n^5} + \ldots , \tag{14d}$$

so that for large n,

$$\sqrt{\beta_1}(h) \sim - \left[\frac{8}{k-1} \{1 + \frac{11k^2-2k-1}{24n^2} + 0(\frac{1}{n^3})\} \right]^{\frac{1}{2}} , \tag{15a}$$

$$\beta_2(h) \sim \frac{3(k+3)}{(k-1)} \left[1 + 0(\frac{1}{n^2}) \right] , \tag{15b}$$

and there is no term in n^{-1} in either skewness or kurtosis. These asymptotic formulae must be used with caution and n>k is a minimum requirement.

Actually it can be proved that all terms in the mean (14a) apart from the first, are negative for k>1 and increase without limit for all sample sizes, so that the series is totally divergent. However, for n/k sufficiently large it can be used in the usual asymptotic sense. Put another way, if the series is continued far enough it will always lead to a large negative value of μ_1'; on the other hand, limited accuracy is theoretically available from the series for given values of n,k (in fact Basharin [1] has proved that $|\mu' - \ln k + (k-1)/2n|$ is bounded), and if $n>2k$ then the error in using the n^{-1} term is about 5%. Further details and discussion on this aspect of the problem are to be found in [17] and [2].

Similar remarks apply to the higher moments, and an indication of the behavior of these and the mean is given in Tables 9 and 10.

Before looking at some aspects of the exact distribution of h, we record the exact values of the first two moments. We find

$$Eh = \ln n - \binom{n-1}{1} P_2(\ln 2 - \ln 1) - \binom{n-1}{2} P_3(\ln 3 - 2\ln 2 + \ln 1)$$

$$+ \ldots + -\binom{n-1}{n-1} P_n\left(\ln n - \binom{n-1}{n}\ln(n-1) + \ldots + (-1)^{n-1}\ln 1\right)$$
(16a)

where

$$P_r = \sum_{j=1}^{k} \pi_j^r .$$

For example,

$$n=2 \quad Eh = (1-P_2)\ln 2 \tag{16b}$$

$$n=3 \quad Eh = -2(P_2-P_3)\ln 2 + (1-P_3)\ln 3 \tag{16c}$$

$$n=4 \quad Eh = -3(P_2-2P_3+P_4)\ln 2 - 3(P_3-P_4)\ln 3 + (1-P_4)\ln 4 . \tag{16d}$$

Similarly, after lengthy algebra, Hutcheson [17] has derived

$$\text{Var } h = \sum_{a=0}^{n-2} \binom{n-1}{a} \sum_{i=1}^{k} p_i^{n-a} q_i^a \left[\sum_{b=a+1}^{n-1} \binom{n-1}{b} \sum_{i=1}^{k} p_i^{n-b} q_i^b \ln^2 \frac{n-a}{n-b} \right]$$

$$- (1-\tfrac{1}{n}) \sum_{b=0}^{n-3} \binom{n-2}{b} \left[\sum_{a=0}^{B} \binom{n-b-2}{a} \sum_{i\neq j} \sum p_i^{n-a-b-1} p_j^{a+1} \right.$$

$$\left. (1-p_i-p_j)^b \ln^2 \frac{n-a-b-1}{a+1} \right] . \tag{17}$$

(B=integer part of $(n-b-2)/2$).

Previous work by Miller [11] and Rogers and Green [16] was concerned with the mean and variance in the equiprobable case; the expression (17) is new, and the form of (16a) probably new. These expressions are easily computerized, and can be used as long as n and k are not excessively large (say n,k up to about 200).

2.3 The inadequacy of these asymptotic results lies in the largeness of "large" description, and we can only resolve this difficulty by attempting to describe the exact distribution (or adequate approximations) for selected values of n,k. At the outset it must be remembered that h takes a discrete set of values and that for small n the probability figure is far from smooth.

For example:

	Partition	Diversity h'	$Pr(h=h')$
	(3)	0.000	P_3
n=3	(2,1)	0.637	$3(P_2-P_3)$
	(1,1,1)	1.099	$1 - 3P_2 + 2P_3$
	(4)	0.000	P_4
	(3,1)	0.562	$4(P_3-P_4)$
n=4	(2,2)	0.693	$3(P_2^2-P_4)$
	(2,1,1)	1.040	$12P_4 - 12P_3 + 6P_2 - 6P_2^2$
	(1,1,1,1)	1.386	$1 - 6P_2 + 8P_3 + 3P_2^2 - 6P_4$

(here, $P_r = \sum p_i^r$)

and similar tables can be set up for n=5(1)12 for general category probabilities by using Table 1.1.4 from [4]. Some illustrations are given in Figures 1m, 1e, and 2. For the MacArthur model the category probabilities are

$$P_i = \frac{1}{k} \sum_{i=1}^{k} \frac{1}{k-i+1} \qquad \cdot \qquad \left(\begin{matrix} i=1,2,\ldots,k \\ k=2,3,\ldots \end{matrix} \right)$$

The unevenness of the distribution is evident, as is the fact that for n fixed the distribution tends to be a unit mass at ln n as k→∞.

2.4 There are several approaches possible in the evaluation of the exact distribution. Briefly

(i) Monte-Carlo simulation is quite suitable and we merely draw random samples of n from the categorized data and compute h and print out either moments, percentage points or whatever else is of interest. Of course, the computing time necessary may become an important factor and 50000 cycles for k=15, n=200 may run into a fair portion of an hour.

(ii) By using the tables of symmetric functions [4] we can set up (see section 2.3) the exact distribution of h for practically any value of the number of categories (k) for small samples n=2(1)12. This has been done for n=2(1)10 for the MacArthur model and the equiprobable case. We make no pretence that this is far from practicality, but nonetheless, as we shall relate in the sequel, some interesting facets of the situation emerge.

(iii) By constructing all sample configurations it is possible to set up the distribution using a computer provided n, and nk are not too large. This method was used for a few cases as a check on the Monte-Carlo work. However, it becomes uneconomical to use this method for nk>200 since the number of partitions of n increase as rapidly as $\phi(n)$, where

$$\phi(n) = \frac{1}{4n\sqrt{3}} \ \exp \ \left\{ \pi \left(\frac{2n}{3}\right)^{\frac{1}{2}} \right\}$$

and probabilities associated with a partition of many parts increase almost as fast as that power of k. The number of terms involved in the set of all sample configurations is given in Table 1, Thus, half a million terms approximately would be involved for the (n,k) values (1000,3), (55,5), (23,7), (12,10), (8,15), (6,20). Values of the moments and the skewness parameters $\sqrt{\beta}_1, \beta_2$ are given for a few selected (n,k) in Table 2.

(iv) Another approach is to use the first four asymptotic moments, and hope to come up with a rule to determine for what region in the (n,k) plane it is safe to rely on them. This is a very complicated problem because the third and fourth moments have complex algebraic structures, and again because the possible configurations of the category structure $(p_1, p_2, \ldots, p_k)$ are so varied.

3. DETAILED CONSIDERATION OF THE DISTRIBUTION OF $\hat{h}$ FOR MACARTHUR'S MODEL

3.1 Mean and Variance

Using the exact formulae (16a) and (17) the mean and variance of $\hat{h}$ have been calculated for values of n, lying between 3 and 100. For fixed k the mean increases rapidly as n increases through small values but soon asymptotes towards its ultimate value $-\sum p_j \ln p_j$. The variance however for given k goes through a maximum value prior to decreasing steadily to zero. The results are illustrated in Figures 3 and 4e,m.

3.2 Higher Moments and Skewness Parameter

Values of $-\sqrt{\beta_1}, \beta_2$ for the distribution are shown in Figures 5e,m and contours of constant β_2 for varying n,k are shown in Figure 6. It is remarkable how near to normality the distribution is provided $10 \le k \le 100$ and $10 \le n \le 100$ (the upper limit is inserted because this is the limit of our numerical investigation). Numerical assessments of the mean, standard deivation (s.d.), and β_1, β_2 are given in Table 3 for n,k between 2 and 200. It is of particular interest to note that for a fixed sample size n, the distribution of $\hat{h}$ is far from normal for small k, approaches a best near-normal form as k increases and finally departs without limit from the normal (as the distribution approaches a unit mass at k=ln n). Actually for large k the s.d. of k is always quite small (and approaches zero in a general class of p's) even for small samples. This somewhat unexpected result is plausible because if all p's→0 as k→∞, then for large k it is rather unlikely to get two or more observations in the same category. An illustration of this is given in Table 4 which shows the result of drawing 20 random samples of three from MacArthur's model with k=25; 17 out of the 20 had identical values of k and in fact the mean=1.03 and s.d.=0.16. The result would not be true if the probabilities p_i tended to non-zero constants as k→∞; for example the property would break down for

$$p_i = \frac{c}{i^2} \qquad (i=1,2,\ldots,k) \quad .$$

Out of interest, and for comparison, the moment parameters μ_1, σ, β_1, β_2 are given for the diversity index $\hat{h}$ for the equiprobable case in Table 5; the last column shows the values of β_1, β_2 for the asymptotic assessments given in (15). It is noted again that the limiting distribution (n>>0) is now that of a χ^2-variate (see section 2.2 and in particular (9)).

4. DISTRIBUTION OF McINTOSH'S[*] AND BRILLOUIN'S INDICES OF DIVERSITY

For comparison purposes a selection of values of the moment para-meters μ_1', σ, β_1, β_2 for these diversity indices, with the population probabilities following MacArthur's model, are given in Table 6. These were evaluated by Monte-Carlo methods in part, the cycle length being such that in each case a total of 10^5 random numbers were used, and the sample configuration method described in section 2.4. In addition, checks were carried out on McIntosh's statistic by using the computerized approach to the asymptotic moment expansions described in [2]. In this connection briefly it can be said that the series expansions are more stable than those for the entropy values of the coefficients for k=4(1)10 for μ_1', μ_2, μ_3, μ_4 are given in Table 8.

As might be expected the distributions tend to normality as n→∞ (k fixed) and to a unit probability mass as k→∞ (n fixed). As long as n and k are both reasonably large (say ≥15) the distributions of $\hat{h}_b$ and $\hat{h}_m$ are not far from normal (of course, this remark only applies to 5≤n, k≤100). However, in general the entropy statistic $\hat{h}$ is nearer to normality than McIntosh's and Brillouin's.

5. PIELOU'S 'SEQUENTIAL' APPROACH TO THE ASSESSMENT OF h

To estimate the diversity (h) of a population, Pielou [15] has suggested drawing cumulative samples and evaluating the successive values $h_1, h_2, \ldots, h_r$ of the diversity, the stopping rule being based on some such rule as sampling until

$$|h_r - h_{r-1}| \leq \theta h_{r-1} \qquad (0<\theta<1)$$

in certain sample units. Actually Pielou's method is rather more

---

[*] McIntosh actually refers to the modified statistic $h_m = 1 - (\sqrt{\Sigma} n_i^2)/n$ suggested to us by Dr. Wiegert.

complicated than this, but we have simplified her approach with the object of gaining some little insight into what the process involves. As an illustration we have drawn sample of five from MacArthur's model with k=5, 10, 15, 25, 50, 100 and set up the usual moment parameters (Table 11), (for the case θ =0.05) for the ultimate sample size n* and the assessed value h* of h. It is of considerable interest to note that the final sample size has a mean which changes from 19 to 28 as k goes from 5 to 100; moreover the s.d. of this sample size is around 6 and its distribution is nearly normal; by contrast, and as we should expect, the mean value of h* increases from 1.2 to nearly 3 and the distribution of h* deviates more from normality. However, the small s.d. of h* is noteworthy.

There is thus at least a suggestion here that if MacArthur's model holds then the 'stopping' rule would not involve a large sample and it seems conceivable that one might quit with a value of h* much lower than the true value.

6. CONCLUSION

Distributional properties of several indices of diversity (Shannon, McIntosh, Brillouin) under a categorized population following MacArthur's model and the 'equiprobable' model are studied, using (a) sample configuration procedures, (b) asymptotic series developments, (c) Monte-Carlo simulations. The asymptotic series development makes use of a computerized approach to expectations of products of homogeneous forms in category frequencies (described in detail elsewhere).

As a loose general remark it can be said that the distribution of Shannon's h-statistic is nearer to normality (provided the number of categories is not large) than Brillouin's or McIntosh's measures, the category probability structure following that of MacArthur. Comment is also directed at the corresponding situation under equally-probable categories. Again it is found that contrary to first impressions, the variance of Shannon's index of diversity is unusually small where categories are large in number and vanishingly small in probability.

Finally some attention is given to the distribution of h when a 'sequential' approach (similar to a procedure suggested by Pielou) is followed, sampling being terminated when successive values of h are sufficiently close.

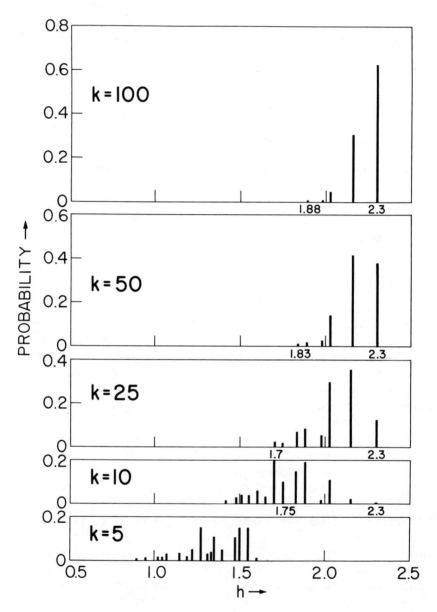

Figure 1e. Distribution of the index of diversity (h) for
            equiprobably case for sample of n=10 and varying
            categories (k)

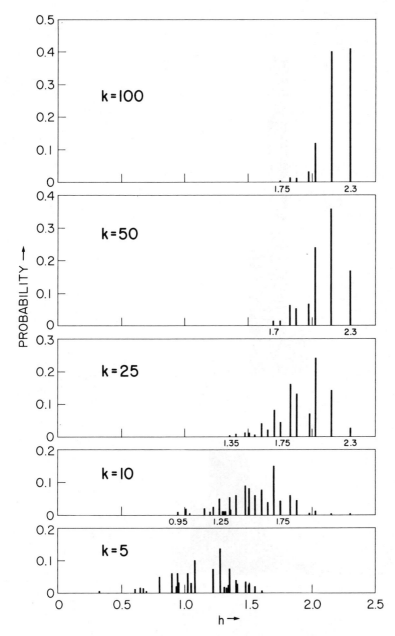

Figure 1m.   Distribution of the index of diversity (h) for
              MacArthur's model for sample of n=10 and varying
              categories (k)

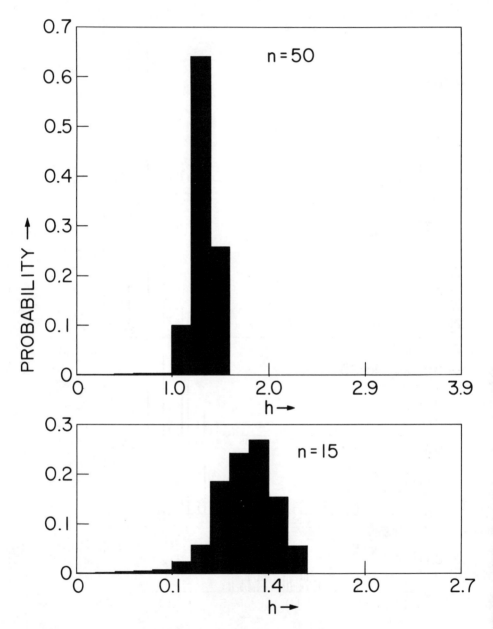

Figure 2.   Distribution of the index of diversity (h) for
            MacArthur's model for five categories and samples
            of n=15 and 50

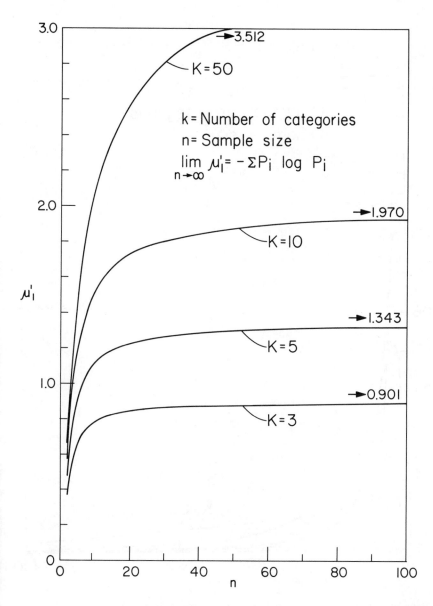

Figure 3.   Mean of the index of diversity (h) for MacArthur's
            model

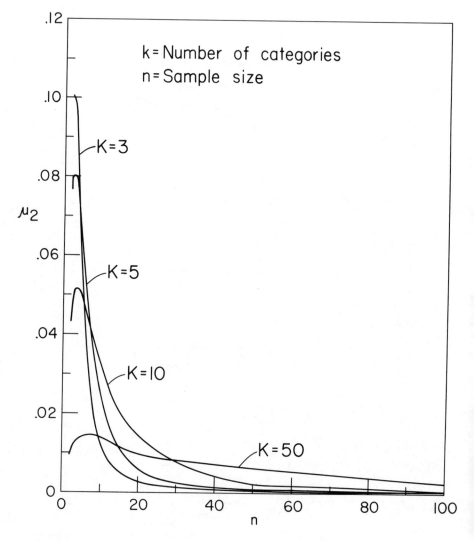

Figure 4e.   Variance of the index of diversity (h) for
             equiprobable case

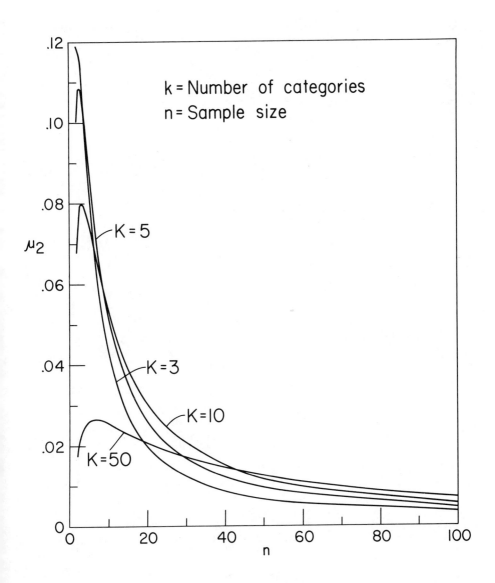

Figure 4m.   Variance of the index of diversity (h) for
MacArthur's model

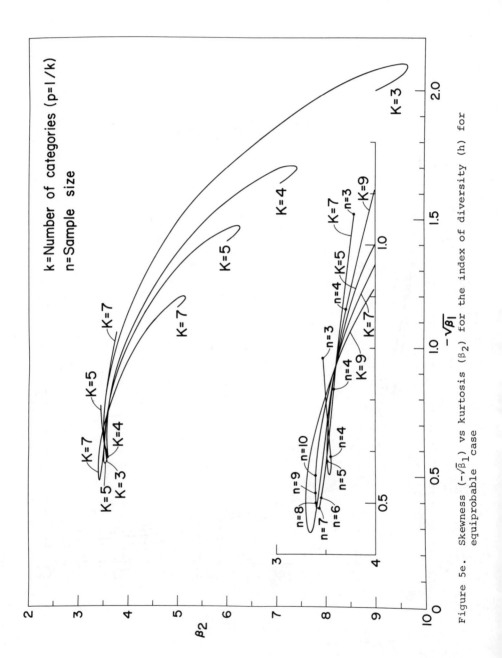

Figure 5e.   Skewness $(-\sqrt{\beta_1})$ vs kurtosis (β_2) for the index of diversity (h) for equiprobable case

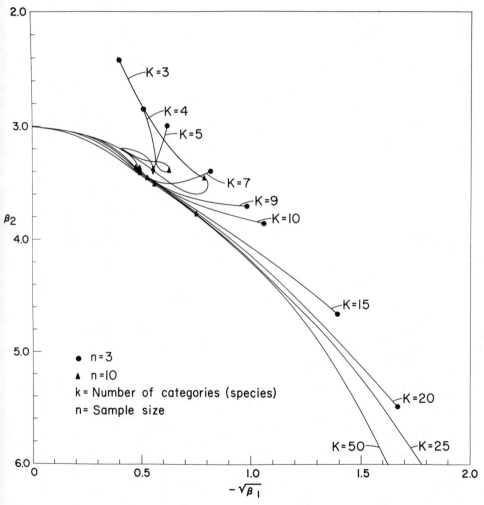

Figure 5m.   Skewness $(-\sqrt{\beta}_1)$ vs kurtosis (β_2) for the index of
diversity (h) for MacArthur's model

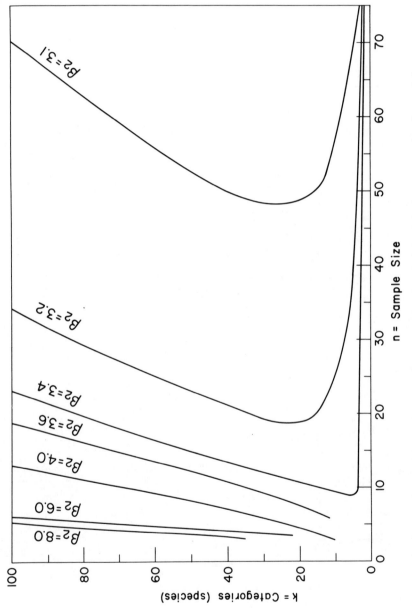

Figure 6.   Kurtosis (β_2) for the index of diversity (h) for MacArthur's model

Table 1. Number of terms in multinomials $A_{k,n}$

Index*	Categories						
	3	5	7	10	15	20	25
3	10	35	84	220	680	1,540	2,925
5	21	126	462	2,002	11,628	42,504	118,755
7	36	330	1,716	11,440	116,280	657,800	262,958^{1}
10	66	1,001	8,008	92,378	196,126^{1}	200,300^{2}	131,128^{3}
15	136	3,876	54,264	130,750^{1}	775,588^{2}	185,597^{4}	251,408^{5}
20	231	10,626	230,230	100,150^{2}	139,198^{4}	689,233^{5}	176,104^{7}
25	351	23,751	736,281	524,513^{2}	150,845^{5}	140,883^{7}	632,053^{8}
50	1,326	316,251	324,684^{2}	125,657^{5}	478,557^{8}	462,527^{11}	175,295^{14}
100	5,151·	459,813^{1}	170,590^{4}	426,342^{7}	312,629^{12}	491,037^{16}	260,110^{20}

* The index represents the power of ten required as a multiplier.

Table 2. $\sqrt{\beta_1}$, β_2 for the index of diversity for MacArthur's model

Categories (k)	Sample Size (n)	$\sqrt{\beta_1}$	β_2	Categories (k)	Sample Size (n)	$\sqrt{\beta_1}$	β_2
3	10	-0.7820	3.4580	5	10	-0.5523	3.3820
	15	-0.8068	3.5207		15	-0.5482	3.3321
	20	-0.7905	3.5785		20	-0.5321	3.2987
	25	-0.7547	3.6009		25	-0.5232	3.2747
	40	-0.6299	3.5012		40	-0.4779	3.2307
	50	-0.5652	3.4112		50	-0.4492	3.2123
	75	-0.4623	3.2703		75	-0.3879	3.1764
	100	-0.4018	3.2018	9	10	-0.4711	3.3742
7	10	-0.4921	3.3710		15	-0.4356	3.2873
	15	-0.4721	3.3002		20	-0.4183	3.2424
	20	-0.4583	3.2606	11	10	-0.4672	3.3833
	25	-0.4453	3.2341	13	10	-0.4721	3.3954

Table 3. Distribution of the index of diversity for MacArthur's model[†]

k→	n→	2	3	4	5	6	7	8	9	10
2	2	0.260	0.358	0.410	0.442	0.464	0.479	0.491	0.499	0.506
	3	0.336	0.316	0.286	0.258	0.235	0.215	0.199	0.185	0.173
	4	0.267	0.063	0.446	0.876	1.243	1.519	1.702	1.804	1.842
	5	1.267	1.063	1.598	2.267	2.921	3.499	3.977	4.348	4.618
3	2	0.372	0.529	0.616	0.670	0.708	0.736	0.757	0.774	0.787
	3	0.346	0.338	0.314	0.289	0.268	0.249	0.233	0.220	0.208
	4	0.022	0.160	0.291	0.391	0.465	0.519	0.560	0.589	0.611
	5	1.022	2.423	2.921	3.169	3.302	3.375	3.417	3.441	3.458
4	2	0.437	0.633	0.746	0.819	0.870	0.908	0.937	0.960	0.979
	3	0.335	0.337	0.318	0.298	0.279	0.261	0.247	0.233	0.222
	4	0.291	0.264	0.290	0.318	0.340	0.358	0.371	0.380	0.387
	5	1.291	2.852	3.208	3.332	3.382	3.401	3.406	3.404	3.400
5	2	0.479	0.704	0.837	0.925	0.987	1.034	1.070	1.099	1.122
	3	0.320	0.329	0.316	0.298	0.282	0.266	0.252	0.240	0.228
	4	0.686	0.384	0.323	0.307	0.303	0.303	0.304	0.305	0.305
	5	1.686	3.080	3.342	3.408	3.421	3.416	3.400	3.394	3.382
6	2	0.509	0.756	0.905	1.005	1.077	1.131	1.173	1.207	1.235
	3	0.306	0.320	0.310	0.296	0.281	0.267	0.254	0.242	0.232
	4	1.131	0.517	0.372	0.318	0.294	0.281	0.273	0.268	0.265
	5	2.131	3.252	3.434	3.461	3.450	3.431	3.410	3.391	3.374

[†] The entries in the columns are as follows: (1) Mean value of index of diversity; (2) S.D.; (3) β_1; (4) β_2. Those marked * and ** are derived from Monte-Carlo simulations of 10^5 to 5.10^4 respectively.

Table 3 (continued).

7	0.532	0.795	0.957	1.067	1.147	1.208	1.256	1.295	1.327
	0.293	0.310	0.303	0.291	0.278	0.266	0.254	0.243	0.233
	1.599	0.660	0.431	0.341	0.298	0.274	0.259	0.249	0.242
	2.599	3.407	3.510	3.504	3.476	3.445	3.417	3.392	3.371
8	0.549	0.827	0.999	1.118	1.205	1.272	1.325	1.367	1.403
	0.281	0.300	0.269	0.286	0.275	0.263	0.254	0.243	0.239
	0.080	0.808	0.495	0.371	0.310	0.275	0.254	0.239	0.229
	3.080	3.559	3.582	3.545	3.500	3.460	3.425	3.396	3.371
9	0.563	0.852	1.034	1.160	1.253	1.325	1.382	1.429	1.467
	0.270	0.291	0.289	0.281	0.271	0.261	0.251	0.241	0.233
	2.570	0.961	0.564	0.406	0.327	0.282	0.254	0.235	0.222
	3.570	3.711	3.653	3.584	3.524	3.475	3.425	3.402	3.378
10	0.575	0.873	1.062	1.195	1.294	1.370	1.431	1.481	1.523
	0.261	0.282	0.282	0.275	0.266	0.257	0.248	0.239	0.231
	3.064	1.118	0.636	0.443	0.347	0.292	0.258	0.235	0.219
	4.064	3.865	3.724	3.624	3.549	3.491	3.445	3.409	3.378
15	0.611	0.940	1.155	1.311	1.420	1.523	1.599	1.662	1.716
	0.224	0.247	0.251	0.249	0.245	0.239	0.233	0.227	0.221
	5.567	1.929	1.022	0.659	0.477	0.373	0.307	0.263	0.232
	6.567	4.660	4.098	3.835	3.681	3.579	3.507	3.452	3.409
20	0.630	0.976	1.206	1.375	1.506	1.610	1.697	1.769	1.831
	0.199	0.222	0.228	0.228	0.220	0.222	0.218	0.214	0.210
	8.088	2.760	1.428	0.895	0.627	0.474	0.278	0.313	0.268
	9.088	5.482	4.496	4.064	3.827	3.679	3.579	3.505	3.450

Table 3 (continued).

25	0.642	0.998	1.239	1.417	1.556	1.668	1.761	1.840	1.907
	0.181	0.204	0.210	0.212	0.210	0.208	0.205	0.202	0.199
	10.612	3.599	1.843	1.139	0.786	0.584	0.457	0.372	0.313
	11.613	6.316	4.905	4.303	3.983	3.787	3.657	3.564	3.495
50	0.667	1.046	1.309	1.507	1.665	1.795	1.906	2.001	2.084
	0.133	0.151	0.158	0.161	0.162	0.162	0.162	0.161	0.160
	23.218	7.806	3.943	2.393	1.616	1.171	0.891	0.705	0.574
	24.218	10.523	7.001	5.552	4.806	4.368	4.087	3.894	3.755
100	0.680	1.072	1.346	1.556	1.726	1.867	1.988	2.093	2.186
	0.096	0.110	0.116	0.119	0.120	0.121	0.122	0.122	0.122
	48.351	16.201	8.145	4.916	3.297	2.370	1.790	1.402	1.130
	49.351	18.930	11.213	8.080	6.491	5.570	4.986	4.592	4.311
200	0.686	1.085	1.366	1.582	1.758	1.906	2.032	2.144	2.243
	0.068	0.079	0.083	0.086	0.087	0.088	0.089	0.089	0.089
	98.501	32.936	16.521	9.945	6.652	4.768	3.588	2.801	2.249
	99.501	35.681	19.601	13.120	9.854	7.794	6.789	5.994	5.433
∞ Mean	0.693	1.099	1.386	1.609	1.792	1.946	2.079	2.197	2.303

Table 3 (continued).

k↓ \ n→	15	20	25	50	100	∞
2	0.527	0.536	0.542	0.552	0.557	0.562
	0.134	0.113	0.100	0.069	0.048	0.000
	1.575	1.220	0.979	0.502	0.258	0.000
	4.932	4.604	4.282	3.631	3.321	3.000
3	0.827	0.847	0.858	0.880	0.891	0.901
	0.166	0.140	0.124	0.084	0.058	0.000
	0.651	0.625	0.570	0.319	0.161	0.000
	3.521	3.578	3.601	3.411	3.202	3.000
5	1.195	1.232	1.255	1.301	1.322	1.343
	0.187	0.161	0.143	0.098	0.067	0.000
	0.301	0.289	0.274	0.202	0.126	0.000
	3.332	3.299	3.275	3.212	3.171	3.000
10	1.658	1.731	1.777*	1.874*	1.923**	1.970
	0.198	0.173	0.156	0.109	0.075	0.000
	0.183	0.165	0.156	0.127	0.068	0.000
	3.300	3.251	3.195	3.171	3.056	3.000
15	1.983	1.995	2.061*	2.201*	2.275**	2.350
	0.194	0.175	0.158	0.113	0.078	0.000
	0.168	0.132	0.115	0.094	0.063	0.000
	3.312	3.226	3.162	3.102	3.078	3.000

Table 3 (continued).

20	2.040	2.164	2.244*	2.424*	2.523*	2.623
	0.189	0.171	0.158	0.115	0.080	0.000
	0.169	0.123	0.107	0.077	0.056	0.000
	3.287	3.212	3.161	3.129	3.090	3.000
25	2.142	2.283	2.375*	2.589*	2.711*	2.838
	0.182	0.167	0.155	0.115	0.081	0.000
	0.167	0.126	0.107	0.071	0.050	0.000
	3.292	3.214	3.191	3.118	3.067	3.000
50	2.384	2.575	2.711*	3.045*	3.216*	3.512
	0.153	0.146	0.138	0.113	0.084	0.000
	0.264	0.181	0.111	0.056	0.036	0.000
	3.386	3.314	3.164	3.120	3.035	3.000
100	2.532	2.763	2.933*	3.390*	3.725**	4.194
	0.120	0.117	0.114	0.100	0.081	0.000
	0.514	0.307	0.185	0.065	0.027	0.000
	3.690	3.437	3.216	3.123	3.059	3.000
∞ Mean	2.708	2.996	3.219	3.912	4.605	8

Table 4.   Monte-Carlo illustration of small variance of h for small n and large k

Category	Cumulative Probability	1	2	3	4	5	6	7	8	9	10	11	12	13	14	15	16	17	18	19	20
1	0.0000																				
2	0.0016		X																	X	
3	0.0049																				
4	0.0099																			X	
5	0.0167																				
6	0.0254				X																
7	0.0362																				
8	0.0490																				
9	0.0640																				
10	0.0815																		X		
11	0.1014	X																			
12	0.1239						X	X													
13	0.1494							X													
14	0.1779																				
15	0.2097																X				
16	0.2452	X														X					
17	0.2847				XX									X	X						
18	0.3286	X	X										X								
19	0.3775					X			X						X					X	X
20	0.4322						X			X			X	X				X			X
21	0.4935			X				X													
22	0.5628					X	X			X											
23	0.6421											XX	X		X	X	X				
24	0.7347			X		X			X		X	X					X	X	X		
25	0.8474		X	X					X		XX			X		X		X	X		X

Table 5. Distribution of index of diversity, equiprobable case[†]

k↓ \ n→	3	5	10	15	25	50	100	∞
3	0.668	0.853	0.989	1.028	1.057	1.078	1.098	1.099
	0.302	0.223	0.113	0.072	0.041	0.020	0.010	0.000
	0.315	1.033	3.708	4.416	4.133	4.024	4.004	4.000
	3.563	3.777	7.436	9.492	9.318	9.048	9.007	9.000
5	0.833	1.119	1.375	1.462	1.525	1.569	1.589	1.609
	0.283	0.246	0.156	0.106	0.061	0.029	0.014	0.000
	0.611	0.306	0.780	1.517	2.165	2.030	2.006	2.000
	3.480	3.544	3.759	4.679	6.105	6.072	6.013	6.000
10	0.963	1.349	1.769	1.948	2.100	2.208	2.257	2.303
	0.226	0.221	0.179	0.142	0.094	0.045	0.022	0.000
	2.019	0.506	0.179	0.227	0.495	0.946	0.935	0.889
	4.523	3.567	3.325	3.299	3.524	4.333	4.426	4.333
15	1.007	1.432	1.929	2.161	2.381	2.556*	2.636**	2.708
	0.192	0.195	0.172	0.148	0.109	0.058	0.028	0.000
	3.601	0.908	0.188	0.124	0.172	0.455	0.612	0.571
	5.994	3.850	3.303	3.225	3.211	3.523	3.842	3.857
25	1.044	1.501	2.069	2.358	2.661	2.944*	3.091**	3.219
	0.153	0.161	0.152	0.139	0.115	0.074	0.038	0.000
	6.868	1.825	0.331	0.139	0.076	0.135	0.318	0.333
	9.176	4.662	3.377	3.212	3.139	3.169	3.381	3.500

Table 5 (continued).

50	1.071	1.555	2.182	2.524	2.915	3.347*	3.632**	3.912
	0.111	0.119	0.119	0.115	0.105	0.082	0.052	0.000
	15.151	4.260	0.820	0.349	0.100	0.035	0.067	0.163
	17.400	7.015	3.791	3.331	3.138	3.082	3.046	3.245
100	1.084	1.583	2.241	2.614	3.060	3.603*	4.036**	4.605
	0.079	0.086	0.088	0.088	0.084	0.074	0.058	0.000
	31.793	9.226	1.894	0.755	0.234	0.048	0.019	0.081
	34.013	11.940	4.822	3.731	3.240	3.095	3.026	3.121
$\mu_1^{\infty} \rightarrow$	1.099	1.609	2.303	2.708	3.219	3.912	4.605	

† The entries in the columns are as follows: (1) Mean value of the index of density (h); (2) S.D.; (3) β_1; (4) β_2. Entries followed by * and ** were evaluated by Monte-Carlo runs of 100,000 and 50,000 samples respectively.

Table 6. Skewness and kurtosis for Brillouin's and McIntosh's
diversity with MacArthur's model[†]

k	n→	5	10	15	50	100
5	(b)	0.589	0.826	0.940	1.171	1.243*
		0.179	0.160	0.143	0.089	0.064
		0.638	0.483	0.387	0.179	0.104
		3.853	3.683	3.518	3.204	3.128
	(m)	0.340	0.390	0.408	0.433	0.439*
		0.105	0.078	0.065	0.037	0.026
		0.684	0.642	0.547	0.243	0.130
		3.788	3.846	3.738	3.337	3.187
10	(b)	0.741	1.080	1.257*	1.643*	1.774*
		0.153	0.146	0.135	0.092	0.069
		0.891	0.481	0.351	0.129	0.076
		4.412	3.824	3.563	3.175	3.105
	(m)	0.427	0.501	0.528*	0.568*	0.578*
		0.089	0.066	0.055	0.031	0.022
		0.958	0.745	0.636	0.288	0.168
		4.493	4.249	4.039	3.463	3.284
15	(b)	0.803	1.194	1.407*	1.896*	2.071*
		0.135	0.132	0.125	0.091	0.069
		1.159	0.528	0.352	0.132	0.067
		4.800	3.941	3.602	3.186	3.090
	(m)	0.463	0.549	0.582*	0.632*	0.643*
		0.079	0.058	0.049	0.027	0.019
		1.211	0.804	0.666	0.337	0.182
		4.909	4.443	4.174	3.579	3.323
50	(b)	0.906	1.398	1.691*	2.480*	2.828*
		0.083	0.086	0.085	0.074	0.062
		3.000	0.937	0.536	0.123	0.068
		6.880	4.517	3.994	3.223	3.133
	(m)	0.522	0.635	0.682*	0.761*	0.782*
		0.049	0.037	0.031	0.017	0.012
		3.005	1.109	0.816	0.362	0.233
		6.965	5.087	4.642	3.734	3.488

Table 6 (continued).

	(b)	0.931	1.451	1.770*	2.688*	3.139**
		0.060	0.063	0.063	0.059	0.053
		5.554	1.512	0.732	0.152	0.068
100		9.520	5.155	4.103	3.287	3.094
	(m)	0.537	0.658	0.710*	0.803*	0.829**
		0.035	0.027	0.023	0.013	0.009
		5.546	1.719	0.944	0.359	0.218
		9.586	5.724	4.630	3.760	3.402

[†] In this table, the entries refer to the mean, standard deviation, β_1, β_2; (b) and (m) refer to Brillouin and McIntosh-Wiegert $[1-\sqrt{\Sigma}(n_2/n)^2]$ respectively. Entires followed by * and ** were evaluated by Monte-Carlo runs of 100,000 and 50,000 samples respectively.

Table 7. Examples of some normed indices of diversity

Distribution	Number of Individuals	Number of Species	Simpson	McIntosh	Shannon	Brillouin
2,2	4	2	1.0000	1.0000	1.0000	1.0000
1(7)	7	7	1.0000	1.0000	1.0000	1.0000
1(4)	4	4	1.0000	1.0000	1.0000	1.0000
7,5,4,2	18	4	0.9465	0.9227	0.9388	0.9342
7,5,4,2,1	19	5	0.9211	0.8810	0.8942	0.8886
7,5,4,2,1,1	20	6	0.9120	0.8620	0.8739	0.8671
7,5,4,2,1(3)	21	7	0.9101	0.8537	0.8647	0.8587
7,5,4,2,1(4)	22	8	0.9115	0.8508	0.8613	0.8554
7,5,4,2,1(5)	23	9	0.9145	0.8511	0.8611	0.8555
7,5,4,2,1(6)	24	10	0.9182	0.8531	0.8627	0.8572
7,5,4,2,1(22)	40	26	0.9646	0.9090	0.9128	0.9258
$5(10)^5$, $5(10)^5$	10^6	2	1.0000	1.0000	1.0000	1.0000
$5(10)^5$, $5(10)^5$,1	10^6+1	3	0.7500	0.6930	0.6300	
$5(10)^5$, $5(10)^5$,1(9)	10^6+9	11	0.5500	0.4193	0.2891	
9,1	10	2	0.3600	0.3225	0.4690	0.4164
4,1	5	2	0.6400	0.5988	0.7319	0.6747
3,2	5	2	0.9600	0.9522	0.9710	0.9652
3,1	4	2	0.7500	0.7150	0.8113	0.7737

Table 8. Moments of index of diversity (McIntosh) for MacArthur's model

	Bias			Variance		
k	n^{-1}	n^{-2}	n^{-3}	n^{-1}	n^{-2}	n^{-3}
4	-0.45	0.045	0.662	0.075	-0.154	0.846
5	-0.56	0.157	0.650	0.070	-0.139	0.898
6	-0.65	0.290	0.523	0.065	-0.125	0.915
7	-0.733	0.440	0.278	0.060	-0.112	0.914
8	-0.808	0.606	-0.085	0.056	-0.102	0.901
9	-0.878	0.785	-0.573	0.053	-0.092	0.883
10	-0.944	0.977	-1.190	0.049	-0.084	0.860

	μ_3			μ_4		
k	n^{-2}	n^{-3}	n^{-4}	n^{-2}	n^{-3}	n^{-4}
4	-0.075	0.569	-3.748	0.0169	0.0234	-0.7778
5	-0.071	0.574	-4.527	0.0146	0.0420	-1.0155
6	-0.066	0.563	-5.024	0.0125	0.0509	-1.1387
7	-0.061	0.545	-5.357	0.0108	0.0547	-1.2064
8	-0.056	0.525	-5.589	0.0094	0.0557	-1.2482
9	-0.052	0.505	-5.758	0.0083	0.0553	-1.2792
10	-0.049	0.487	-5.887	0.0073	0.0541	-1.3073

Table 9. Equiprobable case. Terms in series for first four moments

| | Bias | | | | | μ_4 | |
k	n^{-1}	n^{-2}	n^{-3}	n^{-4}	n^{-5}	n^{-4}	n^{-5}
4	-1.5	-1.250	- 4.000	- 51.750	- 253.60	15.75	52.5
5	-2.0	-2.000	- 8.333	- 69.800	- 883.33	24.00	96.0
6	-2.5	-2.917	-15.000	-154.208	- 2391.00	33.75	157.5
7	-3.0	-4.000	-24.500	-298.500	- 5478.20	45.00	240.0
8	-3.5	-5.250	-37.333	-525.875	-11147.73	57.75	346.5
9	-4.0	-6.667	-54.000	-863.333	-20757.60	72.00	480.0
10	-4.5	-8.250	-75.000	-1341.675	-36075.00	87.75	643.5

| | μ_2 | | | μ_3 | | |
k	n^{-2}	n^{-3}	n^{-4}	n^{-3}	n^{-4}	n^{-5}
4	1.5	2.500	16.000	-3.0	- 7.5	- 72
5	2.0	4.000	33.333	-4.0	-12.0	- 150
6	2.5	5.833	60.000	-5.0	-17.5	- 270
7	3.0	8.000	98.000	-6.0	-24.0	- 441
8	3.5	10.500	149.333	-7.0	-31.5	- 672
9	4.0	13.333	216.000	-8.0	-40.0	- 972
10	4.5	16.500	300.000	-9.0	-49.5	-1350

Table 10.   Variance of index of diversity (h) in equiprobable case
Goodness of approximation of partial sums of series for
varying n and k[†]

	n	1st	1&2	1&2&3	True	Dividing Factor
k = 3	2	2.500*	4.167	7.917	1.068	1
	4	6.250*	8.333	10.677	6.822	2
	5	4.000	5.067*	6.027	4.977	2
	6	2.778	3.395	3.858*	3.653	2
	10	1.000	1.133	1.193*	1.274	2
	200	2.500	2.517	2.517*	2.517	5
k = 4	10	1.500	1.750	1.910*	1.938	2
	15	6.667	7.407	7.723*	8.173	3
	25	2.400	2.560	2.601*	2.646	3
	50	6.000	6.200	6.226*	6.232	4
	100	1.500	1.525	1.527*	1.527	4
k = 5	10	2.000	2.400*	2.733	2.448	2
	15	0.889	1.007	1.073*	1.121	2
	25	3.200	3.456	3.541*	3.667	3
	50	8.000	8.320	8.373*	8.398	4
	100	2.000	2.040	2.043*	2.044	4
k = 10	10	4.500*	6.150	9.150	3.208	2
	15	2.000*	2.489	3.081	2.025	2
	20	1.125	1.331*	1.519⁻	1.312	2
	25	7.200	8.256	9.024*	8.813	3
	50	1.800	1.932	1.980*	2.052	3

[†] The last column gives the power of ten to be used as a divisor.
* Indicates sum closest to true value.

Table 11. Moments of index of diversity (h) and sample size (n) for MacArthur's model under Pielou's 'sequential' approach

Let h_m be the diversity for a sample of m, and $h_{m+m'}$ the diversity for the cumulated sample m+m'. Draw samples of m' up to the stage when

$$\left| h_{m+\lambda m'} - h_{m+\lambda m' - m'} \right| \leq 0.05 h_{m+\lambda m' - m'}$$

occurs for the first time. Let the sample size and diversity at this stage be n^* and h^*.

Monte-Carlo Assessments of the Moments (50,000 Cycles)

$$m = m' = 5$$

Categories	Statistic	Mean	S.D.(σ)	Skewness ($\sqrt{\beta_1}$)	Kurtosis (β_2)
5	h^*	1.214	0.178	-0.809	4.138
	n^*	19.002	7.049	0.743	3.369
10	h^*	1.706	0.204	-0.760	4.071
	n^*	19.887	5.838	0.578	3.483
15	h^*	1.978	0.215	-0.815	4.167
	n^*	20.775	5.602	0.475	3.234
25	h^*	2.301	0.224	-0.770	4.271
	n^*	22.489	5.499	0.353	2.999
50	h^*	2.687	0.225	-0.768	4.210
	n^*	25.427	5.484	0.137	2.818
100	h^*	2.999	0.212	-0.817	4.329
	n^*	28.323	5.228	0.030	2.911

ACKNOWLEDGMENTS

It is a pleasure to record our indebtedness to several colleagues; in particular to Professors D. Wigert and Carl Monk. Of course, any errors or misconceptions implied in the paper are entirely our own responsibility.

REFERENCES

[1]   Basharin, G. P. 1959. On a statistical estimate for the entropy of a sequence of independent random variables, in N. Artin (Ed.), Theory of Probability and Its Applications, Vol. IV, (Translation of Teoriya Veroyatnostei i ee Pvimeneniya), Society for Industrial and Applied Mathematics, Philadelphia :333-336.

[2]   Bowman, K. O. and Shenton, L. R. 1969. Moments of moments and frequencies, CTC-13, Union Carbide Corporation, Oak Ridge, Tennessee.

[3]   Brillouin, L. 1960. Science and Information Theory, 2nd ed., Academic Press, New York.

[4]   David, F. N., Kendall, M. G. and Barton, D. E. 1966. Symmetric function and allied tables, Cambridge University Press, Cambridge.

[5]   Fisher, R. A., Corbet, A. Steven, and Williams, C. B. 1943. The relation between the number of species and the number of individuals in a random sample of an animal population, Journal of Animal Ecology. 12:42-58.

[6]   Good, I. J. 1953. The population frequencies of species and the estimation of population parameters, Biometrika. 40: 237-264.

[7]   Kendall, M. G. and Stuart, A. 1956. The Advanced Theory of Statistics, Vol 1, Charles Griffin and Co., London.

[8]   Lloyd, Monte and Ghelardi, R. J. 1964. A table for calculating the 'equitability' component of species diversity, Journal of Animal Ecology. 33:217-225.

[9]   MacArthur, Robert H. 1957. On the relative abundance of bird species, Proc. Nat. Acad. of Sc. 43:293-295.

[10] McIntosh, Robert P. 1967. An index of diversity and the relation of certain concepts to diversity, Ecology. 48: 392-404.

[11] Miller, George A. 1955. Note on the bias of information estimates, Information Theory in Psychology (edited by H. Ouastler), University of Illinois Press, Urbana.

[12] Odum, Eugene P. 1966. Ecology, Holt, Rinehart and Winston, New York, Toronto and London.

[13]   Pielou, E. C. 1966.   Species-diversity and pattern-diversity
       in the study of ecological succession, J. Theoret. Biol.
       10:370-383.

[14]   ——————————. 1966.   The measurement of diversity in different
       types of biological collections, J. Theoret. Biol. 13:
       131-144.

[15]   ——————————. 1967. The use of information theory in the study
       of the diversity of biological populations, Proc. Fifth
       Berkeley Symposium on Mathematical Statistics and Proba-
       bility, 4:163-177.

[16]   Rogers, M. S. and Green, B. F. 1956.   The moments of sample
       information when the alternatives are equally likely (mss
       received from one of the authors).

[17]   Shenton, L. R. and Hutcheson, K. 1969.   Moments of moments
       and frequencies, Tech. Reprint No. 17, University of
       Georgia, Athens.

[18]   Simpson, E. H. 1949.   Measurement of diversity, Nature.
       163:688.

RECORD OF PREPLANNED AND SPONTANEOUS DISCUSSIONS

L. L. EBERHARDT (Battelle Memorial Institute, Pacific Northwest
                 Laboratory, Richland, Washington)

This paper represents an important effort directed towards a difficult problem, and the authors are to be complimented for their
perseverance. This is a considerable step in the direction of
providing a semblence of order in one of the more disorganized
areas of ecology. To understand diversity, we need to know how
the various supposed measures of diversity behave and how they
are interrelated. Further, there must be some considerable attention paid to sampling variability, both in the sense of its
influence on discrimination between similar models, and in decipher-
ing the underlying structure from analysis of a set of observations.
There have been all too many statements in the ecological literature
in which visual appraisals of data from small samples are said to
confirm "theories." We now have, thanks to the authors, a sequence
of results for appraising the utility of the information-theory
measure of diversity.

Perhaps it should be mentioned that the statistic attributed to
Simpson [7], while so used by ecologists, was not suggested in that
form in the paper referred to. Both Simpson [7] and Good [3]
pointed out the problems in using n_i/n when the n_i may be small,

and suggested unbiased estimators. It might also be noted that
there are a great many situations in which it is very difficult
to be at all sure of the number (k) of species in an area, so one
of the basic requirements for making much analytic progress with
the information-theory measure is often not available.

As Pielou ([5], [6]) has emphasized, use of information-theory
measures may be categorized by whether or not one has a collection
of organisms so-to-speak 'in hand', or is somehow sampling a very
large population. In the former case, the tools are largely
available, while in the virtually-infinite population circumstances,
one comes immediately up against the hard fact that we do not, in
practice, sample individuals at random. The practical (and practicable) methods all collect individuals in groups, and intra-group
correlations seem always to be the order of the day. In fact, most
of what ecology is about has to do with interactions in groups or
clusters of organisms. We thus come to the conclusion that it is
first necessary to study associations in groups, and the idea of a
measure of "the probability that the next randomly-selected individual will be a new species" does not really seem appropriate unless,
perhaps, it is somehow a measure of local conditions.

No small part of the trouble in coming to grips with the question
of what is really needed by way of a measure of diversity has to do
with the fact that most of the theorizing in ecology is limited to
abstract and deterministic models. One is thus faced with, for
example, "species-spaces" with no notion of how those spaces are
supposed to be populated, nor of the possible influence of the
dynamics of such populations on a real-world expression of the
postulated model. In many respects, one is tempted to suppose that
information theory in ecology might well follow the pattern described by Gilbert [2], who concludes: "As an engineering subject,
information theory has flourished for 18 years because of the promise it gave of improved communication systems. The results are
still almost exclusively on paper."

Rather than to struggle with the mathematics of a particularly
difficult measure, I should think we might look rather carefully
into the matter of what really is required of a measure. I think
our real knowledge of ecology at present is largely empirical, so
that it is most important to know how a given measure behaves over
a wide range of circumstances and what its small-sample fluctuations
are like. Simpson's measure thus seems to me to be preferable to
the information theory measures. Its statistical attributes are

known, at least with reference to a much-studied series of frequency
distributions (the Poisson, negative binomial, and logarithmic
series distributions). Also, a natural extension of the popular
MacArthur "broken-stick" model from a fixed interval to the real
line (infinite interval) places that model within the above sequence
(as having a geometric sampling distribution, cf. [1]). Such an
extension from a single randomly-divided habitat to an indefinitely
long series of "niches" drawn from the same parent population has
the further advantage that the number of species per fixed segment
can now comfortably become a random variable for both ecologist and
statistician. The same model may also provide a basis for appraising Lloyd's "equitability" measure [4]. Inasmuch as that index
uses the MacArthur model as a reference point, the prevalence of
success in fitting data to a more extreme kind of distribution--the
logarithmic series--leaves the equitability index based on something
short of maximum diversity. I should think the concept of "evenness" described by Pielou [5], wherein the reference set is one of
equal abundance of species, ought to be easier to handle both
technically and practically. There is in particular the result
that sampling distributions tend to follow the Poisson, giving a
convenient basis for null-hypothesis testing.

Since virtually all of our field sampling procedures are based
on plot sampling or something like it, it is evident that an analysis that adheres to the facts of field practice is faced with two
non-independent random variables--the number of species and the
number of individuals per sampling unit. In such circumstances,
given that it is feasible to do so, it seems one natural recourse
is to use inverse sampling. If we fix the number of species to be
examined at each locale, and tally individuals of each species
encountered in the process, then it would seem the uncertainty-of-
the-next-individual interpretation of information-theory measures
is rather more nearly applicable. Probably a more useful feature
is that there is a simple basis for distribution-free (but conditional) kinds of statements about association of species.
Alternatively, if we fix the number of individuals tallied, the
number of species becomes the relevant random variable, exploitable
perhaps as another's way to look at the "species-area" problem in
terms of recent developments in finite population sampling. Either
approach will evidently lose some information, and may be rejected
as mechanically difficult. However, for plant sampling, I think
the mechanics may readily be handled by using a "variable-area"

plot, i.e., determining the length of the plot by either of the
criteria mentioned above. In practice, one might profitably use
both stopping-points on the same plot, that is, tallying individuals by species and recording the two different plot lengths
required to accumulate a fixed number of species and given total
of individuals. The prospect of loss of information about density
of individuals seems likely to be considerably explored in a number
of papers pertaining to "distance-sampling" at this Symposium.

In summary, it seems to me that one of the main conclusions to
be drawn from the paper by Shenton etc., reinforced by diverse
other sources, is that it seems unwise to try to measure a poorly
defined quantity with a ruler that is so difficult to read.

References

[1]   Eberhardt, L. L. 1969. Some aspects of species diversity
       models. Ecology, In Press.

[2]   Gilbert, E. N. 1966. Information theory after 18 years.
       Science. 152:320-326.

[3]   Good, I. J. 1953. The population frequencies of species
       and the estimation of population parameters. Biometrika.
       40:237-264.

[4]   Lloyd, M. and Ghelardi, R. J. 1964. A table for calculating
       the "equitability" component of species diversity. J. Anim.
       Ecol. 33:217-225.

[5]   Pielou, E. C. 1966. The measurement of diversity in different
       types of biological collections. J. Theoret. Biol. 13:
       131-144.

[6]   ————————. 1966. The use of information theory in the study
       of biological populations. Proc. Fifth Berkeley Symp. Math.
       Statist. Probab.

[7]   Simpson, E. H. 1949. Measurement of diversity. Nature. 163:
       688.

E. C. PIELOU (Queen's University, Canada)

The authors have described an impressively thorough investigation
of the sampling distributions of some of the diversity indices used
by ecologists. I am sure the difficulties will come as a revelation
to those carefree souls who plug numerical data into the celebrated
formula $-\sum(n_i/n)\ln(n_i/n)$ and trust to luck. Even so, the authors
have assumed throughout that a sample from a natural population was
a random sample and that k, the population number of species, was
independently known. All too often neither of these assumptions
holds. In practice one usually has a sample that is non-random from

a population with an unknown number of species. As the authors say, the latter difficulty creates serious problems for the statistical analyst. This is especially true when one wishes to estimate the 'evenness' of a many-species population--what the authors have called the 'normed' diversity. The maximum likelihood estimator of this normed diversity when Shannon's formula is used is

$$\hat{h} = -\sum (n_i/n)\ln(n_i/n)/\ln k \quad .$$

There are various ways of estimating the numerator of this expression and its sampling variance, and consequently if k is known, an estimate of h and its standard error are directly obtainable.

However, k is the number of species in the population, not the sample, and is often unknown. If one were to assume that all the species in the population were represented in the sample and this assumption were mistaken, obviously the normed diversity would be overestimated.

This brings us to the perennial problem of how to estimate the number of species in a population that cannot be fully censused, a problem that has recently been rather neglected. Brian [1] made a promising suggestion: if a negative binomial distribution fits the observed species-abundance frequencies one may estimate the population number of species from the number in the sample and the parameters of the fitted distribution. It seems to me that this method rests on fewer uncertain assumptions, and is likely to be more generally applicable, than Preston's [2] method of fitting a discrete lognormal distribution to the same data. I hope ecologists can be persuaded to try it.

Then, perhaps, Drs. Shenton, Bowman, Hutcheson and Odum will tackle the very difficult problem of exploring the sampling distribution of diversity indices when k itself is subject to sampling variation. In the meanwhile, the results they have already given should make us fully aware of factors that must be taken into consideration when diversities are estimated.

References

[1] Brian, M. V. 1953. Species frequencies in random samples from animal populations. J. Anim. Ecol. 22:57-64.

[2] Preston, F. W. 1948. The commonness and rarity of species. Ecology. 29:254-283.

P. HOLGATE (Birbeck College)

In view of the widespread use of the entropy measure among statistical ecologists, the increase in our knowledge of its sampling distribution provided by the present paper is of the greatest importance. I will confine my comments to questions of general principle.

I think that statistical ecologists need to think thoroughly about the meaning of entropy in the context in which they are working. It is hardly adequate to state, as do MacArthur and Mac-Arthur [3] that "...the bird species diversity is $-\sum p_i \log p_i$. This is a formula used by communication engineers to calculate the information generated, e.g., by a typist who uses the different keys with frequencies p_i." I agree with Hairston [1] who wrote, commenting on another comparison: "The analogy with language is interesting, but unless common or similar underlying causes can be shown for the organization of languages and communities any similarities must remain interesting coincidences and must not be used in interpretation or in the foregoing kind of justification."

Faced with a probability vector $P=(p_1,\ldots,p_k)$ it is certainly tempting to replace it by a single number such as $I(p)= -\sum p_i \log p_i$. It is, moreover, true that the maximum and minimum values of $I(P)$ correspond to the recognizable extreme situations where all frequencies are equal, and where $p_i=1$ for some i, respectively. Furthermore, for any t in (0,1) we have

$$I(tp_1,(1-t)p_1,p_2,\ldots,p_k) = I(p_1,\ldots,p_k) + p_1 I(t,1-t)$$

so that splitting up a category containing p_1 of the specimens into two sub categories, in proportions t, 1-t increases the entropy by a simple quantity depending only on p_1 and t. (This requirement, together with invariance under permutation of the p_i, Lebesgue measurability and the scaling requirement $I(\frac{1}{2},\frac{1}{2})=1$ does in fact define the entropy uniquely, [2].) The above properties do not however hold for the various "normed" versions.

Another interesting approach is through the generalized measures of information introduced by Rényi, namely $I_\alpha(p)=\sum p_i^\alpha$. For $\alpha=2$ this becomes Simpson's measure of diversity, and for integral j, $I_j(p)$ is the chance that j independently sampled specimens will all belong to the same species. If $I(P)$ is considered as the expectation of $p_i^{\alpha-1}$, $I(P)$ itself can be thought of as a bounding case of the $I(P)$ for $\alpha=1$.

Despite this, a careful study of the books of Rényi [5] which contain a 100 page supplement on the statistical aspects of information, and of Onicescu [4] who takes the entropy as a fundamental quantity in his development of elementary probability, leaves me unconvinced that I(P) is actually a better measure of diversity in the present ecological context (I wish to emphasize the last phase) than for instance a more easily comprehensible quantity such as the standard deviation of the abundances of the various species.

References

[1] Hairston, N. G. 1959. Species abundance and community organization. Ecology. 40:404-416.

[2] Lee, P. M. 1964. On the axioms of information theory. Ann. Math. Statist. 35:415-418.

[3] MacArthur, R. H. and MacArthur, J. W. 1961. On bird species diversity. Ecology. 42:594-598.

[4] Onicescu, O. 1964. Nombres et systèmes aleatoires. (Editions Acad. R. P. Roumaine, Bucharest).

[5] Rényi, A. 1963. Wahrscheinlichkeitsrechnung. Mit einem Anhang über Informationstheorie. (VEB Deutscher Verlag der Wissenschaften, Berlin).

W. T. STILLE (Biometrician, Eastman Kodak C., Rochester, N.Y., and Dept. of Prehistoric Medicine, Medical School, University of Rochester)

Dr. Shenton asked me to describe our method of estimating maximum diversity of illness and injury diagnosis (=species) in varying sizes of demographically defined groups of people. Our precedure is a modification of the accumulative method suggested by Dr. Pielou. We have data on illness episodes coded with respect to the International Classification of Diseases (World Health Organization), which contains over 3000 rubrics. In nine years of data, we have encountered 700 kinds of diagnosis. Upon initiation of analysis of this data, using Shannon's H-statistic, we encountered correlations between H and the number of episodes, which vitiated comparisons between unequal size groups.

In an effort to overcome this correlation between H and N, random samples, of 50 to 20,000 episodes of illness were drawn for variously defined groups and Shannon's H computed. The H values and the corresponding N's were used in a series on nonlinear regression analysis of up to 3 parameters and employing such models as: polynomials, logistic and trigometric functions. The best fits tended

to be achieved with the hyperbolic tangent: $y=H=A \; \tanh(BN^C)$.
Values of C tended about .2 with values of B about .33. Thus, for
N=10000, the fourth root is 10 and $\tanh(10/3)$ is about 1.

AUTHORS' COMMENT

Reply to discussants:

We shall consider Dr. Eberhardt's suggestion of giving closer attention to the much simpler idea of diversity due to Simpson. Dr.
Pielou's remarks on varying k, although interesting, would involve
a great deal of further complication; if we understood her stopping
rule idea correctly, then its title would seem to be not very
important.

   Mr. Holgate's remarks were of considerable interest, especially
those on the tie-up with information theoretic studies.

   We look forward to seeing a more detailed account of Dr. Stille's
investigation on industrial diseases and its association with
diversity indices.

SIMULATION OF DISPERSAL IN DESERT RODENTS*

NORMAN R. FRENCH
University of California, Los Angeles

SUMMARY

A computer simulation of rodent dispersal in which movements were random in direction and the distance moved was determined by encountering an unoccupied home range failed to show a bimodal distribution suggested by field data from natural populations and also failed to show a smooth leptokurtic curve resembling a lognormal distribution as depicted by other studies. A bimodal distribution evidently requires selective pressure for the trait that varies with population density or environmental conditions.

* These studies were supported by Contract AT (04-1) GEN-12 between the U.S. Atomic Energy Commission and the University of California

1.   INTRODUCTION

Dispersal as used in this discussion refers to long distance
movements of individual animals that frequently result in the
establishment of a new home range. It is a trait that is generally
considered to be almost universal in the animal (and also the plant)
kingdom. It is important not only in the distribution of individuals for most effective utilization of the habitat, but also
in promoting (or by its absence, preventing) gene flow in the
species and thereby determining the degree of inbreeding or out-
breeding between subunits of the population. Because of the
obvious difficulties of assessment of dispersal movements, there
is not a great deal of data that will permit quantitative evaluation
of this aspect of animal behavior.

It was first pointed out that dispersal of small mammals was
non-normal or non-random in an analysis of deer mouse (Peromyscus
maniculatus) movements by Dice and Howard [5], although a leptokurtic distribution of insect dispersal movements had earlier been
demonstrated by Dobzhansky and Wright ([6], [7]). An evaluation of
available information on dispersal led Bateman [1] to the conclusion
that all gene dispersal is keptokurtic, that is, with a higher
proportion of short and long range movements and a lower proportion
at intermediate distances. In a study of the movements made by
nesting song sparrows (Melospiza melodia), Johnson [11] emphasized
the unusual proportion of long distance movements, and suggested
that these may represent a distinct subgroup of the population.
Howard [10] also concluded that individuals differed in the trait
for dispersal movements. Murray [14] demonstrated that a skewed
frequency distribution of movements can be generated by random
movements of selected individuals in a population of home ranges
where an animal that occupies a home range repels all intruders.
Dispersal within a population may result primarily from young
animals seeking to establish home ranges ([9], [11]), but even
established animals with a strong attachment to their home ranges
may be induced to move if environmental conditions become less
favorable. This was demonstrated experimentally in a population
of the Great Tit, Parus major, by Kluijver [12]. When nest boxes
were closed, thereby reducing the number of nest sites available,
the population was reduced to one quarter of its former density.

One pressing question that remains regarding dispersal is the
existence of a subunit of the population with an unusually strong

trait for dispersal. Murray [14] has shown that a strongly skewed
distribution of dispersal moves can result from movements that are
purely random in direction and distance. The only indications that
remain, then, that seem to support this hypothesis are the bimodal
curves for deer mice [5], for song sparrows [11], for pocket mice
[8], and possibly the more extensive data for lizards assembled by
Blair [2]. Blair's study of dispersal in Sceloporus olivaceus was
based on 227 records. Although only 12 of these records suggest
the bimodality of the dispersal curve, Blair emphasized the problem
of increased area and consequent reduced sampling effort at these
great distances. Movements were also affected by features of the
environment. These problems are common to all studies of dispersal,
and are the reasons why the significance of long-distance dispersal
is still a problem for speculation. In the other cases cited,
bimodal frequency distribution may result from irregularities of
sampling, which may often occur when the sample number is small,
as pointed out by Wright ([15], p. 119).

This paper reports the results of an effort to discover the
conditions that would lead to a bimodal frequency distribution of
dispersal distances by means of computer simulation of the process.

2. METHODS

The model utilized in these studies is an extension of the original
one proposed by Murray [14]. The present model utilized computer
simulation techniques. By using a computer, many modifications of
the original assumptions could be investigated, and many replications
of each set of conditions could be easily conducted. The computer
program set up a grid of 20 by 20 or 40 by 40 animal home ranges.
These were filled on a random basis according to the population
density set by the program. Whether or not an animal dispersed
was also determined randomly, according to the probability set by
the program. If an animal dispersed, the direction of dispersal
was random, and it moved in a straight line until it intercepted a
home range that was vacant, in which case it occupied that home
range. The straight line distance moved by the animal was computed
and stored. The total dispersal movements for ten such populations
were accumulated and summarized as a table and as a histogram depicting the results of one given set of conditions.

A number of assumptions are implicit in the design. An animal

is dominant in its own home range and repells all intruders.   Old
and young animals of both sex make dispersal movements. Habitat is
considered uniform, at least to the extent that the variations
encompassed within one home range are equivalent to those in any
other home range. These assumptions seem valid based upon the field
experience reported in our study of dispersal in desert rodents [8].
    In this study an effort was made to determine the conditions or
assumptions necessary to produce a bimodal frequency distribution of
dispersal movements. To this end, the computer simulation was run
with varying population densities, different probabilities of dispersal, and with the total population divided into subgroups with
different densities and/or different probabilities of dispersal. In
this way the effects of different spatial patterns or dispersion of
animals in the field were examined, and the effects of various
behavioral or genetic attributes that have been proposed to explain
the persistence of the dispersal trait.

3. RESULTS

The results of fifteen different combinations of population density,
dispersal probability, and dispersion pattern (Table 1) indicate
that the frequency distribution of dispersal distances is not
necessarily a smooth leptokurtic curve, even when dispersal is a
completely random event. Divergence from leptokurtosis is in the
form of a shoulder that develops on the curve at a distance of four
or five home range diameters. Development of the shoulder is
dependent upon population density and dispersal probability, becoming more pronounced as each of these variables is increased
(Fig. 1). When the total population is divided into subunits of
different population density, departure from leptokurtosis increases
with increasing size of the subunits. A smooth leptokurtic distribution was produced in populations of low density and populations
of changing density.

Table 1. Combinations of conditions tested by simulation of dispersal

Grid	Grid subunits	Population density (%)	Dispersal probability (%)	Number of computer runs
20 x 20	-	25	25	9
20 x 20	-	25	50	10
20 x 20	-	25	75	10
20 x 20	-.	50	25	10
20 x 20	-	50	75	10
20 x 20	-	75	10	10
20 x 20	-	75	25	10
20 x 20	-	75	50	10
20 x 20	-	75	75	10
40 x 40	(Inside 20x20 (Peripheral 10	90 10	30) 30)	10
40 x 40	(Inside 20x20 (Peripheral 10	10 90	30) 30)	10
40 x 40	-	20→60	30	5 at each density
40 x 40	(Inside 20x20 (plus 10x10 corners (Remainder	80 20	30)) 30)	10
40 x 40	(Alternating 10x10 (	80 20	30) 30)	10
40 x 40	(Alternating 10x10 (	80 20	50) 25)	10

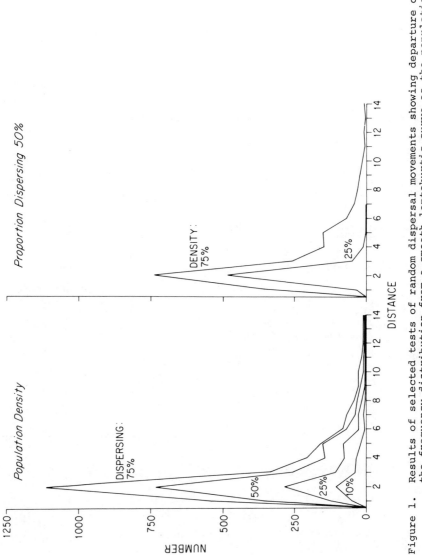

Figure 1.   Results of selected tests of random dispersal movements showing departure of
the frequency distribution from a smooth leptokurtic curve as the population
density increases and as the amount of dispersal increases

4. DISCUSSION

At this stage of the study, it seems uncertain if the departure from leptokurtosis is actually the same as the bimodal frequency distribution suggested by certain field studies. The shoulder of the dispersal curve resulting from simulation studies occurs at a distance of four of five home range diameters, and represents a greater number of simulated animals moving these distances than would be expected in a lognormal distribution. The secondary peak that appears in results of certain field studies, both from desert rodents and birds, which suggests a bimodal curve in the frequency distribution of dispersal movements, is at a much greater distance from the origin, probably twice this distance.

If there is a mechanism operating in the natrual population that permits a separate genotype to persist, a genotype which possesses the behavior pattern for exceptionally long-distance dispersal moves, I have been unable to duplicate it in the model with the assumptions described. The preservation of such a trait would require selection pressure, at least periodically, for the dispersal trait (see [13], p. 212). This could only occur in fluctuating populations, such as desert rodent populations, if the mortality rate among individuals possessing the dispersal trait differed between times of high population density and times of low population density. Under such conditions the degree of selection for the trait would be balanced with the frequency of periods when such selection operates. The simulation procedure described must be modified to allow differential mortality in order to explore this process. Such a process is an essential requisite to the hypothesis of varying behavior mechanisms for population regulation proposed by Chitty ([3], [4]). The existence of a bimodal frequency distribution of dispersal movements in natural populations requires confirmation with extensive field data.

REFERENCES

[1]   Bateman, A. J. 1950. Is gene dispersion normal? Heredity.
       4:353-363.

[2]   Blair, W. F. 1960. The rusty lizard. Univ. of Texas Press,
       Austin. 185 p.

[3]   Chitty, D. 1960. Population processes in the vale and their
       relevance to general theory. Can. J. Zool. 38:99-113.

[4]   Chitty, D. 1964. Animal numbers and behavior. p. 41-53 in
      J. R. Dymond, (Ed.), Fish and Wildlife: a memorial to W. J.
      K. Harkness. Longmans Canada Ltd., Don Mills.

[5]   Dice, L. R. and Howard, W. E. 1951. Distance of dispersal by
      prairie deer mice from birthplaces to breeding sites.
      Contrib. Lab. Vert. Biol., Univ. Michigan, 50:1-15.

[6]   Dobzhansky, T. and Wright, S. 1943. Genetics of natural
      populations. X. Dispersion rates in Drosophila pseudoob-
      scura. Genetics. 28:304-340.

[7]   ──────────. and ──────────. 1947. Genetics of natural popu-
      lations. XV. Rate of diffusion of a mutant gene through a
      population of Drosophila pseudoobscura. Genetics. 32:
      303-324.

[8]   French, N. R., Tagami, T. Y. and Hayden, P. 1968. Dispersal
      in a population of desert rodents. J. Mammalogy. 49:
      272-280.

[9]   Howard, W. E. 1949. Dispersal, amount of inbreeding, and
      longevity in a local population of prairie deer mice on
      the George Reserve, southern Michigan. Contrib. Lab. Vert.
      Biol., Univ. Michigan, 43:1-50.

[10] ──────────. 1960. Innate and environmental dispersal of
      individual vertebrates. Amer. Midland Nat. 63:152-161.

[11] Johnson, R. F. 1956. Population structure in salt marsh
      song sparrows. Part I. Environment and annual cycle.
      Condor. 58:24-44.

[12] Kluijver, H. N. 1951. The population ecology of the Great
      Tit, Parus m. major L. Ardea. 39:1-135.

[13] Mayr, E. 1965. Animal species and evolution. Harvard
      University Press, Cambridge. 797 p.

[14] Murray, B. G., Jr. 1967. Dispersal in vertebrates. Ecology.
      48:975-978.

[15] Wright, S. 1968. Evolution and genetics of populations.
      Volume I. Genetic and biometric foundations. Univ. of
      Chicago Press. 469 p.

RECORD OF PREPLANNED AND SPONTANEOUS DISCUSSIONS

R. M. CORMACK (University of Edinburgh)

A possible mechanism by which the shoulder of the curve might be
found further from the initial state than in the model described
might be the occurrence of home ranges of different sizes, particularly if, as seems reasonable, an occupied home range is larger
than an unoccupied one. Alternatively a hexagonal lattice rather

than a square one might be considered, although intuitively this
would lead to a shorter distance travelled.

N. R. FRENCH

It was considered that the boundaries of an animal's home range
might be expanded in the direction of an unoccupied area, but it
is likely that when an animal settles in that area, the boundaries
of the adjacent home range would be pushed back. So, in effect,
the boundaries would form the more or less regular lattice as
depicted. However, it does seem that if the unoccupied areas are
very much smaller than occupied ones, then greater dispersal
distances might result. It would be interesting to test the effect
of such conditions on the shoulder of the curve. Of course, home
ranges are neither regularly square nor regularly hexagonal. There
is no reason to believe that, for purposes of this model, one is a
better representation of home range than the other.

SOME USES AND LIMITATIONS OF MATHEMATICAL ANALYSIS IN PLANT ECOLOGY AND LAND MANAGEMENT

FREDERICK C. HALL
Forest Service
Oregon

SUMMARY

This investigation has shown (i) continuum tendencies exist in undisturbed vegetation and in seral vegetation; (ii) these continuua can be significantly related to measurable environmental factors by use of step-wise regression analysis; (iii) variation accounted for by regression analysis is often not high; (iv) more variability can be accounted for by limiting analysis to groups of similar plant community types than it can when all communities are combined; and (v) vegetation data means and standard deviations demonstrate considerable variability within community types. Numerical accuracy could be increased by prediction equations utilizing measured environmental factors.

These findings suggest several important factors to consider when ecological studies are designed to investigate land management problems: (a) A continuum tendency in vegetation should be assumed. Samples should be selected to represent natural variability between all types of plant communities. (b) Due to continuum tendencies, vegetation and related environmental factors should be analyzed by both association and regression analysis. (c) The ecologist should understand that a land manager deals with each and every acre under his jurisdiction; he manages real entities. Statistical analysis is based upon inferences about a population. The "population" which is managed must be sampled for existing variability so that statistical inferences will be appropriate for as much land area as possible. The ecologist must present his conclusions in a form useful to the land manager.

1. INTRODUCTION

Statistical inference and mathematical analysis are extremely useful tools for ecological investigations. To use these tools intelli- gently, the population from which inferences are to be drawn must be defined. The major factors influencing this definition are investigation objectives.

An additional factor should be considered for analysis of plant community characteristics. Does the investigator assume a continuum in environment and vegetation or does he classify and work with discreet groups of plant communities or species? Does he sample for variation in environment and vegetation or does he restrict sampling to plant communities representing vegetation groups? Either concept has legitimate uses in meeting management objectives, but it is imperative that the statistical method selected be appro- priate to the concept.

2. THE STUDY

2.1 Concepts and Objectives

A continuum in environment and vegetation was assumed for an eco- logical investigation of the Blue Mountains in Oregon and Southeastern Washington, U.S.A. (5 mill. A./2 mill. ha). The basic objective was to develop guidelines for land management. Specific objectives were the development of (i) guides for eval- uating livestock range condition, trend, production, and site potential for revegetation; (ii) guides for evaluating forest succession, site potential, growth, stocking, and reforestation; (iii) descriptions of plant community characteristics suitable for application in multiple use management; and (iv) a vegetation classification system suitable for mapping purposes which will differentiate sites and/or plant communities differing in management characteristics.

Objectives (i), (ii), and (iii) deal largely with measurable attributes of plants, such as species composition, pounds of forage production, tree site index, and basal area. Land managers use these data for evaluating and predicting livestock range condition, animal stocking, tree growth, allowable cut, and tree effects on water production.

Assumption of a continuum poses serious problems for land

management application. Vegetation continua cannot be effectively
mapped when objectives are both vegetation inventory and characterization. Mapping units must distinguish vegetation and environmental
characteristics significantly different for management. This
requires classification by associations, groups, or types. Land
management guides, such as range condition standards and timber
volume tables, are most easily understood and applied when developed
for vegetation groups or types.

2.2 Methods

Sample locations were selected to encompass variation in environment
and undisturbed vegetation. Plant communities were initially combined into very restricted groups for analysis according to close
similarities in vegetation and environment. Analysis group vegetation characteristics were evaluated using means and standard errors,
simple correlations, and step-wise regressions. Based upon these
evaluations, community types were devised according to three
criteria: similarity in floristics, environmental factors significant in management, and similarities suggested by regression
analysis. These community types are the basis for mapping units
and various vegetation guides. They are the "populations."

3. AN EXAMPLE OF THE ANALYSIS

Table 1 lists six original analysis groups and diagramatically shows
how they were combined for evaluation. Various combinations have
been designated as analysis steps to facilitate interpretation of
Table 4. These six analysis groups were selected for illustration
because all have Pinus ponderosa in common; Carex geyeri is often
a codominant herbaceous species in the mixed conifer/Calamagrostis
groups and is the dominant herb in Pinus ponderosa/Carex geyeri
groups. Analysis groups are named according to the dominant tree,
shrub (if present), and herb species. In Step I, each group was
analyzed individually. The most similar pairs of groups were
combined in Step II, and these pairs were combined to produce groups
of four in Step III. In Step IV, larger combinations of greater
dissimilarity were formed; and in Step V, all Pinus ponderosa
analysis groups were combined to form a single analysis unit.
Table 1 is a greatly simplified example of analysis for 38 groups
from which 22 community types were developed for the Blue Mountains.

Table 1. Step-wise procedure used to progressively combine analysis groups for step-wise regression analysis of selected vegetation characteristics. In step I, each analysis group was analyzed separately. Two of these step I groups were combined for analysis in step II as shown by the connecting lines. Additional combinations were made in steps III, IV, and V.

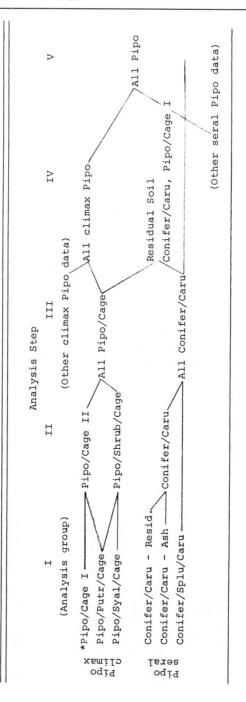

* Abbreviated plant generic names. Communities named for dominant species in tree, shrub (if present) and herb layers. Pipo = Pinus ponderosa, Putr = Purshia tridentata, Cage = Carex geyeri, Syal = Symphoricarpos albus, Conifer = P. ponderosa, Psedudotsuga menziesii and Abies grandis mixture, Caru = Calamagrostis rubescens, Splu = Spirea lucida, Resid. = residual soil, Ash = ash soil.

The following 19 vegetation characteristics were evaluated:
density and composition of Calamagrostis rubescens and Carex geyeri;
composition of Arnica cordifolia, Achillea millefolium, and Hieracium albiflorum; crown cover, site index, and basal area of Pinus
ponderosa, Pseudotsuga menziesii, and Abies grandis; and forage
production, growth basal area, and total basal area. Growth basal
area is defined as that basal area on a specific site which trees
average 15 rings per inch in radial growth.

3.1 Evaluation of a Single Analysis Group

One of the six original analysis groups was the mixed conifer/
Calamagrostis rubescens - ash soil group. This group is dominated
by Pinus ponderosa with Pseudotsuga menziesii and Abies grandis
present to codominant and the rhizomatious grass Calamagrostis
rubescens; shrubs are absent. Soils are derived from fine sandy
volcanic ash overlying bedrock or soil.

Tables 2 and 3 are step-wise regression and simple correlation
analyses. In Table 2, dependent variables were vegetation characteristics. Independent variables were environmental factors
representing topographic, geographic, and soil characteristics.
Tree crown cover was used as an independent variable for shrub and
herb vegetation only.

Of 32 environmental factors tested, 21 were significantly
associated (5% or 1% level) with the 19 vegetation characteristics.
Slope direction was significant in 8 of the 19 vegetation characteristics. Soil C horizon color was significant six times. Soil A
horizon color was most consistent, being significantly related to
site index of all three tree species tested.

On the other hand, eight environmental factors appeared only
once. In only one case was a single factor important, that of
bedrock fracturing for Achillea millefolium composition. The
maximum number of significant environmental factors associated
with a vegetation characteristic was four, a situation occurring
four times.

Table 2 clearly demonstrates multiple factor relationships in
the mixed conifer/Calamagrostis - ash soil analysis group. No single
environmental factor was consistently associated with all vegetation
characteristics. One exception to this statement is the consistency
of soil A horizon color with tree site index. This, however, was
the only consistent factor of the five associated with tree site
index.

Table 2. Variation accounted for by step-wise regression analysis of selected vegetation characteristics in the conifer/Calamagrostis – ash soil analysis group†

Increase in R^2, Rank and
Variation Accounted for by Environmental Factor

Vegetation Characteristic → Environmental Factor ↓	Caru Den Com	Cage Den Com	Arco Com Com Com	Acmi Com Com Com	H1al Com Com Com	Forage Production	Pipo Cover	Fir Cover	Total Cover	Pipo S. I.	Psme S. I.	Abgr S. I.	Pipo B. A.	Psme B. A.	Abar B. A.	Growth B. A.	Total B. A.
Vegetation																	
Pipo Cover																	
Fir Cover																	
Total Cover	*18^3		25^1		*10^3	*43^1											
Herb Specie										37^1		66^1	38^1				
Slope Position																	
Micro Topog.															63^1		
Elevation			11^3						12^2							16^2	17^3

Table 2 (continued).

Slope Direction	*23[1]	*20[2]	9[3] 20[1]	*19[2]	12	9[3] 24[1]
% Slope			13[2] *12[4]			
Township					10[2]	
Range						
Total Soil Depth	23[1]				23[2]	
Eff. Soil Depth	*22[2]	*17[2]				
% Soil Stone						
Ash Depth						
Buried Soil				19[2]		
A pH			*18[1]	*9[3] *57[1] *27[2]		*19[2]
A Texture						
A Structure						
A Color						
A Stone						
A Depth						
B pH	12[3]				28[2]	*9[4]
B Texture				14[3]		*17[4]
B Structure	*12[1]					
B Color					64[1]	
B Stone						
B Depth						

Table 2 (continued).

C pH	25^1																	25^1	
C Texture		14^2		$*11^2$													11^3	$*10^2$	
C Structure																		$*12^1$	
C Color	22^2			$*10^3$			28^1	$*12^3$	12^2										
C Depth	9^4																		
C Stone		$*9^3$																	
Bedrock Fracture					$*22^1$														
TOTAL VARIABILITY Accounted for (R^2)	73	46	58	33	53	22	58	67	53	31	32	65	90	93	48	92	98	59	77
Data Mean	7.6	46	4.1	26	7.3	2.7	1.8	307	40	24	61	71	72	46	125	42	49	90	157
Data S. D.	4.3	19	3.8	18	11.0	3.4	2.1	90	14	24	15	6.5	7.9	4.2	71	44	40	25	68
Standard Error of Estimate	2.4	15	2.6	16	8.0	3.1	1.4	55	11	21	13	4.1	3.2	1.3	54	14	8	12	36

† An asterisk indicates negative correlation. Increase in R^2 is shown by the number (i.e. 18). Order in which each environmental factor entered the equation is shown as a power of the number (i.e. 3rd to enter).

Table 3. Variation accounted for by simple correlations in the conifer/Calamagrostis - ash soil analysis group

Environmental Factor \ Vegetation Characteristic	Pipo Cover	Pipo B. A.	Fir Cover	Fir B. A.
No. fire scars		29		*38
Years since last fire		*19		21
Fir cover	*56			
Total cover			67	
Fir B. A.		*32		

Table 3 lists simple correlations used to evaluate forest succession in this mixed conifer/Calamagrostis - ash soil analysis group. Fire scars on Pinus ponderosa were used to count the number of ground fires which damaged the tree and estimate the years since the last ground fire. These correlations suggest that Pinus ponderosa basal area increases with increasing number of fire scars, whereas fir basal area decreases. Fir basal area increases and ponderosa basal area decreases with increasing number of years since fire. Pinus ponderosa crown cover decreases as fir crown cover increases; a similar but less accurate relationship is found between Pinus ponderosa basal area and fir basal area. Accuracy is less because fir dominates reproduction where it contributes little to basal area. These data suggest that fir is increasing and Pinus ponderosa is decreasing in this analysis group. Apparently, this change is being caused by fire control which is preventing the formerly natural ground fires.

Simple correlation and step-wise regression analysis of the mixed conifer/Calamagrostis - ash soil analysis group may be summarized as follows:

---

* An asterisk denotes a negative correlation.

(a)   A successional continuum is suggested by simple correlation analysis.

(b)   An environmental continuum tendency is suggested within this analysis group by highly significant step-wise regressions associating vegetation characteristics with environmental factors.

(c)   Each plant species and each characteristic of a species is associated with different environmental factors. This implies an immensely complex relationship in plant communities.

3.2   Comparison of Several Analysis Groups

Two characteristics, Pinus ponderosa site index and growth basal area, were selected to compare six analysis groups and illustrate progressive analysis of group combinations (Table 1). Table 4 shows step-wise regression results for each analysis step.

In Step I, 65-98% of Pinus ponderosa site index variability is accounted for. Thirteen environmental factors appear in the six analysis groups. Soil B horizon texture is most consistent, occurring in three of the six groups. Six of the thirteen environmental factors appeared only once. Single factor relationships occurred twice. This suggests that a plant species characteristic is not associated with the same environmental factors throughout its ecological range. The concept of a "primary controlling factor" is not supported by this data. Relationships of environmental factors to species characteristics vary with the kind of plant community.

In Step I, 64-89% of growth basal area variation is accounted for by environmental factors. Only bedrock fracturing was reasonably consistent, occurring in three of the six analysis groups. Eight of the twelve environmental factors appeared only once. As with site index, no single environmental factor was consistently associated with growth basal area.

Only four of the 21 environmental factors are significantly associated with both site index and growth basal area in Step I. These observations suggest the following:

(i)   Continuum tendencies are suggested in all original analysis groups by significant associations of vegetation characteristics with environmental factors.

(ii)   Different environmental factors are associated with Pinus ponderosa site index than with growth basal area.

(iii)   Significant environmental factors will differ by analysis groups for the same species characteristic, which suggests that

Table 4. Variability accounted for by step-wise regression analysis for Pinus ponderosa site index and growth basal area. Analysis groups combined in each analysis step are shown in Table 1.†

Increase in R^2, Rank and Variation Accounted for by Environmental Factor

Site Index Analysis Step

Environmental Factor	I						II			III		IV		V
Analysis Group	Pipo/Putt Cage	Pipo/ Cage I	Pipo/Syal/ Cage	Pipo/Caru- Resid.	Pipo/Caru- Ash	Pipo/Splu/ Caru	Pipo/ Cage II	Pipo/Shrub/ Cage	Pipo/ Caru	All Pipo/ Cage	All Pipo/ Caru	Residual Cage/Caru	Pipo Climax	All Pipo
Subord. Specie				60[1]	37[1]	20[1]								
Slope Position														
Micro Topog.														
Elevation														
Slope Direct.						14[2]								
% Slope							30[1]							
Township														
Range						9[5]								

Table 4 (continued).

	1	2	3	4	5	6
Total Soil Dpt.						
Eff. Soil Dpt.						
% Soil Stone	10^4			23^2		$*21^2$
Ash Depth			19^2	11^4		
Buried Soil			12^3			
A pH						
A Texture				18^2	21^2	
A Structure			$*14^3$	56^1	47^1	$*10^2$
A Color			$*9^3$	$*17^2$	$*9^2$	
A Stone						
A Depth		13^3				9^3
B pH	$*12^3$					
B Texture		$*14^2$	$*8^6$	$*14^4$	$*10^3$	
B Structure	$*24^2$	91^1	$*81^1$			
B Color						
B Stone						
B Depth						
C pH						
C Texture						
C Structure						
C Color						
C Depth	$*52^1$		30^1	16^1	47^1	34^1

Table 4 (continued).

	98	91	81	87	65	74	90	77	49	80	39	35	68	34
C Stone Bedrock Fracture									32[1]					
TOTAL VARIABILITY ACCOUNTED FOR														
Data Mean	61	63	69	72	71	70	62	66	72	64	71	60	68	65
Data S. D.	4.9	3.4	9.0	7.8	6.5	12.0	4.4	7.7	7.1	6.7	8.3	9.1	8.2	10.0
Standard Error of estimate	1.0	1.2	4.5	3.1	4.1	6.1	1.9	4.0	5.3	3.3	6.6	7.6	4.8	8.2

Growth Basal Area Analysis Step

	98	91	81	87	65	74	90	77	49	80	39	35	68	34
Subord. Specie														
Slope Position														
Micro Topog.														
Elevation					16[2]	18[2]			8[4]		8[3]	*20[2]	27[1]	
Slope Direct.					9				12[3]		12[2]		12[3]	
% Slope	33[2]													
Township						10[4]		86		70[1]				
Range							32[2]							44[1]
Total Soil Dpt.							10[3]							
Eff. Soil Dpt.														

Table 4 (continued).

Variable				
% Soil Stone				
Ash Depth				
Buried Soil				
A pH		*47[1]		
A Texture	8[3]			
A Structure				
A Color				
A Stone				
A Depth				
B pH	*9[4]			
B Texture				*10[3]
B Structure	*14[2]			
B Color	*36[1]			
B Stone	20[3]			
B Depth	38[1]	14[2]	18[1]	
C pH	79[1]		10[2]	
C Texture	24[1]			
C Structure	25[1]	12[3]		
C Color	8[4]	20[1]		
C Depth	84[1]			12[2]
C Stone				34[1]
Bedrock Fracture				

Table 4 (continued).

TOTAL VARIABILITY ACCOUNTED FOR	89	79	84	68	59	64	89	86	54	80	39	64	52	44
Data Mean	44	72	96	88	90	130	55	64	90	66	93	73	53	63
Data S. D.	5.2	8.4	11	30	25	20	15	27	27	24	28	29	22	30.1
Standard Error of estimate	2.3	4.4	5.3	13	12	9	6.0	11	20	11	23	19	16	23.9

† An asterisk indicates a negative correlation. Increase in R^2 is indicated by the number (i.e. 12). Order in which the factor entered the equation is shown as a power of the number (i.e. 3rd to enter).

original analysis groups may represent unique sets of environmental conditions.

(iv)   The analysis group designation appears to be a useful means for evaluating site index and growth basal area. Each of these groups can be defined as a population. If they are, biological or numerical criteria for assigning plant communities to an analysis group must be stipulated.

3.3 Analysis of Combined Groups

Steps II through V illustrate step-wise regression results obtained by combining similar analysis groups. In general, as increasingly dissimilar groups are combined, variation accounted for becomes less and standard error of estimate becomes greater. If all the samples came from the same population, we would expect the standard error of estimate to decrease with increasing degrees of freedom--but the opposite occurs with this data.

As analysis progresses to Step V, the kind and importance of environmental factors associated with site index and growth basal area change. Of 21 factors in Step I, only 12 appear in Steps II through V, and 11 other environmental factors become significant. These data and observations suggest:

(i)   Continuum tendencies in vegetation occur within initial analysis groups and also within combined groups.

(ii)   Environmental factors associated with these vegetation continuua change and become more obscure as increasingly dissimilar analysis groups are combined.

(iii)   Very seldom are specific environmental factors consistently associated with particular vegetation characteristics within either single or combined analysis groups.

(iv)   Analysis groups, when classified on the basis of undisturbed vegetation similarity, seem to reflect rather restricted and unique environments with apparently limited vegetation characteristics. This tends to support some ecologist's contention that undisturbed vegetation reflects the sum of the effective environment.

(v)   These unique environments may be an important factor in the greater variations accounted for and lower standard errors of estimate found in original analysis groups. This seems particularly significant since degrees of freedom are lowest in the original community groups and highest in combinations thereof.

(vi)   In general, these data suggest a series of continuum gradients rather than a single, universal continuum. Classification into

groups can be justified as an aid to analysis of vegetation--
environmental relationships. Designation of a group or type need
not be restricted to floristics; other objectives such as specific
land management problems can and often should influence categori-
zation.

4. INTERPRETATION

This study aimed to develop broad inclusive types, easily used and
interpreted by land managers, without sacrificing precision. Progressive combination and evaluation of analysis groups were used
to assess accuracy. It was hoped that larger combined groups could
serve as working classification types. Unfortunately, these few,
broad types did not materialize. In some cases, no combination was
possible. In other instances, results permit combination of analysis groups into community types. Of the 38 original analysis groups,
22 community types had to be developed.

4.1 Development of Community Types from Analysis Groups

   Analysis groups listed in Table 4 are used to illustrate development of community types. Resultant combinations are shown in Table
5. Vegetation data from the combination of conifer/Calamagrostis--
ash and residual soil groups met accuracy criteria. Floristic
considerations would permit combination. However, these soils
differ greatly in their erodability characteristics. Ash soils,
due to low bulk density and poor structural characteristics, are
extremely susceptible to erosion after logging. Therefore, mixed
conifer/Calamagrostis - ash soil was established as a community type.
Conifer/Calamagrostis - residual soil numerical data and soil characteristics were similar to Pinus ponderosa/Symphoricarpos albus/
Carex geyeri. Symphoricarpos was less common in conifer/Calama-
grostis and Calamagrostis was less common in Pinus/Symphoricarpos.
However, herbaceous dominants in both analysis groups react similarly under livestock grazing and the shrubby Symphoricarpos does not
tend to increase with overgrazing (downward range trend). These
analysis groups were combined for the conifer/Calamagrostis-Sym-
phoricarpos community type.
   Mixed conifer/Spiraea lucida/Calamagrostis had similar site index
values to other Calamagrostis groups; however, growth basal area was
significantly higher and shrub reaction to overgrazing was

Table 5. Analysis groups combined to form community types

Analysis Group	Community Type
Pipo/Putr/Cage	
*Pipo/Putr/Feid	Pipo/Putr/Cage-Feid
Pipo/Cage	Pipo/Cage
Pipo/Syal/Cage	
Conifer/Caru-residual	Conifer/Caru-Syal
Conifer/Caru-Ash	Conifer/Caru-Ash
Conifer/Splu/Caru	Conifer/Splu/Caru

dramatically different. Spiraea and rhizomatious Symphoricarpos
increase tenfold with range deterioration and clearly dominate
poor condition ranges. These shrubs greatly retard a return to the
former grass dominance (upward range trend) when livestock utilization is properly controlled. Conifer/Calamagrostis - ash and
residual soil groups respond well to proper livestock management.
Conifer/Spiraea/Calamagrostis was established as a community type.

   Pinus ponderosa/Purshia tridentata/Carex geyeri and P. ponderosa/
Carex geyeri groups are similar in herbaceous data, forage production and Pinus ponderosa site index; however, growth basal area is
highly significantly different. Pinus/Carex was selected as a
community type. Pinus/Purshia/Carex was combined with Pinus/Purshia/
Festuca idahoensis because all numerical data were similar, shrubs
and their characteristics were nearly identical, and herbaceous
dominants react similarly under grazing. This combination was
designated the Pinus ponderosa/Purshia tridentata/Carex-Festuca
community type.

4.2 Community Types

Each community type has management characteristics which differ
significantly from other types. Differences are based upon many

----

*Not shown in Table 4.

factors such as species dominance, community successional characteristics, productivity, numerical values like average basal area
or regression relationships, and environmental limitations such as
soil erodability and suitability for range revegetation or reforestation.

Community types are described with photographs, narrative
explanations of vegetation and environmental characteristics, and
management opportunities and limitations. With this material, the
land manager can grasp the nature and complexities of his vegetation
resource. He has a type "standard" by which he can interpret
variability. He has a mapping unit with which to inventory kinds
of significantly different vegetation--environmental situations.
He has the basic biological tools with which he can intelligently
and accurately refine his management.

4.3 Management Considerations

The land manager is responsible for each and every acre under his
jurisdiction. He is concerned with real entities because his
business is treatment of specific stands of vegetation growing
upon defined pieces of ground.

Statistical analysis, on the other hand, deals with inferences
about a population. The land manager's "population" is all of his
land and its associated vegetation and environmental factors. To
infer something about this population, we must sample the existing
variability and evaluate this variability with statistical methods.
Prediction equations derived from this analysis will hopefully
apply to most areas and will offer the land manager a means to
refine his interpretations.

Prediction equations raise a question. How well can standards
be established so different observers will obtain similarity in
measurement or estimation of environmental factors? For example,
guidelines must be given for estimating percent soil stone and soil
texture and structure. Unless these standards can be established,
formulas cannot be effectively applied. Training sessions help
alleviate this problem.

SOME ASPECTS OF HETEROGENEITY IN GRASSLANDS OF CANTAL (FRANCE)

M. GODRON
PH. DAGET
J. POISSONET
P. POISSONET
Centre d'Etudes Phytosociologiques et Ecologiques
Montpellier 34, France

SUMMARY

The distributions of frequencies of the species of a vegetal community can be expressed by different ways (curves of Raunkiaer, of Henry, of Estoup and Zipf, of Gini-Lorenz, etc., and indexes of diversity), but they seem always to reveal a very general rule. This rule can be connected with the distribution of the frequencies of one species in several sites.

The relative importance of the different species in a plant community can be expressed by the "frequencies" (absolute or relative) of these species. It is possible to connect the distributions of the frequencies of the species with several models and to compare these models before sketching possible extensions of the current research.

1. SOME EXPRESSIONS OF THE HIERARCHY OF THE SPECIES OF A PLANT
    COMMUNITY

1.1 Review of Some Previous Works

It was shown by Ph. Daget [2] that the lognormal distribution used
by zoologists (Preston, [17] seems also applicable to the frequencies of species in certain plant communities. J. Poissonet [14]
confirmed that the frequencies of the species in a great number of
dense herbaceous formations can be expressed often by a concentration curve of the "20-80" type. P. Poissonet [15], when comparing
several methods of studies of grassland vegetation, also used
several distributions of the frequencies of the species. The notion
of frequency itself has been discussed previously (Godron, [7]) and
an estimation of the information given by the observation of the
frequency of species had been proposed (Godron; [4], [5] and [6]).

   In order to see how these expressions relative to the frequencies
of species are bound together, the simplest way is to employ them
simultaneously in a case for which we dispose of a very complete set
of observations:

   Nine different methods of vegetation analysis were used in a
Grassland community of Cantal, at an altitude of 950 m. [8]. The
most efficient method to obtain a distribution of the frequencies
for almost all species is to observe their presence under consecu-
tive segments. This method has provided, for 1024 segments of
.25 m., the following results:

Table 1.

	Absolute frequency	Relative frequency
Festuca rubra L.	1006	98.6
Agrostis vulgaris With.	1003	98.3
Trifolium repens L.	963	94.4
Galium verum L.	690	67.6
Luzula campestris L.	634	62.2
Poa pratensis L.	509	49.9
Trisetum flavescens (L.) Ry	421	41.3
Ranunculus bulbosus L.	411	40.3

Table 1 (continued).

Anthoxanthum odoratum L.	286	28.0
Stellaria graminea L.	280	27.5
Achillea millefolium L.	178	17.5
Phleum pratense L.	164	16.1
Trifolium pratense L.	146	14.3
Conopodium majus (Gouan) Loret et B.	145	14.2
Cerastium caespitosum Gilib.	138	13.5
Potentilla verna L.	107	10.5
Cynosurus cristatus L.	93	9.1
Lolium perenne L.	89	8.7
Ajuga reptans L.	67	6.6
Holcus lanatus L.	66	6.5
Plantago lanceolata L.	65	6.4
Lotus corniculatus L.	55	5.4
Campanula rotundifolia L.	52	5.1
Carex caryophyllea Latourr.	49	4.8
Rumex acetosa L.	48	4.7
Veronica chamaedrys L.	41	4.0
Campanula glomerata L.	16	1.6
Pseudoscleropodium purum (Hedw.) Fleisch.	14	1.4
Campanula rapunculus L.	12	1.2
Ranunculus acer L.	11	1.1
Hypochoeris radicata L.	10	1.0
Genistella sagittalis (L.) Gams	8	0.8
Taraxacum officinale Weber	7	0.7
Poa tirivalis L.	7	0.7
Rhytidiadelphus squarrosus (Hedw.) Warnst.	7	0.7
Alchimilla vulgaris L.	6	0.6
Hieracium pilosella L.	6	0.6

Table 1 (continued).

Thymus serpyllum L.	5	0.5
Veronica officinalis L.	5	0.5
Galium asperum Schreb.	4	0.4
Centaurea pratensis Thuill.	4	0.4
Deschampsia caespitosa (L.) P.B.	4	0.4
Avena pubescens Huds.	3	0.3
Veronica serpyllifolia L.	3	0.3
Viola tricolor L.	2	0.2
Hieracium auricula L.	1	0.1
Sarothamnus scoparius (L.) Wimmer	1	0.1
Cirsium eriophorum (L.) Scop.	1	0.1
Dactylis glomerata L.	1	0.1
Hypnum cupressiforme Hedw.	1	0.1
Plantago media L.	1	0.1

1.2   Main Methods of Graphical Expression of the Distribution of Specific Frequencies

1.2.1   From the Table 1, we can deduce the classic histogram of Raunkiaer, simply be regrouping the relative frequencies in five classes (Fig. 1). Raunkiaer's criterion of homogeneity is very empiric and the interpretation of the histogram could proceed as proposed by Guinochet [11] or Dagnelie [3], but in any case, the regrouping of the frequencies in "linear" classes is arbitrary (Greig-Smith, [10]).

1.2.2   It is rather convenient to start from the logarithms of frequencies (Preston, [16]; J. Daget, [1]; Ph. Daget, [2]) which leads to Figure 2.*

1.2.3   Preston [16] observed that the distribution of logarithms of frequencies seems often normal. In order to test the normality of the distribution presented in Figure 2, Ph. Daget uses the Henry's graphical method (Fig. 3). The curve thus obtained can be deducted of the preceding curve by cumulating the ordinates--which is equiva-

lent to calculating an integral function--and by carrying out an
anamorphosis. This curve could be transformed in an "S" curve on
semilogarithmic paper.

1.2.4 It is also possible to express the frequencies according to
the method of Estoup--which gives a straight line--or to the method
of Estoup and Zipf (Mandelbrot, [12]) which suggests a notion
similar to that of "temperature," in order to give an idea of the
"freedom" of the species, in the observed samples (Fig. 4*).

1.2.5 The straight line of Estoup is expressed by the relation

$$\log y = \log a - b \log x .$$

According to this, the frequency would be an homographical function
of a power of the rank, the species being arranged in the decreasing
order of their frequencies. In the present case, this relation
would be acceptable at 1% significance level.

It is equivalent to arrange the species in an increasing order
of the frequencies or to carry out on Figure 5 a symmetry with
respect to a parallel to ordinate axis, which gives the curve B.*
This latter, designed with bilogarithmic coordinates, would become
a curve which would give indications on the distribution of the
frequencies of the more frequent species.

1.2.6 Taking for the abscissae a Gaussian-logarithmic scale, the
curve B lends itself to another graphical expression of Henry's
method (Fig. 6). This latter shows also that the distribution of
the frequencies does not deviate significantly from a lognormal
distribution.

1.2.7 It is possible to obtain a very simple expression of the
hierarchy of the species by plotting the ranks on the abscissae
and the logarithms of the frequencies on the ordinates (Fig. 7).
When the size of the samples increases, the curve is deformed in
the direction shown by the arrow. The size for which the relation
is linear seems to present remarkable properties.

---

* When the size of the samples increases, the curve is displaced
  in the direction of the arrow.

1.2.8 The cumulation of the frequencies and their expression in percentage of the sum of the frequencies transform the curve B in a "concentration curve" of Gini-Lorenz (Fig. 8). The coordinates of the point of maximal curvature (close to 20-80%), are characteristical parameters of this "concentration curve."

1.3 Other Expressions

Moreover, the frequencies of the species allow one to calculate an index of heterogeneity deduced from the information theory (Godron, [5]) and, as far as they can be assimilated to probabilities (Godron, [7]), to calculate several diversity indexes (Margalef, [13]). Compared to the preceding curves, such indexes are the equivalents or climatic indexes with respect to climagrams. They lead to classifications which are simpler but more schematic than those deducted from the curves.

2. DISTRIBUTION OF THE FREQUENCIES OF A SPECIES IN SEVERAL SITES*

The graphical expressions of the preceding paragraph present various advantages but they all present a number of common characters sufficient to make us try, notwithstanding the difficulty, to draw a general rule: the form of all these curves confirms always that the distribution of frequencies is not gaussian and that the frequent species are more frequent than would give a gaussian distribution.

All that reflects, in a way, a multiplicative effect, which may proceed from the fact that a species is able, in a station, to multiply as much as it is already frequent in this station and to reduce proportionally the place free for the rare species.

In order to get a nearer view of this hypothesis, we examined the distribution of the frequencies of some species in 76 "relevés" collected in the grasslands of Cantal; these distributions seem also to be lognormal distributions (Fig. 9).

In another way it is possible to tell that, when a species is established in a station, its frequency is distinctly higher than it were to be expected if the frequencies of this species were randomly distributed around the average value. This catches up

---

* For the definition of "Site" see "Station" in Godron et al [9].

with the aphorism of botanists: "The rare species are frequent in the sites where they do exist."

It seems even possible to go further: the higher frequencies are often located above the Henry's straight line, and the lower frequencies below. For certain species, the logarithm of the frequency is nearly proportional to the rank (Figure 10, where the scales on the coordinate axis are "analogue" to those of Figure 7). These curves would perhaps allow one to detect a certain heterogeneity in the whole of the "relevés"; this has been observed firstly in forests of Sologne (France) but our examples, however, are not yet sufficiently numerous to bring more than a working hypothesis.

3. CONCLUSION

Finally, it seems useful to lay stress upon two points:

(i) It is generally useful to compare several expressions of distributions of frequencies of the species.

(ii) Some distributions of frequencies can be rapidly used in the field as a guide for the study of the heterogeneity of the vegetation.

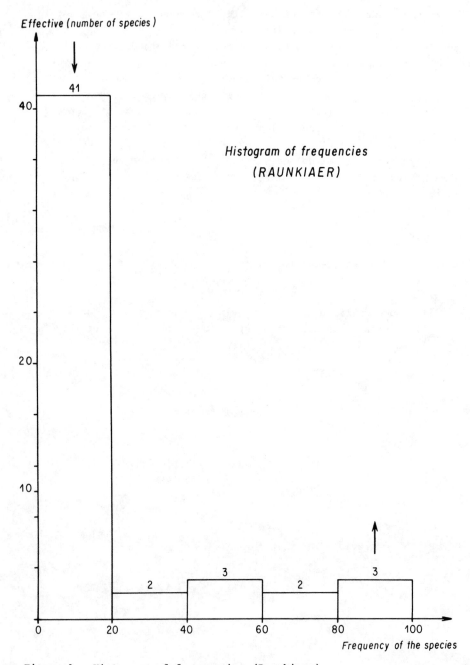

Figure 1.   Histogram of frequencies (Raunkiaer)

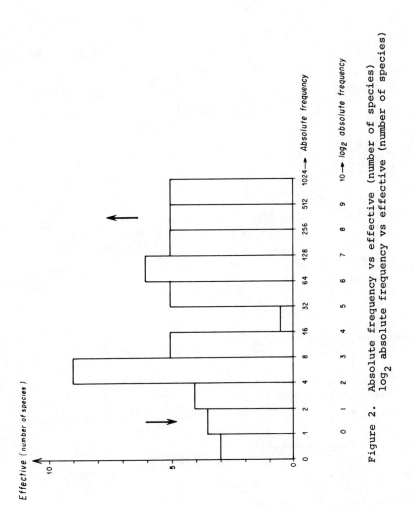

Figure 2.   Absolute frequency vs effective (number of species)
log$_2$ absolute frequency vs effective (number of species)

Cumulated number of species (%oo)

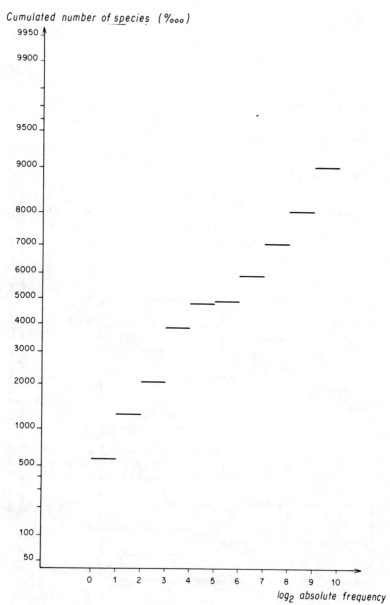

Figure 3.   Log₂ absolute frequency vs cumulated number of
            species (0/000)

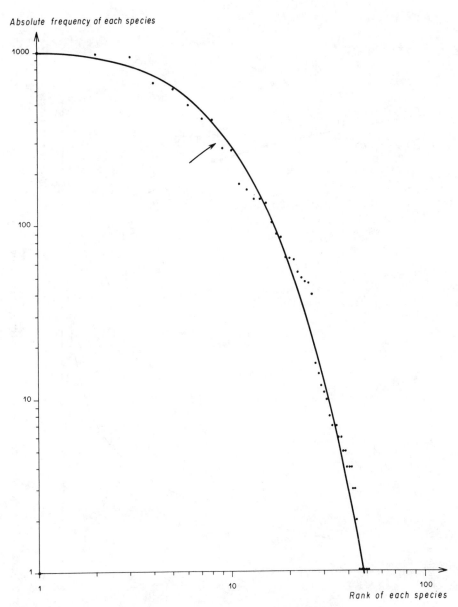

Figure 4. Rank of each species vs absolute frequency of each species

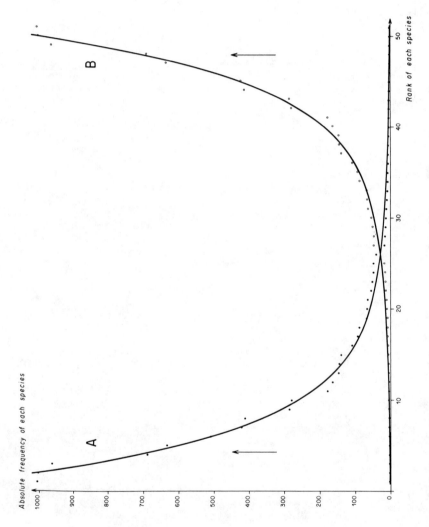

Figure 5.   Rank of each species vs absolute frequency of each species

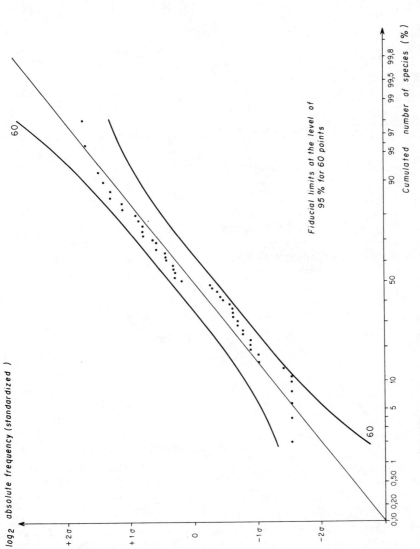

Figure 6.   Cumulated number of species (%) vs log$_2$ absolute frequency (standardized)

log. absolute frequency of each species

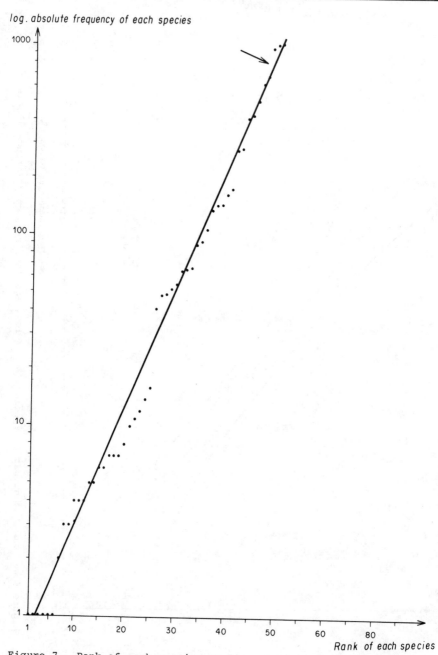

Figure 7.   Rank of each species vs $\log_e$ absolute frequency of
each species

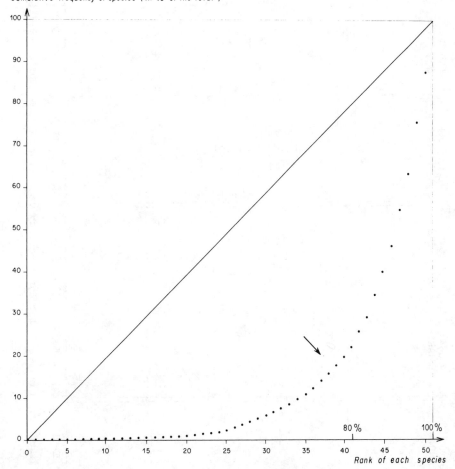

Figure 8.   Rank of each species vs cumulative frequency of species
             (in % of the total)

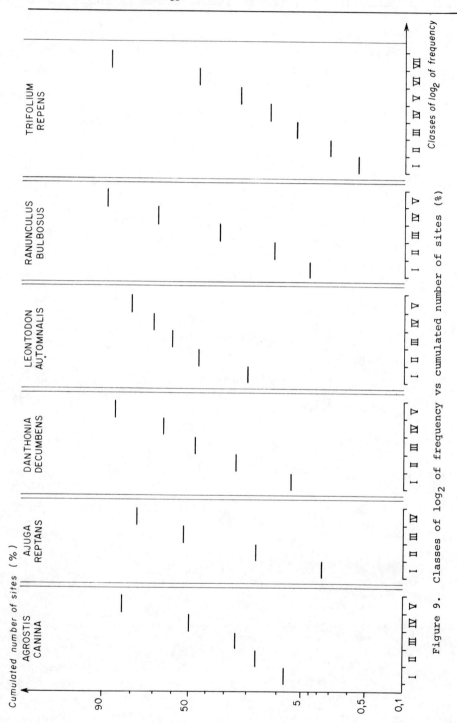

Figure 9.   Classes of $\log_2$ of frequency vs cumulated number of sites (%)

Number of sites (ranked by increasing frequency
for each species)

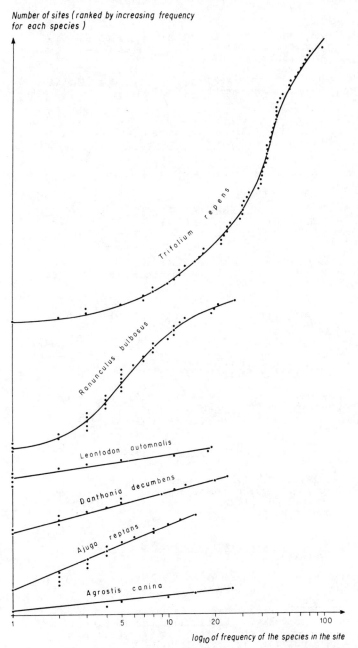

log₁₀ of frequency of the species in the site

Figure 10. Log_{10} frequency of the species in the site vs number of sites (ranked by increasing frequency for each species)

REFERENCES

[1]   Daget, J. 1966. Abondance relative des poissons dans les
       plaines inondées par le Benoué à hauteur de Garoua (Cameroun).
       Bull. I.F.A.N. 28(A1):245-258.

[2]   Daget, Ph. 1968. Quelques remarques sur les distributions de
       fréquences spécifiques dans les phytocénoses. Oecol. Plant.
       3:299-312.

[3]   Dagnelie, P. 1962. Etude statistique d'une pelouse à Brachy-
       podium ramosum. Les liaisons interspécifiques: présence
       et absence des espèces. Bull. Serv. Carte Phytog., série B,
       7, 1:85-97.

[4]   Godron, M. 1965. Actes du Colloque de Copenhague. UNESCO:
       484-485.

[5]   _____. 1966. Essai d'application de quelques éléments
       simples de la théorie de l'information à l'étude de la
       structure et de l'homogénéité de la végétation. Thèse,
       C.E.P.E., C.N.R.S., Montpellier (France), 67 p.

[6]   _____. 1966. Application de la théorie de l'information
       à l'étude de l'homogénéité et de la structure de la végéta-
       tion. Oecol. Plant. 1:187-197.

[7]   _____. 1968. Quelques applications de la notion de
       fréquence en écologie végétale. Oecol. Plant. 3:185-212.

[8]   _____, Poissonet, J. and Poissonet, P. 1967. Méthodes
       d'étude des formations herbacées denses. Essais d'appli-
       cation à l'étude du dynamisme de la végétation. C.E.P.E.,
       C.N.R.S., Montpellier (France), 28 p.

[9]   _____, _____ and _____. 1968. Code pour
       le relevé méthodique de la végétation et du milieu (Prin-
       cipes it transcription sur cartes perforées). Ed. C.N.R.S.,
       42 fig., 292 p.

[10]   Greig-Smith, P. 1954. Quantitative plant ecology. Butter-
       worths scientific publications, London, 198 p.

[11]   Guinochet, M. 1954. Sur les fondements statistiques de la
       phytosociologie et quelques-unes de leurs conséquences.
       Aktuelle Probleme des Pflanzensoziologie. Veröffenclichun-
       gen des Geobotanischen Institutes Rübel in Zürich, Heft.
       29:41-67.

[12]   Mandelbrot, B. 1957. Linguistique statistique macroscopique.
       Presses universitaires de France, Paris, 80 p.

[13]   Margalef, R. 1957. La teoria de la informaciòn en ecologia.
       Mem. R. Acad. Cienc, Barcelone, 32(13):373-449.

[14]   Poissonet, J. 1968. Recherche sur les lois générales d'équi-
       libre dans la composition floristique des formations
       herbacées denses. Premiers résultats et hypothèses.
       C.E.P.E., Montpellier (France), 18 p.

[15] Poissonet, P. en prép. Méthodes d'analyse des formations
      herbacées denses permanentes. Leur comparaison, leur
      intérêt agronomique. (C.E.P.E., env. 100 p.)

[16] Preston, F. 1948. The commonness and rarity of species.
      Ecology. 29:254-283.

[17] ─────────. 1962. The cannonical distribution of commonness
      and rarity. Ecology. 43(2):185-215; (3):410-432.

A BIBLIOGRAPHY OF SELECTED PUBLICATIONS ON POPULATION DYNAMICS, MATHEMATICS AND STATISTICS IN ECOLOGY

VINCENT SCHULTZ
Washington State University
Pullman, Washington

SUMMARY

This article is a bibliography of selected books and monographs on Population Dynamics, and Mathematics and Statistics in Ecology which play important roles in the study of Mathematical and Statistical Ecology.

1. INTRODUCTION

At the suggestion of Dr. G. P. Patil I have prepared the following
bibliographies which I hope will be of interest at this stage of
development of the subject of statistical ecology. For rather obvious reasons, I have restricted entries to books and monographs and
those not dealing with individual species.

The periodical literature contains many excellent articles but to
have included them was impossible in the time available. A bibliography of this literature is in preparation.

Initially I attempted to prepare a bibliography on mathematics
and statistics for biologists but because of the rather broad audience to whom such publications are directed and the extensive literature, I decided to discontinue the task and restrict my efforts to
ecology. The listings do not include books on population genetics
and include only limited reference to human ecology.

The bibliographies necessarily reflect my interests and material
available to me.

I am grateful for the assistance of W. Baltensweiler, E. Batschelet, R. R. Giese, M. Godron, C. S. Holling, D. B. Mertz, G. P. Patil,
O. Persson, E. C. Pielou, T. R. E. Southwood and W. G. Warren.

2. SELECTED BOOK REFERENCES ON POPULATION SYNAMICS

[1]   Andrewartha, H. G. 1961. Introduction to the Study of Animal
         Populations. Methuen, London. xvii, 281 pp.

[2]   ───────. and Birch, L. C. 1954. The Distribution and
         Abundance of Animals. Univ. of Chicago Press, Chicago,
         Illinois. xv, 782 pp.

[3]   Beverton, R. J. H. and Holt, S. J. 1957. On the Dynamics of
         Exploited Fish Populations. Fishery Investigations Series
         II, Volume XIX. Her Majesty's Stationary Office, London.
         533 pp.

[4]   Bodenheimer, F. S. 1958. Animal Ecology To-Day. Dr. W. Junk,
         The Hague. 276 pp.

[5]   Boughey, A. S. 1967. Population and Environmental Biology.
         Dickenson, Belmont, California. x, 108 pp.

[6]   ───────. 1968. Ecology of Populations. Macmillan, New
         York. viii, 135 pp.

[7]   Browning, T. O. 1963. Animal Populations. Harper and Row,
         New York. 127 pp.

[8]   Clark, L. R., Geier, P. W., Hughes, R. D. and Morris, R. F.
      1967. The Ecology of Insect Populations in Theory and
      Practice. Methuen, London. xiii, 232 pp.

[9]   Cragg, J. B. and Pirie, N. W. (eds.). 1955. The Numbers of
      Man and Animals. Oliver and Boyd, Edinburgh. viii, 152 pp.

[10] Cushing, D. H. 1968. Fisheries Biology: A Study in Popula-
      tion Dynamics. Univ. of Wisconsin Press, Madison, Wisconsin.
      xii, 200 pp.

[11] Elton, C. 1942. Voles, Mice and Lemmings. Oxford Univ. Press,
      London. 496 pp.

[12] Elton, C. S. 1958. The Ecology of Invasions by Animals and
      Plants. Methuen, London. 181 pp.

[13] Gause, G. F. 1934. The Struggle for Existence. Williams and
      Wilkins, Baltimore. ix, 163 pp.

[14] Gerking, S. D. (ed.). 1967. The Biological Basis of Fresh-
      water Fish Production. John Wiley, New York. xiv, 495 pp.

[15] Hazen, W. E. 1964. Readings in Population and Community
      Ecology. W. B. Saunders, Philadelphia. x, 388 pp.

[16] Hewitt, O. H. (ed.). 1954. A Symposium on Cycles in Animal
      Populations. Journ. Wildlife Management. 18(1):1-112.

[17] Keith, L. B. 1963. Wildlife's Ten-Year Cycle. Univ. of
      Wisconsin Press, Madison, Wisconsin. xvi, 201 pp.

[18] Lack, D. 1954. The Natural Regulation of Animal Numbers.
      Oxford Univ. Press, London. viii, 343 pp.

[19] ──────────. 1966. Population Studies of Birds. Oxford Univ.
      Press, London. v, 341 pp.

[20] LeCren, E. D. and Holdgate, M. W. (eds.). 1962. The Exploita-
      tion of Natural Animal Populations. Blackwell, Oxford.
      xiv, 399 pp.

[21] MacArthur, R. H. and Connell, J. H. 1966. The Biology of
      Populations. John Wiley, New York. xv, 200 pp.

[22] ──────────. and Wilson, E. O. 1967. The Theory of Island
      Biogeography. Princeton Univ. Press, Princeton, New
      Jersey. 203 pp.

[23] Malthus, T. R. 1890. An Essay on the Principle of Population.
      Ward, Lock, London. xiii, 614 pp.

[24] Pearl, R. 1925. The Biology of Population Growth. Alfred A.
      Knopf, New York. xiv, 260 pp.

[25] Schwerdtfeger, F. 1968. Ökologie der Tiere. Band II.
      Demökologie: Struktur und Dynamik tierischer Populationen.
      Paul Parey, Berlin. 460 pp.

[26] Slobodkin, L. B. 1961. Growth and Regulation of Animal Popu-
      lations. Holt, Rinehart and Winston, New York. viii, 184 pp.

[27] Southwood, T. R. E. (ed.). 1968. Insect Abundance. Blackwell,
      Oxford. viii, 160 pp.

[28] Sweeney, J. S. 1926. The Natural Increase of Mankind.
      Williams and Wilkins, Baltimore, Maryland. 185 pp.

[29] Universtetets Biologiske Laboratorium. (ed.). 1933. Essays
      on Population. Scientific Results of Marine Biological
      Research No. 7. Det Norske Videnskaps-Akademi I Oslo, Oslo.
      159 pp.

[30] Uvarov, B. P. 1931. Insects and Climate. Trans. Entomologi-
      cal Society of London. 79(1):1-247.

[31] Warren, K. B. (ed.). 1957. Population Studies: Animal
      Ecology and Demography. Cold Spring Harbor Sumposium on
      Quantitative Biology, Vol. XXII. Biological Laboratory,
      Cold Spring Harbor, L. I., New York. xiv, 437 pp.

Added in the proof
[32] Solomon, M. E. 1969. Population Dynamics. St. Martin's Press,
      New York. 60 pp.

3. SELECTED REFERENCES ON MATHEMATICS AND STATISTICS IN ECOLOGY

[1]   Anderson, K. P. 1965. Manual of Sampling and Statistical
      Methods for Fisheries Biology. Part II. Statistical
      Methods, Chp. 5. Computations. FAO Fisheries Technical
      Paper No. 26, Supplement 1, Food and Agriculture Organi-
      zation of the United Nations, Rome. iii, 25 pp.

[2]   Andrewartha, H. G. 1961 Introduction to the Study of Animal
      Populations. Methuen, London. xvii, 281 pp.

[3]   Anonymous. 1959. Techniques and Methods of Measuring Under-
      story Vegetation. Proceedings of a Symposium at Tifton,
      Georgia, October, 1958. Southern Forest Experiment Station,
      New Orleans, Louisiana and Southeastern Forest Experiment
      Station, Asheville, North Carolina (U. S. Forest Service).
      174 pp.

[4]   Bailey, N. T. J. 1957. The Mathematical Theory of Epidemics.
      Charles Griffin, London. xiii, 194 pp.

[5]   —————————. 1964. The Elements of Stochastic Processes with
      Applications to the Natural Sciences. John Wiley, New York.
      xi, 249 pp.

[6]   —————————. 1967. The Mathematical Approach to Biology and
      Medicine. John Wiley, New York. xiii, 296 pp.

[7]   Bartlett, M. S. 1960. Stochastic Population Models in Ecology
      and Epidemiology. Methuen, London. x, 90 pp.

[8]   Batschelet, E. 1965. Statistical Methods for the Analysis of
      Problems in Animal Orientation and Certain Biological
      Rhythms. The American Institute of Biological Sciences,
      Washington, D. C. 57 pp.

[9]   Beverton, R. J. H. 1954. Notes on the Use of Theoretical
      Models in the Study of the Dynamics of Exploited Fish
      Populations. U. S. Fishery Laboratory. Beaufort, North
      Carolina. Misc. Contr. No. 2. 186 pp.

[10] —————————. and Holt, S. J. 1957. On the Dynamics of Exploi-
      ted Fish Populations. Fishery Investigations Series II,
      Volume XIX, Her Majesty's Stationary Office, London. 533 pp.

[11] —————————. and Parrish, B. B. (eds.). 1956. Problems and
      Methods of Sampling Fish Populations. (Contributions to
      Special Scientific Meetings, 1954). Conseil Permanent
      International pour l'Exploration de la Mer. Rapports et
      Procès-Verbaux des Réunions, Vol. 140, Part I. 111 pp.

[12] Brooks, C. E. P. and Carruthers, N. 1953. Handbook of Statis-
      tical Methods in Meteorology. Meteorological Office
      Publication M. O. 538, Her Majesty's Stationary Office,
      London. viii, 412 pp.

[13] Brown, D. 1954. Methods of Surveying and Measuring Vegetation.
      Commonwealth Bureau of Pastures and Field Crops, Hurley,
      Berks, England. Bulletin 42. xv, 223 pp.

[14] Buchanan-Wollaston, H. J. 1945. On Statistical Treatment of
      the Results of Parallel Trials with Special Reference to
      Fishery Research. Freshwater Biological Association of the
      British Empire, Scientific Publication No. 10. 55 pp.

[15] Cain, S. A. and Castro, G. M. de O. 1959. Manual of Vege-
      tation Analysis. Harper and Row, New York. xvii, 325 pp.

[16] Cohen, J. E. 1967. A Model of Simple Competition. Harvard
      Univ. Press, Cambridge, Massachusetts. x, 138 pp.

[17] D'Ancona, U. 1954. The Struggle for Existence. Series D
      Bibliotheca Biotheoretica Volume VI. E. J. Brill, Leiden.
      xii, 274 pp.

[18] Eberhardt. L., Peterle, T. J. and Schofield, R. 1963. Probl
      Problems in a Rabbit Population Study. Wildlife Monographs
      No. 10. 51 pp.

[19] Gani, J. 1966. Stochastic Models for Bacteriophage. Methuen,
      London. 46 pp.

[20] Gause, G. F. 1934. The Struggle for Existence. Williams and
      Wilkins, Baltimore. ix, 163 pp.

[21] Godron, M. et al. 1968. Code pour le Relève Methodique de la
      Végétation et du Milieu (principes et transcription sur
      cartes perforées). Ed. du Centre National de la Recherche
      Scientifique, Paris. 292 pp.

[22] Grieb, J. R. 1958. Wildlife Statistics. Federal Aid Division,
      Colorado Game and Fish Department, Denver, Colorado. 96 pp.

[23] Greig-Smith, P. 1964. Quantitative Plant Ecology. (2nd
      Edition). Butterworths, London. xii, 256 pp.

[24] Griffith, A. L. and Ram, B. S. 1947. The Silviculture Re-
      search Code. Vol. 2. The Statistical Manual. The Manager
      of Publications, Delhi. viii, 214 pp.

[25] Gulland, J. A. 1965. Manual of Methods for Fish Stock Assess-
      ment. Part I. Fish Population Analysis. FAO Fisheries
      Technical Paper No. 40 (revision 1), Food and Agriculture
      Organization of the United Nations, Rome. misc. pages.

[26] Gulland, J. A. 1966. Manual of Sampling and Statistical
      Methods for Fisheries Biology. Part I. Sampling Methods.
      FAO Manuals in Fisheries Science No. 3. Food and Agricul-
      ture Organization of the United Nations, Rome. misc. pages.

[27] Hanson, W. R. 1963. Calculation of Productivity, Survival,
      and Abundance of Selected Vertebrates from Sex and Age
      Ratios. Wildlife Monographs N0. 9. 60 pp.

[28] Holling, C. S. 1965. The Functional Response of Predators to
      Prey Density and its Role in Mimicry and Population Regula-
      tion. Memoirs of the Entomological Society of Canada,
      No. 45. 60 pp.

[29] Kurten, B. and Rausch, R. 1959. Biometric Comparisons Between
      North American and European Mammals. Acta Arctica, Fasc.
      XI. 45 pp.

[30] Jeffers, J. N. R. 1959. Experimental Design and Analysis in
      Forest Research. Almqvist and Wiksell, Stockholm. 172 pp.

[31] Jones, R. 1966. Manual of Methods for Fish Stock Assessment.
      Part IV. Marking. FAO Fisheries Technical Paper No. 51,
      Supplement 1. Food and Agricultural Organization of the
      United Nations, Rome. misc. pages.

[32] Kershaw, K. A. 1964. Quantitative and Dynamic Ecology.
      Edward Arnold, London. viii, 183 pp.

[33] Keyfitz, N. 1968. Introduction to the Mathematics of Popula-
      tion. Addison-Wesley, Reading, Massachusetts. xiv, 450 pp.

[34] Kinashi, K. 1954. Forest Inventory by Sampling Methods.
      Bulletin of the Kyúshú University Forests No. 23. 153 pp.

[35] Kostitzin, V. A. 1939. Mathematical Biology. George G.
      Harrap, London. 238 pp.

[36] Lewis, T. and Taylor, L. R. 1967. Introduction to Experi-
      mental Ecology. Academic Press, New York. xi, 401 pp.

[37] Lotka, A. J. 1925. Elements of Physical Biology. Williams
      and Wilkins, Baltimore, Maryland. xxx, 460 pp.

[38] Matérn, B. 1960. Spatial Variation. Meddelanden från Statens
      Skogsforskningsinstitut 45(5):1-144.

[39] Miller, R. L. and Kahn, J. S. 1962. Statistical Analysis in
      the Geological Science. John Wiley, New York. xiii, 483 pp.

[40] Morris, M. J. 1967. An Abstract Bibliography of Statistical Methods in Grassland Research. U. S. Forest Service, Miscellaneous Publication No. 1030, Washington, D. C. 222 pp.

[41] Patil, G. P. 1965. Classical and Contagious Discrete Distributions. Statistical Publishing Society, Calcutta, India; Pergamon Press, New York. Proc. of a Symposium. 552 pp.

[42] ――――――. and Joshi, S. W. 1968. A Dictionary and Bibliography of Discrete Distributions. Oliver and Boyd, Edinburgh. 279 pp.

[43] Patil, G. P., et al. 1964. Certain Studies of the Structure and Statistics of the Logarithmic Series Distribution and Related Talks. Office of Technical Services, U. S. Dept. of Commerce, Washington, D. C. 389 pp.

[44] Phillips, E. A. 1959. Methods of Vegetation Study. Holt, Rinehart and Winston, New York. xvi, 107 pp.

[45] Pielou, E. C. 1969. An Introduction to Mathematical Ecology. John Wiley, New York. 286 pp.

[46] Prodan, M. 1968. Forest Biometrics. Pergamon Press, New York. xi, 447 pp.

[47] Ricker, W. E. 1958. Handbook of Computations for Biological Statistics of Fish Populations. Fisheries Research Board of Canada, Bulletin No. 119. Queen's Printer and Controller of Stationary, Ottawa. 300 pp.

[48] Schultz, V. 1954. Wildlife Surveys--a Discussion of a Sampling Procedure and a Survey Design. Tennessee Game and Fish Commission, Nashville, Tennessee. 153 pp.

[49] ――――――. 1961. An Annotated Bibliography on the Uses of Statistics in Ecology--A Search of 31 Periodicals, U. S. Atomic Energy Commission Report TID-3908. Washington, D. C. vii, 315 pp.

[50] Schumacher, F. X. and Chapman, R. A. 1948. Sampling Methods in Forestry and Range Management. School of Forestry, Duke Univ., Durham, North Carolina. Bulletin 7 (revised). 222 pp.

[51] Simpson, G. G., Roe, A. and Lewontin, R C. 1960. Quantitative Zoology. Harcourt, Brace, New York. vii, 440 pp.

[52] Smith, J. M. 1968. Mathematical Ideas in Biology. Cambridge Univ. Press, London. vii, 152 pp.

[53] Sokal, R. R. and Sneath, P. H. A. 1963. Principles of Numerical Taxonomy. W. H. Freeman, San Francisco, California. xvi, 359 pp.

[54] Southwood, T. R. E. 1966. Ecological Methods with Particular Reference to the Study of Insect Populations. Methuen, London. xviii, 391 pp.

[55] Toril, T. 1956. The Stochastic Approach in Field Population
      Ecology: with Special Reference to Field Insect Populations.
      Japan Society for the Promotion of Science, Uéno, Tokyo.
      277 pp., plus chart.

[56] Ullyett, G. C. 1953. Biomathematics and Insect Population
      Problem--a Critical Review. Memoir of the Entomological
      Society of South Africa, Pretoria, No. 2. 89 pp.

[57] Various authors. 1966. Statistics in the Study of Bird
      Populations. The Statistician 16(2):119-202.

[58] Wadley, F. M. 1967. Experimental Statistics in Entomology.
      Graduate School Press, U. S. Department of Agriculture,
      Washington D. C. viii, 133 pp.

[59] Watt, K. E. F. 1961. Mathematical Models for Use in Insect
      Pest Control. Canadian Entomologist, Supplement 19. 62 pp.

[60] ——————————. (ed.). 1966. Systems Analysis in Ecology. Aca-
      demic Press, New York. xiii, 276 pp.

[61] ——————————. 1968. Ecology and Resource Management: A Quanti-
      tative Approach. McGraw-Hill, New York. xii, 450 pp.

[62] Wilimovsky, N. J. and Wicklund, E. C. 1963. Tables of the
      incomplete bata function for the calculation of fish popu-
      lation yield. Inst. of Fisheries, Univ. of British
      Columbia, Vancouver. iii, 291 pp.

[63] Williams, C. B. 1964. Patterns in the Balance of Nature and
      Related Problems in Quantitative Ecology. Academic Press,
      New York. vii, 324 pp.

Added in the proof
[64] Cole, A. J. (ed.). 1969. Numerical Taxonomy. Academic Press,
      New York. 320 pp.

[65] Freese, F. 1962. Elementary Forest Sampling. U. S. Dept of
      Agriculture, Forest Service, Washington D. C., Agriculture
      Handbook No. 232. iv, 91 pp.

[66] ——————————. 1967. Elementary Statistical Methods for Fores-
      ters. U. S. Dept. of Agriculture, Forest Service, Washing-
      ton D. C., Agriculture Handbook No. 317. iv, 87 pp.

[67] Giles, R. H., Jr. (ed.). 1969. Wildlife Management Techniques
      (Third edition: revised). The Wildlife Society, Washington,
      D. C. vii, 623 pp.

[68] Khil'mi, G. F. 1962. Theoretical Forest Biogeophysics.
      Office of Technical Services, U. S. Dept. of Commerce,
      Washington, D. C. 155 pp. (translated from: 1957.
      Teoreticheskaya Biogeofizika Lesa. Izadatel'stvo Akademii
      Nauk SSR, Moskva).

[69] Kuczynski, R. R. 1969. The Measurement of Population Growth:
      Methods and Results. Demographic Monographs, Vol. 6.
      Gordon and Breach, New York. 255 pp.

[70] Levins, R. 1968. <u>Evolution in Changing Environments: Some
      Theoretical Explorations</u>. Princeton Univ. Press, Princeton,
      New Jersey. ix, 120 pp.

PANEL DISCUSSION: TRAINING AND RESEARCH PROBLEMS IN QUANTITATIVE ECOLOGY

DOUGLAS G. CHAPMAN
GERALD J. PAULIK
Center for Quantitative Science in Forestry,
Fisheries and Wildlife
University of Washington

GEORGE M. VAN DYNE
Departments of Range Science and Fishery and Wildlife Biology
College of Forestry and Natural Resources
Colorado State University

WILLIAM E. WATERS
Chief, Forest Insect Research
U. S. Forest Service
Washington, D. C.

JOHN WILSON
Graduate Student
University of California, Davis

SUMMARY

An evening panel discussion was conducted during the Symposium.
Drs. Chapman, Paulik, Van Dyne and Waters and Mr. J. W. Wilson, a
graduate student, served on the panel. Professor Patil was chairman. The contribution of each panelist is given in the following
section.

1. MATHEMATICS AND ECOLOGY (D. G. CHAPMAN)

1.1 Introduction

There have been in the past decade a large number of conferences
devoted in part or whole to some interweaving of mathematics and
biology. Much of the emphasis of these conferences is on the use
of mathematics in that section of biology oriented towards physiological and medical problems, but more recently the mathematical
role in ecology and the role of ecology in the world today have
both received more emphasis. What is the role of mathematics in
ecology and how does this bear on the level and type of mathematical
training that should be recommended for ecology students?

Let me begin by noting that I obviously bring a biased viewpoint
to this question. By background training, I am primarily a mathe-
matician and while since that time I have delved into some aspects
of ecology, particularly in fisheries biology and fisheries management, many ecologists would perhaps rightfully question my
qualifications to speak on these subjects. I recently participated
in a site visit to a biomathematics program. A prime concern of
one of the site visitors was whether and to what extent the trainees
would get "blood on their hands." While I occasionally get "blood
on my hands" it has not been too frequently, and I therefore admit
the bias that comes from being more of a pencil pusher than a
laboratory or field researcher.

This is an important aspect of our discussion because most of us
will express subjective views rather than objective analyses, for
the simple reason that almost no careful studies have been made to
compare different curricula, or methods of teaching, or types of
courses particularly at the University level. The difficulties of
making such studies are obvious. For example, some years ago I
participated in a study to determine whether there was any difference between calculus grades of students in large lecture sections
from a regular faculty member and those in small lecture sections
from a graduate teaching assistant. The outcome of the experiment
was as follows: there was no significant difference between methods
but there was a significant difference between students in morning
and afternoon classes. The latter were poorer to begin with and
also at the end--regardless of teaching method, which seems to
support the sometimes voiced idea that the students with innate
ability will do well regardless of the curriculum and/or the method
of teaching. All such students need is motivation. In addition,

it may be suggested that any program that appears to offer something new will at least for a while appear to have great potential because it attracts and motivates good students. Thus, ecology may at the moment be receiving benefits from this effect--it is relatively new and it has the advantage of its having received much publicity and of having appeal because of social needs and social benefits.

Despite or perhaps because of these problems, let me begin with certain assumptions that you may wish to challenge:

(i) that use of mathematical language and mathematical models introduces and forces a clarity of ideas that may be otherwise lacking.

(ii) that ecology necessarily involves a considerable amount of data; the more realistic the study the larger and more complex the amount of data.

(iii) that ecology both in its research and in its management phases is becoming more system oriented and that increasingly sophisticated mathematical tools (including the computer) make possible more sophisticated models encompassing larger and complex systems.

(iv) that the important and exciting research of tomorrow will be done by researchers who today form a small fraction of the student body (not all of whom will have Ph.D.'s), and that this select group requires stimulation and direction rather than teaching.

(v) a much larger group of students remains who will receive some kind of training in ecology and who will go on to play lesser roles on research teams or be in management positions that have great influence. Of course, it will probably still be true in the 1970's as it is now that many of those in a position to make decisions affecting our resources and environment have had little or no training in ecology.

1.2 Needs for Mathematical Sciences in Ecology

For convenience I will divide mathematics into three areas--analysis, statistics and computing. By analysis I refer to what otherwise might be called classical applied mathematics--the part using the tools of the calculus, differential and difference equations, integral equations, operator theory, etc.--and which has been most associated with physics and engineering, mechanics, thermodynamics, and hydrodynamics. In fact, it is well known that it was problems of physics and astronomy that gave this large area of mathematics its great impetus. The classical theory might be said to have begun with the work of Newton and the recent developments in mathematical physics grew out of relativity and quantum mechanics stemming from Einstein's work.

While these are the three main areas that I wish to treat, I will make some comments on matrix theory, operations research and systems analysis. I will also subsume under the heading analysis some of the more recent developments in mathematics that have found applications in various fields, e.g., topics in convexity, combinatorics, etc.

Let us turn first, however, to statistics. Management of fisheries, forestry and wildlife resources has for the most part passed beyond seat-of-the-pants decisions. Few managers will operate without an adequate inventory of the resources--an inventory usually obtained through sample procedures. But an inventory alone is insufficient to properly manage a renewable resource. Particularly in fisheries it has become clear that management requires quantitative knowledge about the processes involved. In fact the less we understand the processes in detail, the more we have been forced to use a "black box" approach that inevitably is statistical. Let me illustrate with the management of our salmon resource. One of the most important pieces of information for management of salmon is the relationship between parent stock and their progeny. Obviously a lot of steps are involved in this process--the competition between spawning females for egg deposition sites, the male-female ratio and the effectiveness of fertilization, the competition between the salmon fingerlings after they hatch, the relationship of numbers of juvenile salmon to their predators among others. In the absence of detailed knowledge about all of the processes--which might give rise to intra-specific competition and compensatory mortality processes--and in the even more obvious absence of precise information about the exogeneous variables which affect young salmon survival, Ricker set up a simple mathematical model for the spawner-progeny relationship. This model has been widely accepted and used so that for many salmon management agencies, this phase of the problem reduces to data collection and parameter estimation.

That management of our resources requires statistical competence-- particularly in sampling, in experimental design, in estimation-- should be a matter of agreement. It is perhaps not so clear that basic ecological research requires such an input of statistical methods. For example, one might argue that Darwin and Mendel

achieved quite a lot without any technical statistical competence.
I agree that substantial research breakthroughs will be achieved
which will not necessarily be dependent on statistical competence,
but it is also true that a large fraction of the research being done
will involve statistical methods and concepts. The details of broad
ideas will be based on sample studies and statistical tests, and the
student of ecology who is conversant in statistics will find his way
through the literature much more easily than one who is not so conversant. More importantly he can screen the wheat from the chaff
and disregard those results that are based on poor design or involve
outright errors of analysis.

To avoid heavy and tedious calculations, statisticians have in
the past acted as if the many variables in nature could each be
examined separately and treated with univariate methods. This
simplification which was not desirable from the point of view of
modelling the real world is no longer necessary. Therefore, the
student of statistics must become reasonably proficient in the
language of multivariate analysis--matrix algebra. How much matrix
algebra he requires and where he should obtain this knowledge I will
return to later.

Next, I turn to computing. I think no one would dispute that
those who now work with data must learn the elements of computing.
Of course, most standard statistical analyses are now available in
packaged programs that can be used without computer training. Simulation studies are now increasingly in the vogue and to undertake
these requires a little more computer knowledge than merely putting
in a canned program. Returning now to the area of mathematical
analysis, there will be controversy over the amount of mathematics
necessary to an ecologist. An ecologist will probably think of the
work of Volterra, Lotka and Gause as applications of mathematical
analysis to ecology, and many ecologists will assert that this has
been a rather sterile exercise in mathematics and the models constructed have been too far from any possible reality to be even
remotely useful. On the other hand, recent work of MacArthur and
Holling, to name only two examples, is closely related to some of
the ideas and some of the models originally constructed by Volterra
and Lotka. Moreover, the construction of models for ecosystems
involves transfer functions that are usually expressed as differential equations. The International Biological Program (IBP) is
placing emphasis on the study of ecosystems and more particularly
on developing models, primarily mathematical for such ecosystems.

1.3 Training in the Mathematical Sciences for an Ecologist

It becomes fairly clear from what I have said above that I would
recommend some training in statistics, in matrix algebra, in computing and in analysis for the potential ecologist. Note that I
say potential ecologist, for I anticipate a point that may be made
against this proposal--how does the student squeeze it all in and
still find time to study biology, and obviously some genetics, and
chemistry, and still not be totally illiterate in the social
sciences and humanities?

One way this problem can be met is to start the program early.
The average junior high school student now learns much of the
algebra that I took in Grade XII and yet the average biologist does
not encounter calculus, if at all, until his upper division or
graduate college years. By this time he has forgotten his algebra
and finds the mathematics rather tough and unrelated to his chosen
specialty. He also dislikes having to compete with bright freshmen
and sophomores specializing in mathematics or a closely allied field.

So I suggest that much of the required mathematics can be inserted into the high school curriculum. Nor do I mean only the
typical calculus now to be found for a very select group of students
in better senior high schools but also computing and statistics.
It has now been well demonstrated that grade school students can
learn to compute and are fascinated by it--who would expect our
children in this machine and gadget-oriented society to be otherwise.
From the computer we can progress to statistics, using the computer
to assist in statistical analyses and to generate random data for
sample experiments that are so important in elementary statistical
analysis.

Up to now I have felt that a first statistics course must begin
with basic probability and our first course for a variety of biological majors includes a substantial dose of probability model
building, though not of combinatorics which is so difficult, confusing and misleading for many beginning students. I would still
insist that all students of statistics need to get a foundation in
probability but perhaps a computer-oriented introduction will be
more motivating at least for less experienced students.

If the student completes his mathematics early as I am suggesting,
then we are faced with two problems--he does not have it related to
applications in ecology and he does not have an opportunity to use
it for several years. Ideally, of course, both these problems could

be met if the biology courses he takes are sufficiently quantitative--
but for this to happen it is necessary that the appropriate mathematical sciences be prerequisite to the course. This may occur to
a limited degree but its general acceptance is perhaps some time off.

In this interim it will be necessary to develop integrated
courses, applying the mathematical methods acquired, to ecological
situations particularly in model building. Moreover, we will still
have to provide for the student of ecology who does not get his
mathematical science at the late high school or early college stage.
For such students I suggest that they not be sent back to compete
with physics and mathematics majors in freshman calculus but rather
that special courses be constructed, tailored to their interests and
background. Such courses should be oriented not towards formal
mathematical training but towards problem solving and model building.
In view of the time demands, perhaps they can be compressed from the
usual calculus sequence. We are attempting this at the University
of Washington but are not now in a position to report on its outcome.

It may be asserted that this represents a step backwards--mathe-
matics departments used to have special courses for non-majors but
have tended to eliminate these in recent years. However, I believe
that a transitional period is required. At the present time a
physicist or engineer, for example, sees many applications of mathematics in courses in physics, mechanics, etc., and since he starts
his mathematical training early it is reasonable that he take a
standard mathematical course. Of course it is true that there may
be problems in arranging for such courses--I have no particular
suggestions concerning this and believe that each university must
handle this according to the local climate, individuals available
to handle special courses, etc. How soon such courses should be
eliminated is also a matter that will vary from place to place,
depending on both the quality of the courses, the rapidity with
which ecological and other biological courses become more mathematical.

Earlier I made brief reference to the need for some study of
matrices as the language of multivariate statistics and perhaps
also of multivariate calculus. Many universities include some
matrix algebra in courses in statistics where multivariate analysis
is treated. In view of the fact that multivariate statistical
analysis represents one of the most important applications of matrix
algebra for students of ecology (though by no means the only one),
this seems to be a logical procedure. In view of the pressure on

students because of the increasing number of subjects they must study, it seems reasonable to try to compress some in this manner. Of course matrix algebra is another subject that could well be added to the high school curriculum--again provided that the student sees it used reasonably soon after he learns the language of matrix algebra.

Finally, let me comment on my views on the role that courses in operations research or systems analysis may play in the training of an ecologist in the next decade. I regard these as specialized options which may well be taken up but only after the student has taken the more basic courses in statistics, computing and analysis. In this sense I would put them in the same category as various other specialized options--courses in mathematical statistics, in advanced analysis and perhaps in other areas. What priority I would suggest for the student who wants to go beyond the basic mathematical sciences would depend on his interest and his mathematical ability.

In summary then, I believe that the student of ecology who is pre-paring to do ecological research or to participate in a unit making management decisions that affect our environment and our resources will be poorly trained unless he has a substantial component of mathematical sciences. Personally, I do not see how the topics that are essential can be adequately treated in less than 12 semester or 18 quarter units: At the University of Washington we do not try to squeeze it this much but in fact suggest courses of 21 quarter units plus any work the student does on acquiring computing skill. Of course, at the present time not all students in broad ecological areas take all this, though most students in fisheries and forestry do, and this much will be required of a new wildlife program. Ecology students in botany and zoology tend to take less than this as undergraduates, though there are few graduate students who do not complete at least this much. I also urge that consideration be given to development of courses that emphasize model building that follow the basic mathematics, particularly for those students who have early mathematical training. Finally, let me say that all such courses should be oriented towards the formulation and solution of problems and/or the building of models. I have emphasized these points of view for many years--I wish that I could say that I could present objective evidence that the approach is an unqualified success and that it is better than other approaches. That I cannot do; I will listen with interest to any others who have objective data or subjective views on the role or the methods of teaching mathematical sciences to students of ecology.

2. QUANTITATIVE ECOLOGY AND RESOURCE MANAGEMENT AT THE UNIVERSITY OF WASHINGTON (G. J. PAULIK)

The opportunities for training at the University of Washington in the area of natural resources can be understood best by first examining some of the powerful external and internal forces acting on large public universities today that affect training programs concerned with environmental biology. These forces have acted in a variety of ways to shape current programs and will probably continue to exert an even stronger influence in the future. An understanding of these forces is also important because any description of an existing program is merely a blurred snapshot. Programs in resource management are evolving so rapidly that it is almost impossible to stop the action long enough to take a close look.

The factors that seem to be exerting some subtle and some not so subtle influences on our training program include:

(i)   Slowly increasing public awareness of the Malthusian dilemma.

(ii) Scientific awareness of the power and general applicability of the collection of techniques known as the "system approach."

(iii) Widespread availability and rapid development of large electronic computers.

(iv) Developments in environmental monitoring and sensor devices.

(v)   Students' quest for relevance in their studies and their demands for academic reforms.

Many of these forces are exerting pressure on universities to offer new types of interdisciplinary programs. The traditional structure of the mature university with administrative entities representing various academic disciplines is finding these new pressures difficult to handle. Although the university community is most sympathetic to any programs, interdisciplinary or not, devoted to wise use of natural resources and to improving the quality of human existence, the structural ridigity of the university into academic specialties is embarrassingly opposed to some of the more imaginative and innovative programs being proposed in the resources area.

In spite of considerable personal frustration over day-to-day problems brought about by the tug and pull of these opposing forces, I feel that the traditional structure of the university serves as a valuable dampener preventing over-reactions to the highly visible crises in population density, natural resources, and pollution. A strong departmental structure almost automatically insures that a

student will master at least one field and minimizes the danger that
broadness will be accompanied by superficiality. The production by
our educational system of shallow generalists cannot be condoned
irregardless of needs for the multidisciplinary training. Of course,
a strong disciplinary structure puts considerable pressure on the
student and his graduate committee to properly select those courses
outside of his major which will make his training broad enough so
that he will be able to work effectively with other specialists.
This type of arrangement may encourage dilettantism and some students
will carefully pick easy introductory courses in other fields and
thus avoid any serious intellectual involvement outside of the fields
in which they are majoring.

The physical setting of the University of Washington in the Northwest where a relative abundance of many important natural resources
still exists is somewhat unique. There is no College of Agriculture
on our campus and consequently we have a quite different sort of
resource program than those programs associated with land grant
schools where interdisciplinary studies related to agriculture have
a long history. Training in some aspect of natural resources is
found in the following groupings at the University of Washington:
the Colleges of Engineering, Fisheries, Forest Resources, and Arts
and Sciences which includes the Departments of Atmospheric Sciences,
Civil Engineering, Economics, Geography, Oceanography, and Political
Science; and the Schools of Business Administration, Law, and Public
Affairs. The diversity of this arrangement gives our resource
training program a broad base and accompanying stability.

Public concern about environmental problems has begun to manifest
itself in the form of financial support for broad interdisciplinary
studies. Such programs as the Ford Foundation Ecology Program, the
Sea Grant Program, the International Biological Program, and the
Water Resources Research Program are shaking the traditional academic groupings at the University of Washington as well as elsewhere.
The impact of these programs is amazing to behold and difficult to
comprehend.

The students' quest for relevance is causing some of the bright-
est to consider resource management as an exciting, challenging,
and respectable field. Their demands for academic reforms are
responsible for the scrapping or revision of such time-honored
requirements as foreign languages and grades. A recent change in
regulations at the University of Washington of interest to inter-
disciplinarians is the optional pass-fail system that a graduate

student may request in place of the usual grading system when
taking courses outside of his major field. This policy encourages
broader interests and training.

The Center for Quantitative Science in Forestry, Fisheries, and
Wildlife is a new organization on the University of Washington
campus. It has been in existence for just one year and I think
it is fair to say that it is a product of the forces mentioned
earlier. The Ford Foundation was the midwife in this birth. They
helped create the Center by financing some permanent faculty positions for an existing Biomathematics group. These new positions
allowed the group to establish the Quantitative Center and to begin
offering new types of courses. Faculty appointments in the Center
are joint appointments in the Colleges of Fisheries and of Forest
Resources. Dr. Douglas Chapman, who is on our panel tonight, left
the Department of Mathematics to become its director.

During the first year of the Quantitative Center's operation,
five students have been supported by Ford Foundation Fellowships
in Quantitative Ecology and Resource Management. The number of
fellowships available increases to 12 per year by the fourth year.
It happens that each of these five students is taking his Ph.D. in
a different area. The five areas represented are: Biomathematics,
Business Administration, Economics, Fisheries, and Forest Resources.
However, in addition to the students formally labeled as having a
special interest in resources, a number of other students pursuing
advanced degrees in fields ranging from Public Affairs to Zoology
have graduate programs which look very much like those of the five
resource scholars. For example, several other students in the
Biomathematics Program have taken strong minors in ecology and some
will spend their professional lives in this field.

Lists of course offerings tend to make dull reading and even
duller listening. For a detailed list of offerings, I suggest that
you write to Dr. Chapman. However, I would like to sketch the
general outline of the course program. We offer a more or less
standard series of courses in statistics covering statistical
inference and experimental design with special emphasis on biological problems, a series in operations research with special
applications in forestry and resource management, and several
new mathematics courses designed especially for biologists. These
include matrix algebra, techniques of applied mathematics and
differential equations, systems analysis, and an experimental

calculus course. I am not sure whether the proliferation of
courses in quantitative methods dealing with specific areas of
application is the best way to offer quantitative training at a
large public univeristy. Courses in systems analysis, for example,
are nearly as ubiquitous on our campus as courses in statistics.

Although there is no official core program in resources, most of
our students take Advanced Ecology in the Zoology Department and
Economics of Natural Resources in the Department of Economics as
well as a variety of offerings in Computer Science. International
law, advanced statistical methods, and bioengineering are also
fairly common parts of our graduate students' programs.

As is true for most large universities, the University of Washington provides an extremely rich intellectual environment. A wide
variety of seminars constitute a continuing public dialog between
faculty members from all University divisions as well as outside
visitors from all segments of society.

One special training project at the University of Washington is
the development of computer simulation games as Link trainers to
provide students with the opportunity to exercise their management
skills under simulated real-life management conditions. Although
I find it difficult to be objective about these games since a fair
evaluation requires that I crawl out of my own psychic skin, I think
they may revolutionize the training of potential resource managers.
Such games also have potential for use in short courses for practicing managers and for research purposes.

A sometimes overlooked method of training resource managers is
to involve them in a real life resource problem. Last spring
quarter our class in population dynamics undertook as a special
project the application of a "total systems approach" to the
problem of controlling the abundance of dogfish (Squalus acanthus)
in Puget Sound. This small shark is a nuisance to sports and
commerical fishermen and is thought to be a competitor with some
life stages of salmon and a predator on other life stages. The
class tagged a number of dogfish to estimate population parameters
and has built computer models of the dogfish population to test
the economic feasibility of various control strategies as well as
to engage in some ecological speculation as to the long-range
effects of a successful control program on the Puget Sound ecosystem
and its value as a recreational and commerical resource. It is
rather surprising that one of our main difficulties has been to

restrict our study to a small-scale class exercise. So far we have resisted the temptation of offers of financing a large-scale research study on this problem. However, the offers themselves provide some assurance that we are working on a problem of real importance to the local population.

To summarize briefly, the University of Washington's program in resource management and quantitative ecology is an extremely flexible and diverse program imbedded in a university environment characterized by traditional academic disciplines organized into discrete and administratively separate departments. Many of these individual departments have their own programs concerned with various aspects of managing natural resources. Most students in the resources area major in one of the existing academic fields and tailor the parts of their programs involving other fields to suit their individual needs and interests. However, all aspects of training in resource management and quantitative ecology are evolving rapidly. Various types of experimental teaching techniques are being used to provide students with more diverse and effective learning opportunities. Academic requirements and arrangements for supervising graduate students are also changing rapidly. The direction of most changes is toward increasing multidisciplinary and interdisciplinary involvement.

3.   ASPECTS OF QUANTITATIVE TRAINING IN THE NATURAL RESOURCE SCIENCES (G. M. VAN DYNE)

3.1   Introduction

Man must manipulate the environment to produce the food, fiber, metals, and power he needs for his existence. But often he has not had adequate understanding of the long-term consequences of his manipulations. As he increasingly takes an ecological view-point, however, and as he increasingly uses quantitative approaches, he will be able to optimize his multiple uses of natural resources. With growing populations, and with dwindling resources, there is a critical need to assemble and activate interdisciplinary teams concerned with research and management of our natural resources and their optimal use. I am concerned here with the training procedures and philosophies to equip multidisciplinarians, for these interdisciplinary teams, who have the quantitative skills to make significant contributions toward the solution of relevant resource problems.

3.1.1   Balance and Source of Training

This paper is concerned primarily with the balance of analytical subject matter in comparison to the biological, physical, and social science areas in the curriculum. Both the balance of training and the source and context in which the training is obtained are impor-tant. For example, as Watt [7] noted, all of the mathematics cannot be taken only in mathematics departments. The natural resource scientist should take some of his biomathematics courses from biologists. Perhaps skill obtained in moving across the gap between the real world and abstraction is as important, or more important, than having specific skills in analysis or a broad knowledge of biology.

3.1.2   Graduate vs. Undergraduate Objectives

The following discussion focuses on both undergraduate and graduate programs, and the objectives of the two must be considered. Using the outlook of Booker [1], we will note here that the undergraduate education is primarily concerned with what is relatively well known. Thus the undergraduate student in natural resource sciences is likely to work in situations wherein he is called upon to draw from existing knowledge to make decisions, to solve problems, or both. He should be able to understand and use much of what already is well

described. With the exponential increase in information, however,
we question whether or not an undergraduate student can amass a
sufficient part of the knowledge in the four-year curricula, unless
his subject matter area is relatively narrow. I do not suggest he
must know everthing in a particular field. But he should have a
good chance of understanding anything that someone else has already
understood about his field. The graduate student, i.e., the Ph.D.
candidate, should develop skills and knowledge in his training so
that when he reaches the real world he will have confidence in his
ability to face the unusual or new. Furthermore, he should be
instilled with the idea of learning, during his entire life. This
reinforces the idea of cintinuing education, for with the rate at
which knowledge is being accumulated, continuing education is a must
for both the undergraduate and the graduate.

3.1.3 Interdisciplinary Team Approach and Division of Labor

Many of today's natural resource scientists, at both the undergraduate and graduate level, will be working in interdisciplinary
teams in research and management when they graduate. These
scientists must be equipped with the tools and techniques of studying natural resource ecosystems and they must have the operational
philosophy for contributing to and operating in such teams (Van Dyne
[5], [6]). In contrast to the natural resource scientists of the
past, there will be less emphasis on complete self-sufficiency in
many phases of both management and research. The scientists must
be trained to participate well in these teams without losing their
own individuality and initiative. Furthermore, they must be trained
for probably one of three levels of contribution: (i) laboratory
or field scientists largely working alone on a specific process or
phase of a problem, (ii) leaders coordinating efforts of a group of
such researchers or managers, (iii) or coordinating the efforts of
the group leaders (Van Dyne [5]). The successful natural resource
scientist must be able to synthesize results of many research and
management experiments into testable theories or practices. He must
have a broad knowledge about the entire resource ecosystem with
which he works, he must be diligent about searching and condensing
the literature, and he must skillfully use analytical tools to
combine and evaluate information from field, laboratory, and library
investigations. He must, in effect, be a systems ecologist (Van Dyne
[4]).

3.1.4 Evolution or Revolution of Training Programs

Much must be done to reshape training programs in natural resource
sciences, and some aspects of revisions and improvements are discussed here. The emphasis here concerns terrestrial ecosystems;
the concepts are extendable to freshwater and marine systems. The
curricula I propose below do not exist, with exception of some
graduate programs approximating those outlined. The examples I show
below are not considered final, the best, or the only approach. But
they are a point of departure. They do not represent my college's
or departments' opinions. Many of the ideas presented in the next
two sections are explained much more completely elsewhere (Van Dyne
[6]).

Table 1. Suggestions for relative timing and course composition of
a 4-year undergraduate program for natural resource
scientists

Curricular Block	Approximate Quarter Credits	Components (examples)	Timing
Basics	90	Chemistry Physics Mathematics Communications Humanities	Primarily in Years 1 and 2
Ecosystem Components	60	Producers Consumers Decomposers Edaphic Climatic	Primarily in Years 2 and 3
Integrative Courses	40	Ecology Watershed Nutrition Genetics Systems management, structure, function	Primarily in Years 3 and 4
Seminars	10	Introductory Analytical Conceptual	All 4 years

3.2   An Approach to Quantitative Training in the Undergraduate Curriculum

Table 1 presents a general outline of a type of curriculum needed in undergraduate natural resource science fields. In converting these general areas of work to specific sourses, however, the comments of the Panel on Natural Resource Science [2] should be considered. Some major problems pinpointed by that panel include: excessive emphasis in narrow vocational training, the departmental barriers making it necessary to offer specialized courses in professional areas, too much emphasis on field practices and too little on principles, and excessive proliferation of curricula and courses.

3.2.1   Assumptions Underlying the Program

The philosophy followed in outlining the courses in curricular blocks in Table 1 is that: (i) the undergraduate student should obtain a clear idea of the principles and concepts about natural resource ecosystems; (ii) he should have an introduction to many, and skills in some, of the tools used in solving these problems; and (iii) with this training he should be prepared for career work or for graduate school. Perhaps it would be unwise for such an individual to major in a specific natural resource management field. Instead, his program would be in natural resource sciences _per se_, with a quantitative aspect.

3.2.2   Emphasis on the Basics

Perhaps some 45% of the credits in the undergraduate program should be in chemistry, physics, mathematics, communications, and humanities. Chemistry training should go through at least a survey organic course; perhaps a total of 20 credits would be sufficient in this area. Physics would take another 10 to 15 credits. Communications, including English, technical writing, and public speaking, and a selected block of humanities would take about 35 to 40 credits.

Some 20 to 25 credits would be devoted to mathematics. Consider both deterministic and stochastic phenomena, each with few or many variables. Few-variable, deterministic problems require analytical geometry-calculus sequences and difference and/or differential equations. Few-variable, stochastic problems require background in probability and statistics, and both applied statistics and statistical theory should be covered. Many-variable, deterministic problems require matrix algebra. Many-variable, stochastic problems require more statistics or perhaps stochastic processes.

To obtain the above type of training in as few credits as mentioned assumes a strong pre-college background in algebra, trigonometry, and perhaps even in introductory calculus. Advanced calculus and stochastic models are not recommended here because there is not sufficient time. The student should obtain a practice in computing and numerical analysis. But perhaps some of this could come in the natural resource fields themselves.

Students obtaining the above amount of analytical training as undergraduates would be rare indeed in today's schools in the natural resource fields. In fact, they would be rare as Ph.D. students in ecology. Watt [7] recommended, but implied that very few graduate students had, analytical training beyond two semesters of calculus and three semesters of applied statistics. He suggested the need for additional training in statistical theory, numerical analysis, differential equations, matrix algebra, and computing. Computers are being used increasingly in basic and applied ecological courses in various universities including Alaska, British Columbia, California-Davis, Colorado State, Cornell, Georgia, Michigan State, San Diego State, Tennessee, Washington, and Washington State.

3.2.3 Understanding Ecosystem Components

Some 30% of undergraduate effort should be in courses on specific ecosystem components covering ecology and physiology of the bio-components and major processes and principles for the abiotic components. Some courses here would overlap in time the group of basics, but generally they would require basics as prerequisites.

3.2.4 Integrative Courses

Taken primarily during the junior and senior year, about 20% of the undergraduate training would be in integrative courses. In natural resource science much less emphasis need be given to separate management courses for each resource field. Principles from the separate fields could be drawn together in cohesive, comparative courses, which would depend more than at present on the prerequisites, i.e., on the basics and on the courses about ecosystem components. Such comparative, integrative courses should include new and innovative approaches to study of natural resource ecosystem structure, function, and management. It is here that analytical

techniques and tools could be used effectively in the classroom.
Many systems analysis techniques, such as simulation and gaming,
could be profitably incorporated. This would help make the needed
transition between the theory and the real world.

3.2.5 Variety and Continuity in Seminars

Perhaps 5% of the undergraduate program should be taken in various
seminars occurring throughout the undergraduate program. Probably
they would include introductory year seminars instilling in the
students a broad ecological sensitivity for the complex problems
of resource management, analytical approaches to solutions of these
problems in the intermediate years, and topical problems treated in
depth and perspective in the senior year. The latter seminars would
test the student's ability to synthesize and integrate information
and to polish his skills in seeking and selecting additional knowledge through independent study.

3.2.6 Resistance to Curricular Change

There are many problems inherent in attempting to implement the type
of scheme outlined above. Many of the land-grant schools, where
most of the natural resource science students are taught, do not
have enough facilities, manpower, or finances to educate students
as well as desired. And implementing the above concepts and ideas
would require major revisions and innovations in undergraduate
programs in the natural resource fields. Perhaps this cannot be
done yet at the undergraduate level. Perhaps it is too early, but
change may be introduced first into graduate programs.

3.3 Example Ph.D. Programs

Example Ph.D. programs outlined in Table 2 presuppose that the
natural resource scientist wants to be a systems ecologist--a
multidisciplinarian to work in interdisciplinary teams. However,
depending upon his interests, his emphasis may be in the experimental area (the first set of credit figures), or in the theoretical-
analytical area (the second set of credit figures).

3.3.1 Increased Emphasis on Formal Work

This program is controversial because it requires more time
(Table 2) in formal classwork than is conventional in many Ph.D.
programs. This is necessary if the objectives of obtaining training

Table 2. Examples of course composition for Ph.D. programs for
          emphasis on quantitative natural resource science training

Curricular Block	Approximate Quarter Credits*	Components (examples)
Ecology and Physiology	36-12	Producers Consumers Decomposers
Abiotic Components	18-12	Climatic Edaphic Hydrologic
Advanced Ecology	22-27	Principles Theory Systems
Analytical Areas	12-39	Mathematics Statistics Logic Engineering
Physical Sciences and Instrumentation	9-6	Radioisotope techniques Biotelemetry Complex data acquisition systems
Seminars	3-4	Topical reviews Conceptual problems Resource system analysis

* The first figure given under credits is for the experimental
  ecologist; the second figure for the theoretical-analytical
  ecologist.

in depth and qualifying as a multidisciplinarian are accepted.
The systems ecologist must be a specialist in generalization;
therefore he is both an interdisciplinarian and a multidisciplinarian. The systems ecologist cannot affort to be superficial in
either breadth or depth. This is emphasized if one examines the
content of recent books useful in advanced ecological courses
(Watt [8], Pielou [3], Van Dyne [6]).

The programs in Table 2 also assume a sound undergraduate
program. Optimally, the graduate student should plan his entire
graduate program very near to its beginning. If one has the
undergraduate background as outlined in Table 1, and if the above
assumptions are met, it should be easy to construct a graduate
course program as in Table 2 which could be completed in a reasonable period of time. This is facilitated if the university has
accepted the modern trend of allowing substitution for the foreign
language requirement. For example, at Colorado State University a
graduate student in many fields, including the natural resource
sciences, has a choice of taking: (i) the conventional reading
knowledge exam for two foreign languages, (ii) one foreign language
in depth, (iii) a reading knowledge in one language plus a "tool,"
or (iv) two "tools." A "tool" here is defined as a 9-credit block
or sequence approved by the Graduate Committee for that particular
student. Thus, some 18 credits of the 100 outlined in Table 2 might
be in the "tool" area. The Ph.D. research, of course, would be in
addition to the material outlined above.

3.3.2 Consideration of Advanced Ecology Courses

The types of courses in Table 2 will not be discussed in detail,
although several are of importance. For example, in the advanced
ecology area there is a great void in most universities in thorough
and challenging graduate-level courses on ecological principles and
theory. There now exist many graduate ecology courses, but they do
not meet the following requirements. Such a course should require
as background plant ecology, animal ecology, or both, and natural
resource management. The course should be team-taught by scientists
from a wide variety of disciplines; each scientist would be assigned
specific principles or theory closely related to his own research or
interest. The course should not, of course, be organism specific.
It should include considerable use of the classic and current literature and require synthesis of topics by the students. Exercising
and developing the students synthesizing ability is important. It

is becoming clearer that more and more scientists should devote much of their time, energy, and imagination to examining, collating, condensing, and synthesizing data and observations provided by others rather than going to the field or bench to add to the existing volumes of experimental results.

3.3.3 Recent Courses in the Systems Ecology Area

Another area of advanced ecology is beginning to be covered in various universities under the title of "systems ecology" and related sequences. Systems ecology work may be either at undergraduate or graduate level; conceptual or analytical approaches may be emphasized. I will discuss briefly only the analytically-oriented, graduate-level coursework. Such courses are now available in several universities, having been taught for several years at such places as Colorado and Washington State, and Universities of Tennessee, Georgia, and California-Davis.

The sequence at Colorado State University is given college-wide numbers in the College of Forestry and Natural Resources, i.e., rather than being departmentally based and biased. It is a team-taught approach involving at least five men from Departments of Forest and Wood Sciences, Fishery and Wildlife Biology, Range Science, Mathematics and Statistics, and Electrical Engineering. The courses are Systems Ecology, Ecological Simulation, and Natural Resource Models, each of which is 5-quarter credits, and each includes both lecture and laboratory.

Students are provided integrated, computer-assisted training to develop and investigate ecological principles in a quantitative manner. Probabilistic and deterministic methods from statistics, operations research, and systems analysis are introduced to, or embellished by, the students. The students gain considerable experience with both remote console and batch processing use of third-generation digital computers and with small analog computers. Prerequisites are a year's background in calculus, statistics, and ecology. Although we do not require differential equations, matrix algebra, and computer programming as prerequisites, these topics are heavily used in the sequence.

Having taught in such systems-oriented courses for several years, I note there are several trends. The models investigated by students have become larger and more realistic. They have more variables and more effects. Although only static systems and models

were once considered, the emphasis now is on dynamic systems, often
with many nonlinear components. There has been increased use of
remote terminals and consoles. Stochastic effects now are included
in the models; previously most were deterministic. When I taught
such materials in Tennessee, we did not require as much background
in mathematics and statistics, nor did the students obtain as much
computational experience as now in the sequence at Colorado State.

3.3.4   Increasing Systems Ecology Interests of Non-biological Science Students

It is probable that many of the Ph.D. systems ecologists will not
come from undergraduate programs in natural sciences, natural resources, or agriculture. Increasingly there are indications of
more physical science and engineering undergraduates becoming
interested in environmental problems. Such motivated and skilled
students are much needed, and the type of Ph.D. program outlined in
Table 2 can accommodate them, especially in theoretical-analytical
emphasis. It is necessary, however, that they acquire sufficient
biological and resource management background. This does not imply
that they need to take all the coursework listed in Table 1. Often
through judicious selection of readings, through separate and
special seminars, and through special problem courses, they can
quickly build an adequate background in these areas. The common-
ality of ecological principles and concepts across fields should be
stressed, such as in Van Dyne [6], rather than emphasize details of
individual practices in each resource management field. The
physical science or engineering undergraduate who switches to
systems ecology at the Ph.D. level generally is a very capable
and highly motivated individual.

Actually, there is a more common problem in graduate ecology
training, e.g., the undergraduate from a small liberal arts school
who has natural science background, but lacks training in natural
resource, agricultural, or some analytical areas. Often it is
necessary for such students (as well as for other undergraduate
biological science students not obtaining analytical background)
to go back through many areas of analytical coursework. For most
graduate students to receive more than a superficial analytical
understanding, they must have the assistance of an instructor.
Self-teaching alone does not provide adequate skill in, and understanding of, these subjects in a reasonable period of time.

It must be emphasized to the graduate student that the systems

ecologist will have to produce in the environment of team research
and team management. Imagination and inventiveness should be
developed and tested. Examination of problems of real-life complexity should not be shunned.

3.4 Some Problems in Quantitative Ecology

3.4.1 Inadequacies of Our Present Systems

As far as I know, neither the undergraduate nor graduate programs
outlined above are being attempted extensively. The undergraduate
program in Table 1 requires much more counselling time than is
normally given to our undergraduate students in natural resources
in many land-grant schools. There is a dearth of solid, advanced
ecological coursework of the nature described above and outlined in
Table 2 for the graduate student. There are even fewer places where
problems of real-life complexity are being investigated in the
classroom. This is due, in part, to the amount of resources required to accomplish adequate evaluation. Such resources include
(i) classroom use of rather sophisticated third-generation computing
systems and (ii) instructors with experience in formulating and
solving real-life problems, yet still having the time to work with
graduate students (or undergraduates) in doing so. It is difficult
to find more than one staff member in a given department with sufficient time, training, or experience to cover in depth basic
ecology, resource management, basic mathematics, applied mathematics,
systems engineering, and computer sciences. It appears necessary
to have a teaching team. Perhaps a teaching team is also necessary
for frequent reciprocal intellectual stimulation.

   The problems of resource management are many, complex, and
becoming more critical. Can we wait for a few university programs
to be modified and to produce the quantitatively-trained resource
managers and scientists needed? Can we wait for even fewer, much-
needed scientists and managers to be accepted quickly into top
management positions in the resource agencies, both state and
federal?

3.4.2 Resource Management Agency Sponsored Training

We need several combined applied research-management training
programs immediately in which quantitative techniques are investigated, adapted, and extended to treatment of real-life problems.
The resource management agencies have many capable young men with

biological backgrounds, and these individuals soon will fill top
management positions. Universities and research agencies have a
few quantitatively-oriented scientists who can converse and interact
with key management people. There are certainly no limitations of
important problems to be studied. The problems have applied research phases of specific and short-term nature which are of
interest to the resource management agencies. A systems analysis
approach is needed to bring together these management and research
specialists to focus upon resource problems. By careful evaluation
of data in the literature (and data gathering dust in the files of
investigators and agencies), by making specific short-term field
measurements, and by translating observations into quantitative
relations, preliminary simulation models can be structured. Computer experiments can be used to "exercise the models" to test
alternative management strategies.

Natural resource management is a complex process, if done in a
knowledgeable manner, and involves simultaneous consideration of
many variables. Implementing the multiple-use concept of wildlands,
and including the probabilistic inputs of uncontrollable factors
such as climate, produces a staggering array of different situations
to be considered. This can be done using modern computational tools
and techniques and simulation models. Obviously, exercising models
often will lead to many new questions; to answer these some participants in the managment-game exercise must go back to the field,
laboratory, or library to gather more information and to restructure
the models. The new model allows different questions to be asked,
or greater precision will be given to answers to the earlier
questions. And so the cycle continues. But spin-off occurs in
this cycling process not only to the resource management agencies,
but also to the scientist-educators. Such exercises would help
define and illustrate the need for improved or new techniques in
systems analysis and operations research.

3.4.3 Criticality of Quantitative Resource Management Approaches

The needs for such tools, techniques, and well-trained people will
not decrease. Increased quantitative approaches, including "PPBS,"
have been initiated in many agencies and are increasingly filtering
down to the local resource manager. As we gain more and more knowledge about resource systems, and as wise resource use becomes more
critical, the need for solid predictions will increase. It will be

important then that most resource scientists and managers become acquainted with these tools, techniques, and approaches. Thus the agencies, and the individuals, should consider regular sessions of continuing education as essential. Why have they not done so? The hardware and software are now available, the problems are still with us, and the trends are clear. Several useful books are available to explain some of these methods and their applications (e.g., Watt [8], Pielou [3], Van Dyne [6]), and more inclusive textbooks are now in advanced stages of preparation.

To consider resource problems of real-life magnitude and complexity requires a level and type of effort that is not easily defended nor easily supported in our conventional university training programs in either the natural resource disciplines, or in the basic ecological or analytical disciplines. One solution is to initiate continuing cooperative agency-funded, workshop-research programs.

3.4.4 Division of Labor and the Reward System

As we continue now to plan coordinated attacks in research and management on natural resource ecology problems, we find that integrated, interdisciplinary teams are essential. We also must note that to accomplish the goals of these programs there must be a division of labor: (i) some team members must focus on collection of samples and data, perhaps without participating in detailed analyses; (ii) some must focus on sample and data analysis and modelling, perhaps foregoing field and laboratory studies; (iii) other members must concentrate on synthesis of results; and (iv) yet others must work in program management, development, and coordination. This division of labor and specialization of effort is well established in many industrial areas in physical sciences and engineering efforts. Perhaps this approach is used sometimes, but is not always successful, in research and management programs of some of the state and federal resource management agencies. But seldom is this cooperation or philosophy evident in university efforts.

The lack of quick acceptance of this division of labor by biologists and resource scientists stems largely from the history of their field. Most of the major biological and resource accomplishments in the past have been made by a scientist working alone, or at most a few scientists working together. But they did not often tackle problems of total-ecosystem complexity. Furthermore, most

of today's university professors in biological and resource areas
have not participated in team efforts. So they often hold
sacrosanct the concept of highly individual contributions to
science. Perhaps even more influential is the fact that the
background and attitudes of the university administrators often
are the same. And they control many of the rewards in the system.
Thus most students who mature in the university environment, either
as undergraduates or as graduates, come to the real world somewhat unprepared to contribute fully to team research or management.
They do not realize that often they can make greater contributions
through services to the group than through entirely individualistic
efforts. Sample and data processing and analysis efforts may often
be the limiting factor in ecosystem research. Team members devoting
their time and energies to such efforts can make significant contributions for the overall effort. Eventually our reward system in
universities will better recognize program efforts, and more
students will learn this and will more effectively participate
during their career in large, integrated research and management
programs.

ACKNOWLEDGMENTS

Many of the ideas, and the Colorado State University courses,
described herein were developed in training and research programs
supported by the National Science Foundation (Grants GZ991, GB7824,
and GB13096) and in interdisciplinary, ecological modelling work-
shops supported by the Ford Foundation.

References

[1] Booker, H. G. 1963. University education and applied science.
     Science. 141:486, 488, 575-576.

[2] Panel on Natural Resource Science. 1967. Undergraduate educa-
     tion in Agriculture and Natural Resources, Agricultural
     Board, Division of Biology and Agriculture, National Academy
     of Sciences-National Research Council Publication 1537. 28 p.

[3] Pielou, E. C. 1969. An introduction to mathematical ecology.
     John Wiley and Sons, Inc., New York. 286 p.

[4] Van Dyne, G. M. 1966. Ecosystems, systems ecology, and systems
     ecologists. Oak Ridge National Laboratory Report 3957.
     31 p.

[5] Van Dyne, G. M. 1969. Grasslands management, research, and
     training viewed in a systems context. Science Series No. 3,
     Range Science Department, Colorado State University. 50 p.

[6] ——————. (ed.). 1969. The ecosystem concept in natural
     resource management. Academic Press, Inc., New York. 383 p.

[7] Watt, K. E. F. (ed.). 1966. Systems analysis in ecology.
     Academic Press, Inc., New York. 276 p.

[8] ——————. 1968. Ecology and resource management--a quanti-
     tative approach. McGraw-Hill Book Co., New York. 450 p.

4. NEEDS AND OPPORTUNITIES IN THE FEDERAL SERVICE (W. E. WATERS)

My role this evening is to speak for the largest single employer of
knowledge and experience in quantitative ecology now and in the
foreseeable future--the United States government. Needs and opportunities for training and research in this field at the federal
level are greater than ever before. The social and political
implications and presumed scientific applications of ecology to
present-day problems such as environmental contamination and the
depletion of vital natural resources have caught the public attention. And our legislators are not immune. Speeches, discussions,
and arguments on these problems are occurring in increasing
frequency in our Congress--and we hear from these, and in the
pronouncements of our governmental agencies, a new phraseology.
The use of terms such as ecosystems, the ecosystems approach, and
eco-systematic management of environmental variables undoubtedly
reflects an awareness and sincerity of intent to view and attempt
solutions to these broad and complex problems in an ecological
context. This is fine, but one quickly gets the feeling that we
had better soon bring effective knowledge and experience to this
effort before the thrust of intentions outweighs reasoned judgment.

By tradition and policy, our government is committed to the
development and maintenance of the Nation's natural resources, to
the physical and spiritual well-being of its people, and to improvement of the quality of the environment for all time. This
brings with it the ecological dilemmas of utilization and preservation, conservation and exploitation, living space and open space,
waste disposal, and many others. Progress toward these ends and
resolution of the dilemmas has been hardly evident at times, and
the development and application of relevant scientific knowledge
and technology has been uneven and inconsistent. But only the
most naturalistic or primitively-oriented individual would take
the view that our lot has not improved from the past, or that no
progress is being made toward happy survival.

We recognize, however, the finite nature of our resources and
the challenge of survival--in a state of well-being or otherwise--
in the face of a population boom with uncertain parameters. Despite
its somewhat nebulous status as a scientific discipline, ecology is
being looked to for enlightenment and guidance. I believe we can
all agree that the problems involved are going to increase in
complexity and importance and that correspondingly greater

sophistication, efficiency, and realism of concepts and technologies will be required to solve them. This is the prime challenge to quantitative ecology today, and the United States and other federal governments must provide the major impetus.

Current federal programs encompass all of Earth's environments, with that of the moon's surface, too, now open to us for study and development. All levels of ecological organization from molecules to populations and higher orders of whole animals and plants are subject to investigation. Some programs are pure research; most are directed toward some operational goal. Many programs are not labelled ecology, nor even ecologically-oriented, but they nevertheless draw from or contribute to knowledge in this field.

To attempt even a partial listing of all federal programs and activities that involve or relate to quantitative ecology would be tedious for me and boring to you. Rather, I will outline briefly the work of just one agency--the U. S. Forest Service--having direct involvement in and responsibility for a wide range of ecological activities. I would like to state the mission and describe the approaches being taken by several Forest Service research projects, also. These typify, I believe, the present needs and opportunities for technical competence, imagination, and initiative to make significant advances in the solution of major problems. By this means, I hope also to indicate what, in my view, constitutes quantitative ecology and the ways in which training and research should be structured and related one to the other.

The U. S. Forest Service activities cover three major areas of operation: (i) management, protection, and development of the National Forests and National Grasslands; (ii) cooperation with State and private landowners in carrying out and improving forest management practices; and (iii) research. It has stewardship over 154 National Forests and 19 National Grasslands located in 44 states and Puerto Rico. Totalling 187 million acres, these encompass nearly all of the vegetation types, or biomes, of North America, with consequent ecological diversity of biological and physical components. Five primary resources are recognized: timber, water, forage, wildlife, and recreation. Management of these lands is aimed at obtaining the maximum benefit from these renewable resources for the greatest number of people. Generally termed multiple use, this concept (and goal) requires the development of optimal strategy for combining the ecological, economic, and human

elements involved. This is an area in which we are just beginning
to apply the principles and methods of operations research and
systems analysis.

Cooperative programs for management and protection of an additional 480 million acres of forest land and associated watershed
lands are carried out through agreements with State and local
governments, forest industries, and private landowners. Major
programs have been developed for cooperative forest management
(silviculture, harvesting, and marketing), forest pest control
(insects and diseases), forest fire control, and watershed development and flood prevention. Here, too, the aim is maximization of
benefits (with understandably some emphasis on dollar returns and
present conditions). If ecological enlightenment is to be built
into this important aspect of forest resource management, then we
must have ecologists trained and willing to work effectively at this
operational level.

The Forest Service's research program is carried out through
eight regional Experiment Stations, the Forest Products Laboratory,
and the Institute of Tropical Forestry. There are 330 individual
Research Work Units at 80 research locations. Of the 3000 total
personnel, approximately one-third are professional scientists.
The annual budget of $45 million represents about 12 percent of
the total budget for the Forest Service. The study programs are
directed to all phases of timber production, including genetic
improvement; protection of forests from fire, insects, diseases,
and animal pests; management of rangelands; improvement and management of wildlife habitat; protection and management of watersheds;
forest recreation; marketing and utilization of forest products;
forest engineering; and forest economics. A continuing forest
survey provides comprehensive information on the current extent,
condition, and productivity of forest lands; trends in timber growth
and harvest; and estimates of future yields and future demands for
timber supplies and other forest resource elements. At present,
about one-third of the research effort has no direct orientation to
action programs. There is a significant trend to multifunctional
planning and organization of research teams to solve problems
holistically defined. This involves the joining together of subject
matter specialists and generalists of diverse sorts, and the cros-
sing of territorial boundaries of the Experiment Stations.

The intramural program is augmented by grants to universities and

other research institutions and through cooperative research by
scientists in foreign countries under terms of the P.L. 480 program.
Training is a vital part of this extramural effort.

Now, very briefly let us look at the missions and current activities of three Forest Service research units dealing with different
aspects of quantitative ecology.

The first is a pioneering research unit in forest mensuration.
Organized around one key scientist, Lewis R. Grosenbaugh, the
mission of this unit is to acquire knowledge and understanding of
mensurational principles as they apply to the form, volume, growth,
spatial arrangement, and quality of forest trees and stands. More
specifically, the research is directed along four lines: (i) the
theory, instrumentation, and techniques of tree measurement; (ii)
analysis of the dynamics of stand structure; (iii) determination
and prediction of forest growth and yield; and (iv) mathematical
concepts for forest management planning and decision-making. Since
its inception in 1960, this unit has in fact made significant contributions in all of these facets, including a conspectus and new
theory on optical dendrometers (for measurement of out-of-reach
diameters on standing trees); the theory of a new variable-probabil-
ity sampling procedure in which the probability of selection of
sample trees is proportional to ocular prediction (now popularly
called 3-P sampling) and two Fortran-4 computer programs for
generating selection criteria and deriving estimates of tree
population parameters from 3-P sample-tree measurements; a Fortran-4
system for combinatorial screening or conventional analysis of
multivariate regressions (this handles up to 50 variables, 8 of
which may be dependent); an analysis and generalization of some
sigmoid and other non-linear functions typical of tree and forest
stand growth; and an analysis of objectives, problems, and methods
of determining annual timber cut for sustained yield management.

I would like to make a special point in regard to this aspect
of quantitative ecology. The development of mensurational and
sampling techniques may have less allure, and may appear to be less
challenging than the modeling and analysis of ecological relationships, processes, and systems. But statistically sound and
biologically valid criteria and methods of sampling real populations
are necessary if hypotheses and principles are to be generalized and
validated. And anyone who has had to deal with the confounding
effects of variation in the spatial and temporal distributions of
any living organism, statistical artifacts imposed by the constraints

of a laboratory situation or by the procedure of data gathering in the field, and other more subtle factors affecting sampling accuracy, precision, and efficiency can testify that sampling problems can be fascinating and challenging indeed.

My second example is a research unit dealing with the impact of insects on the forests of northeastern United States. Composed of 4 scientists located at the Northeastern Forest Experiment Station's Insect and Disease Laboratory here in Hamden, Connecticut, this unit has as its objective to define, measure, predict, and evaluate the full impact of destructive insects on forests of the region and to provide guidelines for forest resource managers in making decisions regarding action against specific pests. The studies in this unit fall within three categories: (i) definition and description of the kinds and degrees of insect-caused impact to different tree species or species groups and the development of methods for measuring or sampling the insect populations and their impact effects; (ii) determination of the factors or processes causing variation in insect numbers and their impact effects, with regard both to differences among places and changes over time; and (iii) integration of the ecological information obtained in phases (i) and (ii) above with management experience, economic inputs, and cost analyses to develop optimum insect control strategy.

Among the contributions generated thus far by this unit is a management guide for evaluating the need for and scheduling control of the white-pine weevil, a major disruptive factor in the management of eastern white pine and other valuable tree species. Combining quantitative data on tree growth rates, weevil attack frequencies and intensities, volume and quality losses in tree values incurred, financial gains from control, and costs of control, working tables and a simple computational exercise permit the selection and scheduling of pine stands for weevil control at specified confidence levels. A computer program has been written for sequential probability ratio tests, with particular reference to inverse sampling procedures applicable to forest insect populations. In cooperation with the Yale School of Forestry, a computerized system has been developed for simultaneous solution of whole sets of equations describing the dynamics of forest stand condition and predicting stand behavior in terms of resource outputs. The analysis of forest stand dynamics has been approached another way-- utilizing life tables--in which the importance of destructive agents (including man) in determining final yield or value can be assessed. This work has considerable significance beyond its regional context in establishing the quantitative concept of impact, or disturbance, in productive forest ecosystems and developing methods of measurement and analysis for evaluating and predicting such effects on a system.

As a final example, our North Central Forest Experiment Station has a research unit located at St. Paul, Minnesota, that is investigating the social and ecological problems of wilderness-type recreation in the Boundary Waters Canoe Area. Its research mission--to improve methods of managing remote areas in the north woods for "pioneer" types of recreational use--is people-oriented, but it has the distinct goal of reconciling the natural ecological factors with human factors in a way that will maintain the wilderness character of the BWCA and other areas similarly used. Staffed with five scientists, the unit thus far has concentrated largely on vegetation inventory and description, with some preliminary studies of community structure, successional changes, and classification. Based on data from 58 natural upland stands, a tentative model of community classification has been developed for the Minnesota wilderness area, using a combination of agglomerative clustering techniques and ordination by principal components analysis. Preliminary analyses of relationships between the provisional community types and some environmental factors, including fire and insect outbreaks, indicate that these disturbances have had a strong influence in determining the composition and structure of the forest community. A conclusion from this is that controlled use of fires and manipulation of insect populations may be required to maintain the existent natural forest ecosystems.

I hope that this rather sketchy outline has given you some idea of the deep-rooted ecological orientation of the Forest Service action and research programs. The range in character and levels of research in progress and the trends indicated for the near future are representative, in general I believe, of many other Federal agencies with similar concerns in environmental problems. The needs and opportunities for people trained in quantitative ecology are clearly evident now, and they will increase--but so too will the demands on technical knowledge, imagination, and initiative.

As a potential employer, I would hope that beyond the necessary courses in concepts and methodology of mathematics, statistics, and ecology--however organized--training in quantitative ecology include

a coming to grips with a real-life situation. And that feedback
from this be worked into the formal study program. Simulation and
controlled laboratory experiments provide useful demonstrations of
concepts and allow practice in both ecological and mathematical
techniques. Field studies of the "controlled experience" type also
provide a certain amount of intellectual exercise. But detailed
study of an ecological process or system in its natural functioning
state, and full mathematical treatment as required, will better
prepare the neophyte to cope with his first assignment and will
condition him for the frustrations of the continual learning process
that he will undergo in his work.

5. GRADUATE WORK IN STATISTICAL ECOLOGY (J. WILSON)

Dr. Patil has asked me to tell you a little of what it is like to
attempt a Ph.D. in the area of Statistical Ecology. Also, he has
asked me to describe briefly the programs available at U. C. Davis.

Probably one of the most difficult tasks that a student has in
Statistical Ecology is bridging gaps between the two areas. In
order to do this successfully, it is required that first the student
have a very good understanding of both fields, and second that he
have the creative ability to link the two areas. I say this because
until just recently, few courses were offered which pointed the way
to bridging the gap between engineering and biology and mathematics
and biology. Particularly as an undergraduate, it was only with the
extreme patience and help of my mathematics professor and my ecology
professors, Drs. Phil Miller and Boyd Collier at San Diego State
that I was at all able to connect the two areas successfully.

At U. C. Davis, Dr. Watt leads the Statistical Ecology group in
the Zoology Department. Working under a $174,000 grant from the
Ford Foundation, he is attempting to determine the feasibility of
using the techniques of the System Analyst, the Engineer, and the
Statistician to build simulation models for the purpose of better
understanding in large-scale interaction of physical and social
processes in the human ecology of California.

Graduate students employed in this effort are learning first-hand
the difficulties of linking mathematics to biological as well as
sociological phenomena.

Secondly, there are the opportunities available in our Institute
of Ecology. Here a student has the opportunity to select not one
graduate professor, but four. Each one can be from a different
department and the professors from our school of Engineering are
favorite candidates for these positions. Still, there remains the
problem for the student of bridging the gaps between the two
disciplines.

The area of Statistical Ecology remains for all involved a challenge, not just for the student, but for the professional
statistician and the Ecologist, due to the demands being placed
on these people today. Hopefully, through the efforts of Dr. Patil
and others, the communication between these two groups will not
always remain a problem.

I would like to take this opportunity to thank all of those
present for accepting that challenge.